Transformation of Organometallics into Common and Exotic Materials: Design and Activation

NATO ASI Series

Advanced Science Institutes Series

A Series presenting the results of activities sponsored by the NATO Science Committee, which aims at the dissemination of advanced scientific and technological knowledge, with a view to strengthening links between scientific communities.

The Series is published by an international board of publishers in conjunction with the NATO Scientific Affairs Division

A Life Sciences	Plenum Publishing Corporation
B Physics	London and New York
C Mathematical	D. Reidel Publishing Company
and Physical Sciences	Dordrecht, Boston, Lancaster and Tokyo
D Behavioural and Social Sciences	Martinus Nijhoff Publishers
E Applied Sciences	Dordrecht, Boston and Lancaster
F Computer and Systems Sciences	Springer-Verlag
G Ecological Sciences	Berlin, Heidelberg, New York, London,
H Cell Biology	Paris and Tokyo

Series E: Applied Sciences – No. 141

Transformation of Organometallics into Common and Exotic Materials: Design and Activation

edited by

Richard M. Laine

Department of Materials Science and Engineering,
Roberts Hall FB-10, University of Washington,
Seattle, Washington, USA

1988 **Martinus Nijhoff Publishers**
Dordrecht / Boston / Lancaster
Published in cooperation with NATO Scientific Affairs Division

Proceedings of the NATO Advanced Research Workshop on
The Design, Activation, and Transformation of Organometallics into Common and
Exotic Materials
Cap D'Agde, France
September 1–5, 1986

NATO Advanced Research Workshop on "The Design, Activation, and
  Transformation of Organometallics into Common and Exotic Materials"
  (1986 : Cap d'Agde, France)
    Transformation of organometallics into common and exotic materials
  : design and activation / [edited by] Richard M. Laine.
      p.   cm. -- (NATO ASI series. Series E, Applied sciences ; no.
  141)
    "Proceedings of the NATO Advanced Research Workshop on 'The
  Design, Activation, and Transformation of Organometallics into
  Common and Exotic Materials,' Cap d'Agde, France, September 1–5,
  1986"--T.p. verso.
    "Published in cooperation with NATO Scientific Affairs Division."
  Includes bibliographies.

    ISBN-13: 978-94-010-7122-2   e-ISBN-13: 978-94-009-1393-6
    DOI: 10.1007/978-94-009-1393-6

    1. Organometallic compounds--Congresses. 2. Materials-
  -Congresses. 3. Chemistry, Technical--Congresses.   I. Laine,
  Richard M., 1947-   . II. North Atlantic Treaty Organization.
  Scientific Affairs Division. III. Title. IV. Series.
  QD410.N35 1986
  547'.05--dc19
                                                    87-34797
                                                        CIP

Distributors for the United States and Canada: Kluwer Academic Publishers, 101
Philip Drive, Norwell, MA 02061, USA

Distributors for the UK and Ireland: Kluwer Academic Publishers, MTP Press Ltd,
Falcon House, Queen Square, Lancaster LA1 1RN, UK

Distributors for all other countries: Kluwer Academic Publishers Group, Distribution
Center, P.O. Box 322, 3300 AH Dordrecht, The Netherlands

ALL ASI SERIES BOOKS PUBLISHED AS A RESULT OF
ACTIVITIES OF THE SPECIAL PROGRAMME ON
SELECTIVE ACTIVATION OF MOLECULES

This book contains the proceedings of a NATO Advanced Research Workshop held within the programme of activities of the NATO Special Programme on Selective Activation of Molecules running from 1983 to 1988 as part of the activities of the NATO Science Committee.

Other books previously published as a result of the activities of the Special Programme are as follows:

BOSNICH, B. (Ed.) – *Asymmetric Catalysis* (E103) 1986

PELIZZETTI, E. and SERPONE, N. (Eds.) – *Homogeneous and Heterogeneous Photocatalysis* (C174) 1986

SCHNEIDER, M. P. (Ed.) – *Enzymes as Catalysts in Organic Synthesis* (C178) 1986

SETTON, R. (Ed.) – *Chemical Reactions in Organic and Inorganic Constrained Systems* (C165) 1986

VIEHE, H. G., JANOUSEK, Z. and MERENYI, R. (Eds.) – *Substituent Effects in Radical Chemistry* (C189) 1986

BALZANI, V. (Ed.) – *Supramolecular Photochemistry* (C214) 1987

FONTANILLE, M. and GUYOT, A. (Eds.) – *Recent Advances in Mechanistic and Synthetic Aspects of Polymerization* (C215) 1987

PREFACE

The design, synthesis, and selective pyrolytic conversion of organo-metallic precursors to materials of high purity or specific morphology (for electronic or optical applications), high strength and/or high-temperature stability (for structural or refractory applications) represents a poten-tial area of extreme growth at the overlap of chemistry and materials science (materials chemistry). Research in this area is likely to have considerable impact at both the academic and societal levels because it will require development of scientific expertise in areas currently not well understood. Examples include: (1) The thermodynamics of molecular rearrangements in organometallic molecules at temperatures above 200°C; (2) The electronic properties of amorphous ceramic materials; (3) The phys-icochemical properties of ceramic molecular composites; and (4) The optical properties of multicomponent glasses made by sol-gel processing.

The opportunity to establish the scientific principles needed to pursue useful research goals in "materials chemistry" requires communica-tion between chemists, ceramists, metallurgists, and physicists. To date, there have been few opportunities to create an environment where such communication might occur. The objective of this NATO Advanced Research Workshop was to promote discussions between experts in the various disci-plines aligned with "materials chemistry." These discussions were intended to identify the scope and potential rewards of research efforts in the development of: Custom-designed precursors to common and exotic materials, methods of selectively transforming these precursors in high yield to the desired material, and methods of characterizing the final products.

The accounts that follow are contributions from the majority of the 30 workshop participants. The first set of contributions, Framework Sciences, represents an attempt to set the stage for the main body of papers. These papers present work from several diverse disciplines that impact on the future directions of three major areas of research in materials chemistry and suggest new areas of investigation. The remaining sections of this volume have been organized to provide the reader with an overview of preceramic polymers, sol-gel processing, and chemical vapor deposition. The intent of this volume is to serve as an introduction to the complex-ities of materials chemistry rather than as a reference work.

The essential financial support for the workshop was provided by the Scientific Affairs Division of NATO and by a generous contribution from the Office of Naval Research, London, England. The organizers would like to acknowledge the valuable guidance of Dr. Craig Sinclair, Director of the Selective Activation of Molecules (SAM) program, and Dr. Igor Tkatchencko of the SAM Advisory Panel in the preliminary planning and organizational details for the Advanced Research Workshop. We would also like to acknow-ledge the help and contributions of Dr. David Venezky, and Professor Robert Vest of the Office of Naval Research, London. The organizers would especially like to thank the exceptional efforts of Mrs. Elaine Adkins and Mrs. Maria Buyco in handling correspondence, paperwork, and manuscript preparation before and after the workshop.

R. M. Laine
Editor

CONTENTS

IX

x

PARTICIPANTS

ALMOND, Eric A.	U.K.
AYLETT, Bernard J.	U.K.
BRINKER, Jeffrey	U.S.A.
CORRIU, Robert	France
DISLICH, Helmut	W. Germany
DUNOGUES, Jacques	France
GEOFFREY, Gregory	U.S.A.
GHERARDI, Paola	Italy
GOROCHOV, Orie	France
HAIGH, John	England
HARROD, John F.	Canada
LAINE, R. M.	U.S.A.
LIVAGE, Jacques	France
MAYA, Leon	U.S.A.
MAURY, Francis	France
MINGOS, D. Michael P.	U.K.
NOLTES, Jan G.	Netherlands
PIERCE, Hugh	U.S.A.
SAPPA, Enrico	Italy
SCHLEICH, Donald	U.S.A.
SIEBERT, Walter	W. Germany
SEYFORTH, Dietmar	U.S.A.
SOULA, Gerard	France
TKATCHENKO, Igor	France
ULRICH, Donald	U.S.A.
VEST, Robert	U.S.A.
WALSH, Robin	U.K.
WILSON, Robert	U.S.A.
WYNNE, Kenneth J.	U.S.A.

LIST OF AUTHORS

ALBANESI, Giancarlo	U.S.A.
ALMOND, Eric A.	U.K.
AYLETT, Bernard J.	U.K.
BACQUE´, Eric	France
BIROT, Marc	France
BRINKER, Jeffrey	U.S.A.
CASTIGLIONI, Mario	Italy
CZEKAJ-KORN, Corinna L.	Italy
DISLICH, Helmut	W. Germany
DUNOGUES, Jacques	France
EZZAOUIA, H.	France
GEOFFREY, Gregory L.	U.S.A.
GHERARDI, Paola	Italy
GIORDANO, Roberto	Italy
GOROCHOV, Orie	France
GUGLIELMINOTTI, Eugenio	Italy
HAIGH, John	England
HARROD, John F.	Canada
HIRSCHON, Albert S.	U.S.A.
LAI, W.Y.F.	U.S.A.
LAINE, R. M.	U.S.A.
LAM, A.	U.S.A.
LIVAGE, Jacques	France
MATIJEVIC, Egon	U.S.A.
MAYA, Leon	U.S.A.
MAURY, Francis	France
MINGOS, D. Michael P.	U.K.
MOGGI, Pietro	Italy
NOLTES, Jan G.	Netherlands
PILLOT, Jean-Paul	France
PREDIERI, Giovanni	Italy

SAPPA, Enrico	Italy
SCHLEICH, Donald	U.S.A.
SEYFERTH, Dietmar	U.S.A.
SIEBERT, Walter	Germany
ULRICH, Donald R.	U.S.A.
WALSH, Robin	U.K.
WYNNE, Kenneth J.	U.S.A.

INTRODUCTION

There is growing interest in the potential utility of chemical synthesis as a means of preparing materials for high technology applications. This interest in "materials chemistry" derives from the thermodynamic advantages offered by chemical synthesis as compared with current metallurgical or ceramic processing methods. These advantages include: (1) low-temperature processing to initial shape, (2) greater control of product selectivity and product phase through closer control of reaction kinetics and thermodynamics, (3) wider range of final shapes (e.g., fibers and coatings), and, (4) potential access to materials or phases that are kinetically or thermodynamically unavailable using accepted industrial materials processing techniques.

One of the primary obstacles to rapid progress in the field of materials chemistry is that fundamental problems often lie at the interface of chemistry, materials science, physics, and electronics where little dialogue between disciplines occurs. For example, preliminary indications are that the complete development of the new 123 ceramic oxide superconductors will require input from all of these disciplines. Until recently, it was difficult to organize meetings where researchers from all four domains were brought together to establish common ground let alone begin to coordinate research efforts. Indeed, as Mingos notes in the first chapter in this book, a common vocabulary must be established as a prerequisite to useful discussion. Although the Mingos chapter correlates concepts and vocabulary common to both theoretical chemists and physicists, similar correlations must also be established at the practical level. This book represents a preliminary effort to cross disciplinary lines to bring a flavor of disparate fields of research to the reader interested in the potential applications of materials chemistry.

The section on framework science contains several chapters designed to familiarize the reader with specific areas of research that draw from "materials chemistry" or that offer direction or provide information necessary for future studies. These chapters include work on the thermodynamics of bond dissociation during the pyrolytic decomposition of organometallic compounds, the problems associated with the development of better hard metals, the use of chemical synthesis to prepare heterogeneous catalysts or to prepare tungsten and molybdenum carbide. Another motivation for the framework chapters is to illustrate the diversity of the fields involved. Inherent within this diversity are the opportunities for new areas of research.

The sections that follow are directed toward specific areas of research where materials chemistry is already having a significant impact; although, in each case the field of study is far from mature.

The section on electronic materials, to date the most advanced/successful area of materials chemistry, focusses on chemical vapor deposition (CVD) of a variety of conductor, semiconductor, and insulator films. The outstanding features of CVD processing are that it permits extremely fine control of film thickness, impurity levels (at the ppb level in some instances), and it can be used to construct multilayer devices. The intent of this section is to illustrate the scope of the CVD process, some of the pitfalls inherent in the process (e.g., in the preparation of boron nitride coatings) and to suggest directions for future study to improve current processes or expand the utility of the process. Related areas, such as PVD and plasma enhanced OM-CVD, where materials chemistry has less of a direct impact have not been addressed in this section.

The following section on ceramic polymer precursors examines the use of materials chemistry to prepare bulk non-oxide ceramics. This area is much less developed than the electronic materials area, and the majority of the work has been directed towards the synthesis of silicon-based preceramic polymers and their pyrolytic transformation to silicon carbide and silicon nitride.

Many commercially important organic polymers are produced through processes requiring at least one metal catalyst-promoted step. With the exception of polysiloxanes (silicones), very few inorganic polymers are recognized as being of potential industrial importance. If inorganic polymers are ever to become commercially useful, especially as materials precursors, then simple methods of preparation need to be developed. Harrod describes one of the first examples of the synthesis of polysilanes using transition-metal catalysts. The chapters by Seyferth and Noltes cover many important aspects of the synthesis of silicon-containing ceramic precursors and their pyrolytic transformation into silicon carbide and silicon nitride.

Given the structural and chemical similarity of boron nitride to carbon, boron nitride fibers are likely to have exceptional properties and represent a useful goal for materials chemistry. Unfortunately, very little work on the synthesis of boron nitride precursors has been reported in the literature. The chapter by Wynne describes some of the first efforts to synthesize polymer precursors to boron nitride.

The final section addresses sol-gel methods of synthesizing oxide ceramics and glasses. Dislich provides an industrial point of view by describing Schott Glasses' application of the sol-gel process to the preparation of commercial specialty coatings and suggests some future directions. Brinker illustrates some of the basic concepts behind sol-gel processing, and the Livage chapters identify some of the basic problems that need to be solved to improve sol-gel processing. Taken in toto, the sol-gel chapters provide an exceptional overview of the present state of the art and opportunities for the future (e.g., non-linear optics).

SECTION A

FRAMEWORK SCIENCE

TOWARDS A COMMON THEORETICAL LANGUAGE FOR MOLECULAR AND SOLID STATE CHEMISTRY

D. MICHAEL P. MINGOS
Inorganic Chemistry Laboratory, University of Oxford, South Parks Road, Oxford OX1 3QR, United Kingdom.

ABSTRACT

The development of a new area of material science at the interface of organometallic chemistry and solid state physics requires the evolution of theoretical and conceptual models which can be appreciated by both chemists and physicists and a common language which will make the communication between scientists with similar goals in this exciting new area more efficient. The continuity of ideas linking molecular and solid state chemistry will be emphasised and illustrated by a number of specific examples.

INTRODUCTION

Creativity and the imaginative leap forward whether it is in the sciences or the arts is often associated with the bringing together of ideas which were previously thought to be unrelated or opposed. Therefore, it is hardly surprising that the rapid development of new and exciting areas of science frequently depend on the exploration of areas lying at the boundaries of two well established and pedagogically distinct scientific disciplines. The spectacular growth of organometallic chemistry during the last thirty years, originally prompted by the discovery of the commercially profitable Zeigler-Natta catalytic process and the novel and rule breaking molecular structure of ferrocene, have resulted in a continuum of new research activities linking the traditional areas of inorganic co-ordination chemistry and synthetic and physical organic chemistry. The continued growth of organometallic chemistry into a mature science has been ensured by the novel valence problems which it has presented, the remarkable ability of low valent transition metal compounds to activate seemingly inert molecules such as methane and N_2, and the development of commercially viable homogeneous catalysts of high stereospecificity.

The mature status of organometallic chemistry makes it timely to consider the new boundaries that it has created with other subjects. Two areas, which are obviously pertinent to this conference, are its interfaces with surface science and solid state physics. The relationships between organometallic cluster chemistry and surface chemistry were particularly well developed by the late Earl Muetterties [1] and will not be discussed in great detail here, instead I shall concentrate on the relationships between molecular and infinite systems in the bulk phase.

New and exciting chemistry does not miraculously appear once the boundary region has been defined and scientists from the different sides of the boundary are brought together. The development of new and significant science requires the definition of real problems, the development of techniques for the rapid recognition and characterisation of new materials and phenomena and finally the emergence of a common language which gives effective communication between those scientists who have been trained in different branches, but now share a common goal.

For the molecular scientists this means coming to grips with some of the mathematical and physical concepts underlying band theory, reciprocal

3

R. M. Laine (ed.), Transformation of Organometallics into Common and Exotic Materials: Design and Activation, 3–7.
© 1988 by Martinus Nijhoff Publishers.

space, Brillouin Zones, charge density waves and Peierl's transitions. For the physicists it will be necessary to begin to develop an essentially non-mathematical and yet powerful three dimensional view of molecules which correlates molecular structure with the number of valence electrons in the system and accounts for reactivity patterns by a subtle interplay of electronic and steric effects. In molecular chemistry it is the relative ease with which these effects can be modified by the synthesis of related compounds which leads to a high degree of selectivity in physical and chemical properties. The great strength of the inorganic chemist is his familiarity with all areas of the periodic table and the skill to synthesise a wide range of materials not available from commercial suppliers.

The purpose of this paper is to demonstrate that the transition from molecular to infinite solid need not be a forbidden one for the synthetic chemist. Indeed many of the ideas and models of solid state physics have their counterparts in molecular chemistry, although they are described in a somewhat different language.

The Table given below summarises the correspondence between some of the theoretical terms commonly used to describe the electronic characteristics of molecular and infinite systems [2].

TABLE

Comparison of theoretical terms commonly used in molecular and infinite systems

MOLECULAR	INFINITE
LCAO Molecular Orbitals	Tight binding approximation
Molecular Orbital	Crystal or band orbital
HOMO	Valence band
LUMO	Conduction band
HOMO-LUMO Gap	Band gap
Jahn Teller Effect	Peierl's distortion
High and intermediate spin states	Magnetic material
Low spin state	Non-magnetic material

In the lecture the relationships between these alternative descriptions will be discussed in some detail and illustrated by examples drawn from the current literature.

Molecular and solid state chemists formally partition the electron density within molecules or infinite arrays in order to emphasise common structural relationships and to rationalise reactivity patterns. At its simplest this methodology uses electronegativity differences in order to establish charge separations and preferred closed shell requirements. In many solid state materials and co-ordination compounds this results in a preference for the attainment of a closed shell inert gas configuration for the more electronegative atom and the definition of a formal oxidation state for the metal atom. The resulting partial filling of the d and f shells for the metals in oxides, fluorides, etc. results in some important magnetic and electronic properties.

In contrast the majority of organometallic compounds attain closed shell inert gas configurations simultaneously at all sites. The resulting diamagnetic molecular complexes have very limited electronic and magnetic properties even in the solid state. Although high nuclearity cluster compounds, which are formally spin paired systems, do show interesting magnetic properties at low temperatures [3] and currently these effects are poorly understood. It has been suggested that these magnetic properties are characteristic of a new meta-metallic state. Similarly it is possible to generate odd electron species from closed organometallic molecules by redox processes and thereby generate some interesting phenomena in the solid state. In particular ferromagnetic and conducting organometallic compounds are being investigated in many academic and industrial laboratories [4,5].

In simple binary systems where the electronegativity difference between the elements is large and the element-element bonding does not play an important role then the construction of a molecular orbital diagram for co-ordination compounds and a band structure for infinite metal oxides or fluorides follow a very similar pattern and the results are strikingly similar. The application of these qualitative ideas to the interpretation of magnetic and electronic conduction properties of solid state materials owes much to the pioneering work of Goodenough [6].

It is interesting that even in solids where there is extensive element-element bonding the two branches of chemistry approach the problem in an analogous fashion. In infinite solids the Zintl-Klemm-Barmann [7] concept recognises that the partititioning of electron density depends not only on the electronegativity difference but also on the number of homo-polar bonds formed by the less electronegative atom. In particular the attainment of the inert gas configuration at the less electronegative atom by a combination of element-element bond formation and formal electron capture leads to the following relationship :

$$\frac{n_e + b_a - b_c}{n_a} = 8$$

Where n_e is the total number of valence electrons,

n_a is the number of anions,

b_a is the number of homopolar bonds formed in the anion, and

b_c is the number of homopolar bonds formed in the cation.

For example, in $CaSi_2$ n_e = 10, n_a = 2, b_c = 0 and therefore b_a = 6 and each Si is co-ordinated to three other silicon atoms. In $CaSi_2$ this is achieved by a two dimensional layer structure, in $SrSi_2$ a three dimensional cross linked structure and in $BaSi_2$ by the formation of isolated Si_4 tetrahedra. This is a most useful principle, but has its limitations when applied to systems where the electronegativity difference is not large, and those compounds which have a high content of either the more or less electronegative element, e.g. Na_8Si_{46} and $Cu_{33}Ge$.

In molecular systems the Polyhedral Skeletal Electron Pair Theory developed primarily by Wade and myself [8,9] utilises the same principle of achieving an inert gas configuration about each vertex atom of a molecular cluster compound to define the number of homopolar bonds and polyhedron associated with the metal skeleton. For ring and three connected compounds, a formula entirely analogous to that given above applies except that the number of electrons associated with the metal atoms is defined by the number of electrons formally transferred to the metal by the ligands co-ordinated to the metal atoms.

$$\frac{(n_e + n_L) \; b_a}{n_a} = 8 \text{ (or 18 for a transition metal)}$$

Where n_e is the total number of valence electrons associated with the vertex atoms of the cluster,

n_L is the total number of electrons donated by the ligands,

b_a number of electrons involved in homopolar bonds,

n_a is the number of vertex atoms in the cluster.

For example, in $[Ir_4(CO)_{12}]$ n_e = 9 (no. of valence electrons of Ir) x 4, n_L = 12 x 2 (since CO is a two electron donor) and n_a = 4 and therefore b_a = 12 corresponding to the formation of six metal-metal bonds along the edges of the tetrahedron defined by the metal atoms.

For more complex systems this formula has to be modified to take into account the fact that for more highly connected clusters the edges of the polyhedron do not correspond to simple localised two-centre two-electron bonds. A similar situation exists in infinite metal boride systems where the octahedral and icosahedral boron cages require the formation of multicentred bonds. The Polyhedral Skeletal Electron Pair Theory has subsequently been extended to clusters which begin to resemble significant proportions of the close packed arrays found in metals and alloys. The evolution of the molecular orbital patterns from discrete clusters to the bulk metals can be traced in some detail in these systems and provides an explanation for the observed stoicheiometries [10].

In addition to providing a general conceptual framework for assisting the development of this new area theoreticians must also prove capable of making specific predictions and parameters for solving particular problems and analysing spectral data. Currently there is no one portmanteau theoretical method [11] which can be used reliably over the whole range of problems of interest to the experimentalist. Therefore it is necessary to develop some appreciation of the strengths and limitations of the alternative theoretical techniques available. The following give a very preliminary indication of the alternatives.

Detailed knowledge of the ground and excited states of small inorganic and organometallic compounds may be required for understanding aspects of chemical vapour deposition techniques, in which case the information could probably most reliably be obtained from <u>ab</u> <u>initio</u> calculations.

Thermodynamic based bond energy terms which might be required for the analysis of the thermal stabilities and physical properties of glasses and ceramics are most accurately derived from the Generalised Valence Bond Method.

Structural problems in infinite solids and a crude one electron view of the band structures in these systems can be conveniently and quickly derived from semi-empirical molecular orbital calculations. However, a more detailed interpretation of their magnetic and electronic properties require more extensive band theory calculations which take into account electron-electron repulsion, exchange energies, and correlation effects.

The weak interactions which result when small inorganic and organic molecules are introduced into cage structures such as zeolites are best approached using molecular mechanics techniques.

In summary, I hope that I have established that the languages of molecular organometallic chemistry and solid state physics are not too dissimilar and that there are a range of theoretical techniques available to assist the experimentalist.

References

1. E.L. Muetterties, T.N. Rhodin, E. Brand, C.F. Bruker and W.R. Pretzer, <u>Chem. Rev.</u>, 1979, <u>99</u>, 91.
2. T.A. Albright, J.K. Burdett and M.H. Whangbo, "Orbital Interactions in Chemistry", John Wiley and Sons, New York, 1985.
3. R.E. Benfield, P.P. Edwards, W.J.H. Nelson, M.D. Vargas, D.C. Johnson and M.J. Sienko, <u>Nature (London)</u>, 1985, <u>314</u>, 231.
4. J.S. Miller, J.C. Calabrese, A.J. Epstein, R.W. Bigelow, J.H. Zhang and W.M. Rieff, <u>J.C.S. Chem. Commun.</u>, 1986, 7026.
5. J.S. Miller, "Extended Linear Chain Compounds", Plenum, New York, 1982.
6. J.B. Goodenough, "Magnetism and the Chemical Bond", Interscience, New York, 1963.
7. H. Schäfer, B. Eisenmann and W. Müller, <u>Angew. Chem. Intern. Ed.</u>, 1973, <u>12</u>, 694.
8. K. Wade, <u>J.C.S. Chem. Commun.</u>, 1971, 792.
9. D.M.P. Mingos, <u>Nature (London) Phys. Sci.</u>, 1972, <u>236</u>, 99.
10. D.M.P. Mingos, <u>J.C.S. Chem. Commun.</u>, 1985, 1352.
11. D.M.P. Mingos, <u>Adv. Organometallic Chemistry</u>, 1977, <u>15</u>, 1.

KINETICS AND THERMODYNAMICS OF GAS PHASE PYROLYSIS OF ORGANOMETALLICS

ROBIN WALSH
Department of Chemistry, University of Reading, Whiteknights, P.O. Box 224, Reading RG6 2AD, U.K.

ABSTRACT

Modern methods of study of the kinetics of gas-phase pyrolysis are described and discussed. These include the VLPP and Laser pyrolysis methods, as applied to metal alkyl and metal carbonyl decomposition. Additionally the more complex pyrolysis of monosilane is analysed mechanistically in the light of most recent studies. This paper refers throughout to the significance of bond dissociation energies as an aid to understanding mechanisms.

INTRODUCTION

The Chemical Vapour Deposition (CVD) process in which solid materials are prepared from decomposing gases is one of established and increasing technological importance. The subject of Gas Phase Kinetics (and to some extent Thermodynamics) which is of the order of a century old, has been applied relatively little to this subject. This undoubtedly derives from the preference of researchers for homogeneous systems. Even the mechanism of decomposition of a simple molecule such as SiH_4, of vital importance to electronic material manufacture, is not well understood. The potential complexity of CVD, in which not only gaseous processes but also gas-surface processes, and even purely surface processes may all be involved is illustrated in Fig. 1. The study of <u>gaseous</u> pyrolysis can clearly contribute

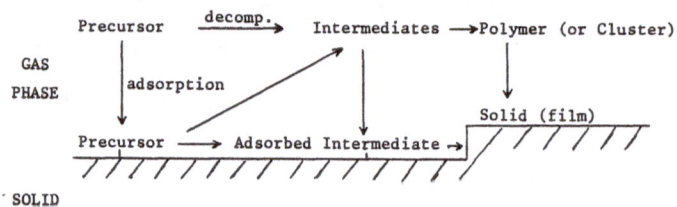

Fig. 1. Schematic diagram of processes involved in CVD

to an understanding of the process although it can by no means provide a complete picture.

This paper will concentrate on <u>pyrolysis</u>, i.e. activation by heat, but it should be borne in mind that other methods of activation viz. photolysis (with or without Lasers) and electric discharge are of great practical importance. Although each method has its own merits, all methods are mechanistically complicated. In terms of ease of interpretation <u>pyrolytic</u> studies have much to recommend them. Furthermore <u>kinetic</u> studies of pyrolysis can, in favourable cases, lead to information on the strengths of chemical bonds broken during the process of decomposition. In a purely homogeneous system, pyrolysis of ditertiary-butyl peroxide, $Me_3COOCMe_3$, offers a good example of this [1]. In this paper, the application of modern pyrolytic techniques is illustrated by discussion of recent studies of metal

8

R. M. Laine (ed.), Transformation of Organometallics into Common and Exotic Materials: Design and Activation, 8–20.
© *1988 by Martinus Nijhoff Publishers.*

alkyls [2] and metal carbonyls [3] carried out by Smith and colleagues. This work is illustrative of straightforward bond fission processes. By contrast the pyrolysis of monosilane appears initiated by a more complex molecular hydrogen extrusion reaction and this system and its complexity offers a third illustrative example. The general importance of a knowledge of bond dissociation energies [†] (BDE) is stressed throughout this paper which is intended to be illustrative rather than a detailed review of the subject.

PYROLYSIS OF METAL ALKYLS

It has been believed for some time that many metal alkyls, MR_n (where $R \equiv CH_3$, C_2H_5) decompose by simple bond fission,

$$MR_n \longrightarrow MR_{n-1} + R \tag{1}$$

The difficulty of study of this process arises from the complexity of subsequent reactions especially the potential attack on the reactant by radical R. In the early work of Price [4], radical R was trapped by toluene. In the recent study of Smith and Patrick (SP) [2] the Very Low Pressure Pyrolysis (VLPP) technique [5] is used in which secondary (bimolecular) reactions are minimised by reducing the possibility of gas-gas collisions.

Application of the VLPP method

A schematic diagram of the apparatus used is shown in Fig. 2. Reactant gases, usually in mixtures with an inert standard (SF_6 or benzene) are flowed at millitorr pressures into a heated Knudsen cell reactor with an

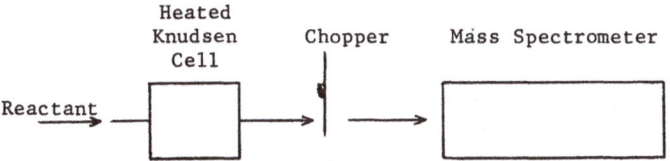

Fig. 2 Schematic diagram of a VLPP apparatus

aperture of calibrated escape rate. Gas-wall collisions activate the gas to temperature and molecules reside in the reactor for a time controlled by the exit aperture. Decomposition will occur and the reaction mixture, emerging from the cell, is analysed by a mass spectrometer using beam-chopping with phase-sensitive detection to enhance sensitivity.

The basic kinetic theory is that of the stirred flow reactor. For a unimolecular decomposition the rate constant is given by

$$k = -\frac{1}{t} \ln(1-f) \tag{2}$$

where t is the residence time measured by comparison with the escape rate constant and f is the fractional decomposition of reactant, measured by the mass spectrometer. Wall catalysis is clearly a potential risk, but may be probed and eliminated by demonstrating k to be independent of cell escape rate (by variation of exit aperture size). Among metal alkyls, SP found,

† More strictly Bond Dissociation Enthalpies

Me₃Al and Et₄Sn to be wall catalysed but others (*vide infra*) not so. Reference [5] may be consulted for further details of the technique.

Rate constants are obtained over as wide a range of temperature as possible and in the study in question, [2], for four compounds viz. GeMe₄, PbEt₃, PEt₃ and SbEt₃. An example of the results for PEt₃ is shown in Fig.3.

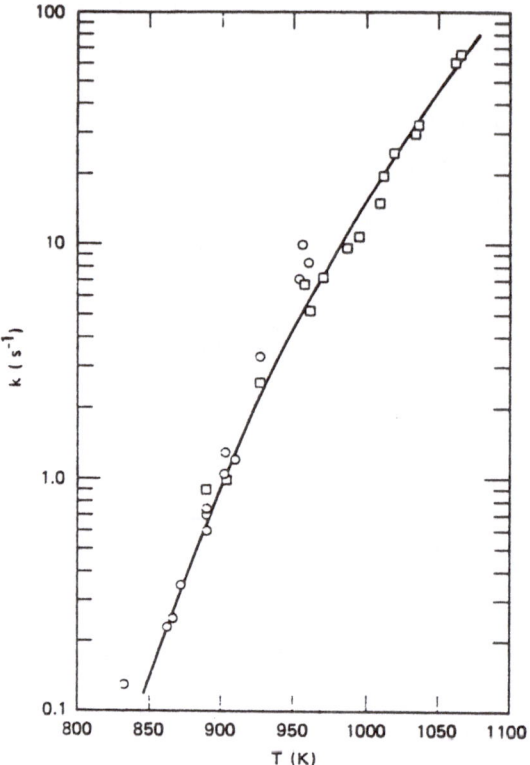

Fig. 3.   VLPP results for triethylphosphorus (reproduced with permission)

The objective is to get the BDE for the initial step (equation (1)) which, provided this step is rate determining, is approximately equal to the activation energy. However, because pressures are low, the rate constants cannot simply be identified with the unimolecular decomposition step (of a Boltzmann energised ensemble of reactant molecules). This is a, by now, standard problem of Unimolecular Reaction theory requiring application of the RRK or RRKM versions of the theory [6]. Details of this are given elsewhere but there is a certain flexibility in the theoretical fitting procedure in choosing an activated complex structure. This leads to some uncertainty (or ambiguity) in the Arrhenius parameters corresponding to the decomposition (high pressure limiting rate constant, k^∞) viz. A^∞ and E_a in equation (3):

$$k^\infty = A^\infty e^{-E_a/RT}$$

(3)

SP [2] argue plausibly for A^∞ values of ca 10^{17} s^{-1} for metal bond breaking processes and choose appropriate model structures to reproduce such values. With a rate constant spread of 10^3, the derived A factors and associated activation energies probably have uncertainties no worse than a power of 10 and ± 3 kcal mol^{-1} respectively. The BDEs are usually close to the activation energies but may be up to 5 kcal mol^{-1} higher depending on details of the activated complex structure. The results obtained for the metal alkyls are shown in Table I.

Molecule	E_a	D
$Me_3Ge - Me$	80.5	83
$Et_3Pb - Et$	53.3	54
$Et_2P - Et$	66.7	68
$Et_2Sb - Et$	56.2	57

Table I. Energetics of metal alkyl decomposition (kcal mol^{-1})

Reinterpretation of Price's work

In addition to their own measurements, SP used unimolecular theoretical modelling to rework the extensive data of Price and his group [4] again by choosing activated complex models to give reasonable A^∞ values (ca 10^{16} s^{-1}). The results of this interpretation and other measurements are included in Table II, which offers close to the current "best available" survey of main

Li ~66					
	B 112	C 83	N 76	O 83	F 109
	Al 87	Si ~72	P 77	S	Cl 84
Zn 66–69	Ga 62–65	Ge 81	As 65–68	Se	Br 71
Cd 58–61	In 49–52	Sn 69	Sb 59–62	Te	I 57
Hg 59–62	Tl 39–41	Pb 55–58	Bi 50–53		

Table II. Summary of metal–methyl bond dissociation energies (kcal mol^{-1})

group metal—methyl BDE values. Metal-ethyl BDEs are typically 3-5 kcal mol^{-1} weaker. The only values suspected by this author to be slightly in error are those for Germanium [7] and Tin (possibly up to 5 kcal mol^{-1} too high), since they imply Ge-Ge and Sn-Sn bonds (in hexa—alkyl compounds) which are stronger than the analogous Si-Si bonds. This would be at odds with normal trends within a group of the periodic table.

Outside the elements of the first row, these BDE values are invariably greater than the mean (thermochemical) bond energies. This offers support for the assumption that initial bond breaking is rate determining in metal alkyl pyrolyses. It also emphasises that average bond energies offer a poor guide to the expected decompositon conditions for such compounds. The best that can be said for average bond energies is that relative values (in a group, or period) usually mirror relative stabilities of the compounds, so that if BDEs are not available, this can be used as a rough guide line to stability. An extreme example of the kind of differences which can occur is illustrated by the first and second dissociation energies in Group II(T) as shown in Table III.

Molecule	D_1	D_2
Zn Me_2	66	22
Cd Me_2	58	13
Hg Me_2	59	3

Table III. Sequential bond dissociation
energies in Group 2(T) (kcal mol^{-1})

In this series the second dissociation energies are extremely low. This weakness arises through operation of the well known "inert pair" effect in which a pair of electrons in the outer shell s orbital are stabilised relative to p and d shell electrons.

PYROLYSIS OF METAL CARBONYLS

These compounds show some similarities to the metal alkyls, in that they are believed to decompose by rate-determining initial bond fission as in equation (1). They are, however, very surface sensitive and so their gas phase pyrolyses have to be investigated by a technique which minimises wall catalysis. Laser pyrolysis is one such technique.

The method of Laser Pyrolysis

A schematic diagram of the apparatus used by Lewis, Golden and Smith (LGS) [3] is shown in Fig. 4. Reactant gases at low pressures (ca 0.1 Torr) are carried in a stream of nitrogen through a thin cylindrical cell. A portion of the cell is irradiated with a uniform pulsed CO_2 laser beam and part of the energy is absorbed by SF_6. This causes rapid heating by

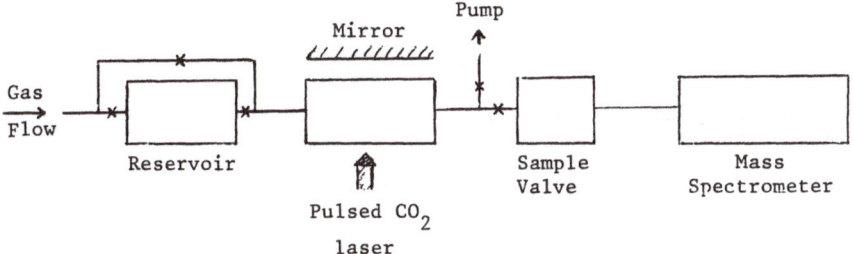

Fig. 4. Schematic diagram of Laser Pyrolysis apparatus

collisional transfer of vibrational energy and an expansion into the cooler surrounding gas occurs. Temperatures of *ca* 850 °C are achieved for reaction times of *ca* 8 μs after which a rarefaction wave causes cooling of *ca* 200 °C which effectively quenches unimolecular reaction. An idealised temperature time profile is shown in Fig. 5.

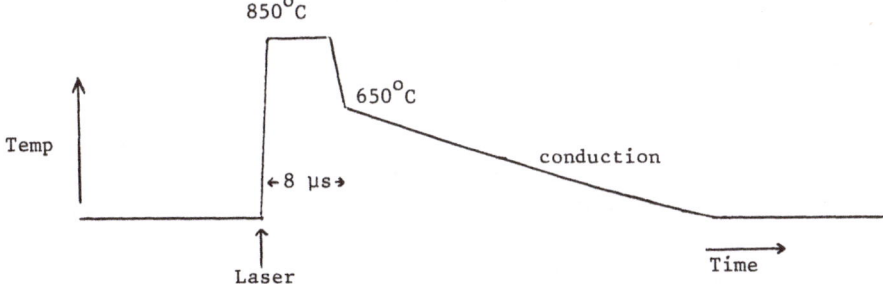

Fig. 5. Idealised temperature-time profile in pulsed laser pyrolysis

The reacted gas stream is sampled and analysed by mass spectrometry. To achieve sufficient decomposition, flow times and pulse frequencies may be appropriately adjusted. The reaction temperature is controlled by the SF_6 pressure, which is varied to achieve the desired temperature range. In addition to the reactant carbonyl, the gas mixture contains an internal standard substance, whose pyrolysis has been studied previously and which decomposes at comparable rates. It is desirable, in order to avoid potential mechanistic complications to choose a substance whose decomposition gives molecular rather than free radical products. In this study dicyclopentadiene was used. The processes occurring are as follows:

$$SF_6 + mh\nu \longrightarrow SF_6^*$$

$$SF_6^* + M(CO)_n \longrightarrow M(CO)_n^* + SF_6$$

$$M(CO)_n^* \xrightarrow{\ 1\ } M(CO)_{n-1} + CO$$

$$\longrightarrow M + (n-1)CO$$

(contd.)

$$SF_6^* + (CPD)_2 \longrightarrow SF_6 + (CPD)_2^*$$

$$(CPD)_2^* \xrightarrow{2} 2CPD$$

The basic kinetic theory is again that of the stirred flow reactor and the rate constant k is given by:

$$k = -\frac{1}{t_r} [\ln 1-((A_o/A)-1)(V_T/V_R)(t_L/t_f)] \tag{4}$$

where A_O and A are the initial and irradiated steady state reactant concentrations, V_T/V_R is the ratio of cell to irradiated volume and t_L/t_f is the ratio of time between laser shots and flow lifetime. t_r, the reaction time is not known precisely, but is common to both reactant and internal standard. Temperatures are varied and a plot is made of $\ln(k_1 t_r)$ versus $\ln(k_2 t_r)$. This plot has a slope of E_1/E_2, the ratio of activation energies. Thus for the known value of E_2, E_1 is obtained. A factors are similarly determined (using the plot intercept). This technique, in which the heated portion of the gas makes minimal contact with the wall, is similar to the Comparative Single Pulse Shock Tube method. The internal standard acts as a kind of gas thermometer and helps avoid difficulties of fluctuations in Laser fluence. Further details of the method may be found in reference [8].

Results and Interpretation

The metal carbonyl compounds studied, their activation energies and resulting BDEs are shown in Table IV.

Molecule	E_a	D
$(CO)_4Fe - CO$	40	42
$(CO)_5Cr - CO$	46	(37)
$(CO)_5Mo - CO$	39	41
$(CO)_5W - CO$	45	46

Table IV. Energetics of metal carbonyl decomposition (kcal mol^{-1})

The BDEs are calculated based on the assumption that there is no recombination activation energy. That this should be so is not immediately obvious, but LGS [3] back this assumption with molecular orbital arguments and more recently direct time-resolved experiments [9] have shown very fast rates of reaction for CO with "unsaturated" carbonyl species viz. $M(CO)_{n-1}$, implying little or no recombination energy barrier [†]. That the initial bond breaking process is rate-determining seems to be supported in three of the four cases studied, the exception being $Cr(CO)_6$. In this case the breakdown product $Cr(CO)_5$ appeared to be stable (from trapping studies with PF_3) so that in the absence of scavenger, the $Cr(CO)_5$ would partially recombine with CO on cooling

[†] In two cases reported I estimate barriers of 0 and ca 4 kcal mol^{-1} for reaction of CO with $Cr(CO)_5$ and $Fe(CO)_4$ respectively.

(between laser pulses). Thus the analyses gave rates which were too low. Nevertheless from the PF_3 scavenged system enough data was obtained by LGS to derive an approximate value for $D((CO)_5Cr-CO)$.

In spite of the difficulties with $Cr(CO)_6$ the measured dissociation energies all turn out, just as for the metal alkyls, to be greater than the mean bond energies, E, which are (kcal mol^{-1}): $Fe(CO)_5$, 28; $Cr(CO)_6$, 26; $Mo(CO)_6$, 36; $W(CO)_6$, 43. LGS [3] note that, in contrast to the first row transition metals, the second and third row examples show D not much greater than E. Maybe in these latter cases the sequential dissociation energies are more closely similar.

Support for complete dissociation of three of these carbonyls was found in the form of metal particles (Mo, W, Fe) and an aerosol was observed in one case (Fe). Obviously these decomposition conditions are unsuitable for the deposition of epitaxial layers ! The mechanism is represented disgramatically in Fig. 6.

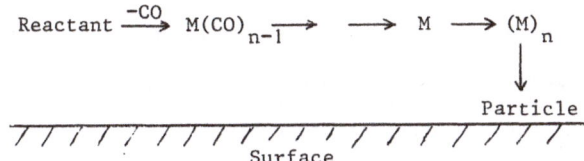

Fig. 6. Schematic diagram of metal (particle)
formation in carbonyl pyrolysis.

These EDEs are very useful, since they may be used to generate enthalpies of formation of the fragment, $M(CO)_{n-1}$. These have been further used by LGS to obtain a number of other ligand binding energies. Thus a few key measurements have yielded much ancillary information.

PYROLYSIS OF MONOSILANE

This system, of great importance to silicon and amorphous silicon hydride (a-Si:H) deposition, has been the subject of much mechanistic controversy. It differs from the pyrolyses of metal alkyls and carbonyls in that simple, rate-determining, bond-breaking is prohibitively high in energy; $D(SiH_3-H) = 90$ kcal mol^{-1} [10]. The observed activation energy (*vide infra*) is considerably lower than this.

The Static System Pyrolysis

This is one of the oldest and time-honoured methods of gas kinetics and was used by Purnell and Walsh (PW) [11] to study SiH_4 decomposition some twenty years ago. The apparatus is shown in Fig. 7. The method simply

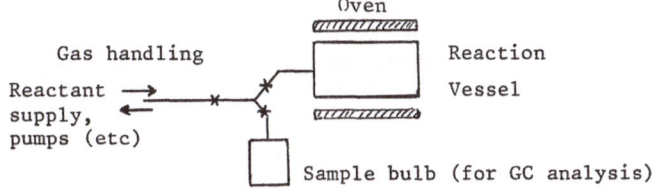

Fig. 7. Schematic diagram of static system pyrolysis
apparatus

involves sharing the reactant into a uniformly thermostatted reaction vessel at the desired pressure. After a known time the vessel contents are shared with a pre-evacuated cold bulb (which quenches the reaction) and the contents analysed, in this case, by chromatography. The point about describing this old work is not that more recent studies have not been made, but that it is still regarded as one of the definitive studies. The main focus of the study was on the _early_ stages of the reaction (before any pressure change occurs). To a good approximation the chemistry may be described by:

$$2SiH_4 \longrightarrow Si_2H_6 + H_2 \qquad (5)$$

The key kinetic observation was the 1.5 order rate law, viz.

$$-\frac{d[SiH_4]}{dt} = k_{3/2}[SiH_4]^{1.5} \qquad (6)$$

From measured values of $k_{3/2}$, the activation energy $E_{3/2} = 56$ kcal mol^{-1} was obtained.

Two mechanistic postulates have been discussed by PW [11] which might account for these results. The preferred one, mechanism (A), which has been substantially borne out by subsequent investigations [12,13] is:

$$SiH_4 \xrightarrow{3} SiH_2 + H_2$$

$$SiH_2 + SiH_4 \xrightarrow{4} Si_2H_6$$

In this mechanism step (3) has to be rate determining, but not at its high-pressure, first-order limit i.e. another example of Unimolecular fall-off [6]. Theory bears this out [12]. The molecular extrusion process, of which step (3) is an example, will probably turn out to be a common decomposition mechanism amongst organometallic compounds when more have been investigated, but it looks at present as if only small hydrogen containing molecules, such as H_2 or HX (X $\equiv$ halogen, alkyl) can be so eliminated. Part of the driving force for step (3) is provided by the lone pair stabilisation (_vide infra_) in SiH_2 which helps lower the activation energy.

It is worth considering briefly, also mechanism (B) which is _not_ supported by the evidence, since this shows up clearly the arguments on the need for reliable values of bond dissociation energies. Mechanism (B) is:

$$SiH_4 \xrightarrow{5} SiH_3 + H$$

$$H + SiH_4 \xrightarrow{6} SiH_3 + H_2$$

$$SiH_3 + SiH_4 \xrightarrow{7} Si_2H_6 + H$$

$$2SiH_3 \xrightarrow{8} Si_2H_6$$

This is a chain reaction, which gains credence in spite of the difficult initiation step (5), by the supposed rapid cycling of intermediates H and SiH_3 in steps (6) and (7). A stationary state treatment (a standard kinetic exercise †) gives:

$$E_{3/2} = E_7 + \tfrac{1}{2}(E_5 - E_8) \qquad (7)$$

†See, for example, ref. [1], chap. 10, p. 390.

where $E_{3/2}$ is the overall activation energy and E_5, E_7 and E_8 are activation energies of the appropriate steps. I estimate [14], $E_7 = 17$ kcal mol^{-1} while $E_5 = 90$ kcal mol^{-1} [10] and $E_8 = 0$ (as for all radical combinations). Thus $E_{3/2} = 62$ kcal mol^{-1} a value too high to fit the experimental data. Lest it be thought that the discrepancy (ca 6 kcal mol^{-1}) is within the error margin, it should be added that the estimated pre-exponential factor, $A_{3/2}$ is at least 10^3 too low [14]. The estimated rate for mechanism (B) is about 10^5 too slow to account for the experimental results.

This analysis has been possible only since ca 1981 when D(SiH$_3$-H) was obtained via a reliable technique [10], although it had, of course, been speculated on earlier by us [11]. The importance of D(SiH$_3$-H) for mechanism (B) lies, not only in the value it gives for E_5, but also for the light it throws on steps (6) and (7). Step (6) is exothermic (D(H-H) > D(SiH$_3$-H)) and therefore fast. Step (3) is endothermic by the difference D(SiH$_3$-H) − D(H$_3$Si-SiH$_3$), the value for which (16 kcal mol^{-1}) lead to my estimate of E_7 [14]. The value for D(H$_3$Si-SiH$_3$) [10] comes indirectly from D(H$_3$Si-H) via a knowledge of ΔH_f^o(SiH$_3$).

In a sense the mechanism for SiH$_4$ pyrolysis is still controversial, since in a recent article, Robertson, Hils and Gallagher [15] claim the process to be a wall-initiated chain reaction. PW had previously excluded this but it appears that under lower pressure conditions [16,17] some surface process can occur (but it seems more likely to involve SiH$_2$ rather than SiH$_3$ generation). As well as others [12], Jasinski and Estes [13] have confirmed PW's original (homogeneous) mechanism (A) by Laser pyrolysis, which should, as explained, avoid wall complications.

Secondary Processes

The processes which lead to solid silicon, do so via polymeric (SiH$_2$)$_n$ formed on the vessel surface. These processes, following the initial SiH$_2$ formation, are complex and not well understood. The difficulty of elucidating the complete mechanism of this decomposition is that so many of the intermediates are themselves unstable. Ring, O'Neal and co-workers [18] have made a valiant attempt to model the whole system based on current kinetic and thermo-dynamic information. The details are too involved to be discussed here but a simplified schematic diagram is shown in Fig. 8. This is undoubtedly a reaction where further progress will require the imaginative investigation of reactions of likely key intermediates such as H$_2$Si = SiH$_2$.

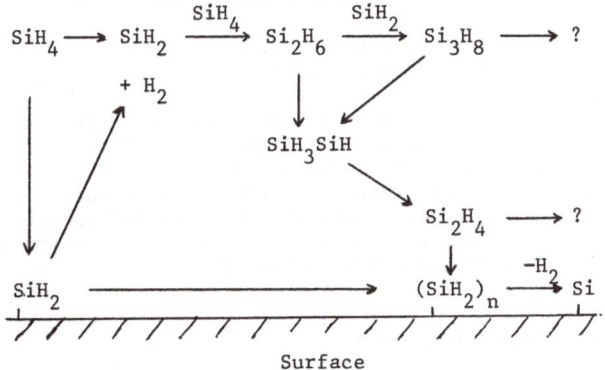

Fig. 8. Schematic representation of mechanism in SiH$_4$ pyrolysis

SOME COMMENTS ON BOND DISSOCIATION ENERGIES

It is clear from the foregoing sections that little critical discussion of pyrolysis mechanisms can go on without a detailed knowledge of BDEs. In some cases pyrolytic studies can lead to their measurement, but in many cases not. In any case alternative methods – if possible more precise – are necessary in order to assess pyrolysis. Average values for bond energies may be obtained from calorimetric determination of heats of formation, but, as we have seen, these are not a reliable guide to actual dissociation energies.

Many methods have been developed. We ourselves have been led, through initial interest in SiH$_4$ pyrolysis, to carry out a long and detailed investigation of Si-H bond dissociation energies [10] which were not, at the start of our work, very reliably known. This was based on a kinetic iodination technique which has been reviewed elsewhere [19]. This is one of the best techniques for gaseous X-H containing molecules, but may not be applicable (unfortunately) for organometallic compounds in general. Mass spectrometric techniques which, in earlier days, gave BDEs of doubtful reliability, have recently developed a whole range of more useful variants, for instance ICR (Ion Cyclotron Resonance) [20] and Guided Ion Beams [21] which look to have much potential in the area of organometallics.

For further, but brief, discussion, I single out two points from our own measurements which seem to have some generality or particular significance for CVD.

Sequential Dissocation Energies in Group IV(M) Hydrides

The most recent values for CH$_4$, SiH$_4$ and GeH$_4$ are shown in Table V. In

Bond	D	Bond	D	Bond	D
H$_3$C–H	105	H$_3$Si–H	90	H$_3$Ge–H	83
H$_2$C–H	120*	H$_2$Si–H	71	H$_2$Ge–H	57
HC–H	91*	HSi–H	≤77	HGe–H	≥60
C–H	81	Si–H	≥70	Ge–H	≤76

* Based on CH$_2$(^{1}A$_1$)

Table V. Sequential bond dissociation energies in Group 4(M) hydrides

contrast to methane, the second dissociation energies for silane and germane show a particular weakening. This demonstrates the particular stability of SiH$_2$ and GeH$_2$ and, as has been explained elsewhere [7,10], may be understood in terms of the stabilised s^2 electron pair, analogous to the Inert Pair effect. It may well be that similar energetic effects may operate to stabilise particular low valent states of transition metals which may therefore show up as (reactive) transients in pyrolysis.

The Impetus to Formation of Si-C Bonds

A comparison of selected BDE values is shown in Table VI. These are for molecules with representative C-C, Si-Si and Si-C bonds. Assuming that these are typical it may be seen that there should be a driving force towards Si-C bond formation in organosilicon compound rearrangements. In particular this explains why, at high temperature, the rearrangement of polydimethylsilane

Bond	D
$MeCH_2 - CH_2Me$	86
$Me_3Si - SiMe_3$	81
$Me_3Si - CH_3$	89

Table VI. Comparison of selected bond dissociation energies (kcal mol^{-1})

(PS) to polycarbosilane (PCS) discovered by Yajima [22] is favoured,

$$\left[\begin{array}{cc} Me & Me \\ | & | \\ Si & Si \\ | & | \\ Me & Me \end{array} \right]_n \longrightarrow \left[\begin{array}{cc} Me & Me \\ | & | \\ Si-CH_2-Si-CH_2 \\ | & | \\ H & H \end{array} \right]_n \qquad (8)$$

A surprisingly low energy kinetic pathway can assist this rearrangement. Once a radical centre has been generated at one of the methyl groups, a silicon-to-carbon migration occurs very readily, viz,

$$\qquad (9)$$

This process, unknown in analogous purely organic free radicals, occurs with the remarkable low activation energy of *ca* 22 kcal mol^{-1} [23]. Thus probably a chain reaction, in which this is one of the propagating steps, is involved in the PS → PCS rearrangement, the important prior step to Silicon Carbide preparation.

References

1. J.W. Moore and R.G. Pearson, Kinetics and Mechanism, 3rd Edn., Wiley, New York (1981) p. 234.
2. G.P. Smith and R. Patrick, Int. J. Chem. Kinet., 15, 167-185 (1983).
3. K.E. Lewis, D.M. Golden and G.P. Smith, J. Am.Chem.Soc., 106, 3905-3912 (1984)
4. S.J.W. Price, Comprehensive Chemical Kinetics, Vol. 4, Eds. C.H. Bamford and C.F.H. Tipper, Elsevier, Amsterdam, 1972, Chap. 4, 197-257.
5. D.M. Golden, G.N. Spokes and S.W. Benson, Angew Chem. Int. Ed., 12, 534-546 (1974).
6. P.J. Robinson and K.A. Holbrook, Unimolecular Reactions, Wiley, London (1972).
7. P.N. Noble and R. Walsh, Int. J. Chem. Kinet., 15, 547-560 (1983).
8. D.M. McMillen, K.E. Lewis, G.P. Smith and D.M. Golden, J. Phys. Chem., 86, 709-718 (1982).
9. M. Poliakoff and E. Weitz, Adv. Organometallic Chem., 25, 277-316 (1986).

10. R. Walsh, Acc. Chem. Res., 14, 246-252 (1981).
11. J.H. Purnell and R. Walsh, Proc. Roy. Soc., A, 293, 543-561 (1966).
12. J.W. Erwin, M.A .Ring and H.E. O'Neal, Int. J. Chem. Kinet., 17, 1067-1083 (1985), and refs. therein.
13. J.M. Jasinski and R.D. Estes, Chem. Phys. Lett., 117, 495-499 (1985).
14. R. Walsh, unpublished.
15. R. Robertson, D. Hils and A. Gallagher, Chem. Phys. Lett., 103, 397-404 (1981).
16. H.E. O'Neal and M.A. Ring, Chem. Phys. Lett., 107, 442-449 (1984).
17. J.H. Purnell and R. Walsh, Chem. Phys. Lett., 110, 330-334 (1984).
18. R.T. White, R.L. Espino-Rios, D.S. Rogers, M.A. Ring and H.E. O'Neal, Int. J. Chem. Kinet., 17, 1029-1065 (1985).
19. D.M. Golden and S.W. Benson, Chem. Rev., 69, 125-134 (1969).
20. A.E. Stevens and J.L. Beauchamp, J. Am. Chem. Soc., 103, 190-192 (1981) and refs. therein.
21. R. Georgiadis and P.B. Armentrout, J. Am. Chem. Soc., 108, 2119-2126 (1986) and refs. therein.
22. S. Yajima, Y. Hasegawa, J. Hayashi and M. Iimura, J. Materials Sci., 13, 2569-2576 (1978).
23. I.M.T. Davidson, P. Potzinger and B. Reimann, Ber. Bunsenges. Phys. Chem., 86, 13-19 (1982).

THE PYROLYTIC TRANSFORMATION OF ORGANOMETALLIC COMPOUNDS INTO REFRACTORY METALS: TUNGSTEN AND MOLYBDENUM CARBIDES

Richard M. Laine and Albert S. Hirschon
Inorganic and Organometallic Chemistry
SRI International, Menlo Park, CA 94025

ABSTRACT

The design of molecular analogs, precursors to tungsten and molybdenum carbides is described. General design criteria for the synthesis of materials precursors are developed. Specific goals were to design nonvolatile, tractable precursors containing multiple metal-carbon bonds to enhance the formation of metal carbides upon pyrolysis. Pyrolysis of $(C_5H_5)_2W_2(CO)_4DMAD$ (DMAD = dimethylacetylene dicarboxylate) at 750°C gives W_2C under appropriate reaction conditions. Pyrolysis of $Mo_2(NMe_2)_6$ at 800°C gives Mo_2C rather than the expected MoN. A rationale for the latter result is presented.

INTRODUCTION

The industrial process for the production of tungsten carbide (WC) powder is a well-studied, mature metallurgical process. Equally mature is the process for forming shapes by sintering mixtures of WC powder with cobalt binder and other refractory or "hard" metal carbides for alloy applications [1,2]. Despite the fact that the hard metal industry is well established, there is considerable room for improvement.

For example, the metallurgical process requires temperatures in excess of 1500°C to transform mixtures of finely divided carbon and particles of tungsten into WC powder [1,2]. This "carburization" step is both energy and equipment intensive. The process of forming shapes (molding and sintering) from the very hard, high-melting (>2800°C) WC powder, also requires specialized equipment and great expenditure of energy. Because the technology for refractory metals powder production and shaping has been optimized over the last 50 years, it is unlikely that revolutionary new processes will derive from the current metallurgical approach. If a revolutionary process is scientifically feasible, it is likely to arise from a totally new approach.

The object of this paper is to develop the concept that designed organometallic compounds can be used as molecular building blocks for the low temperature preparation of refractory metals from organometallic precursors. In particular, fundamental criteria are developed for the macromolecular design of the precursors and for the design of the monomeric

21

R. M. Laine (ed.), Transformation of Organometallics into Common and Exotic Materials: Design and Activation, 21–31.
© *1988 by Martinus Nijhoff Publishers.*

22

unit. Emphasis is on the development of precursors to tungsten and molybdenum carbides.

The basic concept can be stated as follows:

Given the empirical formula of a particular material, it should be possible to synthesize a chemical compound, a monomer, whose molecular formula and structural features approximate the empirical formula and structural features of the material. This monomer then represents a potential precursor to the desired material.

The potential advantages to the use of organometallic precursors are that because they can be shaped at low temperatures, as if they were simple organic materials, processing costs are diminished. Furthermore, because the bonds in most organic and organometallic compounds cleave homolytically at temperatures of 400-500°C, the temperature range avail-able for pyrolytic conversion of the shaped precursors into specific materials can be ~400-2000°C [3-6]. Therefore, one has more opportunity to control the reaction kinetics and thermodynamics of product formation than possible using current metallurgical or ceramic processing techniques.

Figure 1 shows the tungsten-carbon phase diagram [2]. The lower temperature limit shown for Figure 1 is 2200°C. Depending on the mole fraction of tungsten, two stoichiometric tungsten carbides, WC and W_2C and one non-stoichiometric carbide WC_{1-x} are found.

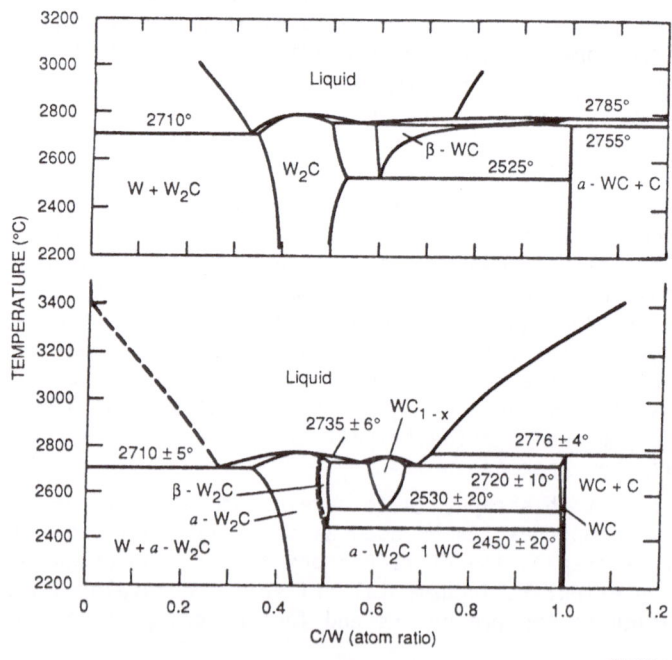

RA-2299-1

FIGURE 1 PHASE DIAGRAM FOR THE W - WC SYSTEM

The literature also reports the existence of W_3C and W_5C [1,2].
Other possible tungsten-carbon materials have not been reported. Because
the goal of materials synthesis via organometallic precursors is to work
at temperatures much lower than are required industrially, there is the
potential to work in temperature regimes that are dominated by different
kinetics and/or thermodynamics than those found in the industrial process.
Thus, ceramic product selectivities obtained using organometallic precursors may not follow those established from phase diagram studies. Indeed,
it may be possible to pyrolytically generate materials that are not thermodynamically stable at industrial process temperatures and therefore
unknown. Alternately, changes in kinetics at lower temperatures may
change not only the type of material formed but also the morphology.

On the surface, the design and synthesis of precursors to refractory
metals appears to be a trivial matter. In the case of tungsten carbides,
organometallic monomer precursors to these carbides will have at least one
carbon atom and one to five tungsten atoms. However, our studies on the
synthesis of silicon nitride preceramics [7-9] and our preliminary work on
tungsten-carbide precursors [10] demonstrates that precursor syntheses
must abide by certain macro- and monomolecular design criteria. For
example, a balance exists between the need to design a shapable or tractable precursor and the need to fix/cure the finished shape (transform or
crosslink to give an intractable material) prior to conversion to ceramic
product. Thus, in addition to the obvious stoichiometric requirements, a
number of other general criteria must be considered. The following set of
criteria discuss macromolecular requirements.

Macromolecular Design Criteria

Volatility--A ceramic precursor must remain in its final shaped form
during pyrolysis. Consequently, the molecular weight of the precursor
must be sufficient to avoid any volatilization during pyrolysis otherwise
ceramic yield (defined as the weight percent of ceramic material produced
following pyrolysis of the precursor) drops precipitously.

Tractability--The utility of a precursor depends on one's ability to
form it. In general, it should be soluble, meltable, or malleable, such
that it can be shaped. Thus, crosslinked, insoluble materials are rarely
useful since they are not readily shaped. Unfortunately, high ceramic
yields are usually obtained from highly crosslinked precursors.

Ceramic Yield--The more extraneous ligands required to make the precursor tractable, the lower the ceramic yields and the higher probability
of impurities. Stated in another way, the greater the difference between
the molecular formula of the monomer unit and the empirical formula of the
desired product, the lower the ceramic yield. This problem is amplified
for low formula weight ceramic materials such as boron nitride (BN formula
wt = 24.8 daltons) where most ligands, except hydrogen, will add considerably to the molecular weight of the precursor.

Latent Reactivity--In order to obtain tractability and high ceramic yields, a precursor must be designed in such a way that it can be cured or crosslinked (made infusible) following the forming process. This design feature is called latent reactivity.

Our studies on precursors to silicon nitride [7-9] suggest macromolecular properties such as molecular weight and extent of crosslinking control the ceramic yield but not product selectivity. In contrast, our preliminary results suggest that design of the monomeric unit strongly affects the product selectivity but not the ceramic yields. We have formulated a set of monomer design criteria, presented below, which offer some general guidelines for the design of precursors to all types of materials. These criteria are based on logic and limited experience. They have not been refined to any significant extent and are open to revision.

Monomolecular Design Criteria

The goal of monomer design is to create the most favorable situation in the precursor such that, during pyrolysis, the last bonds to break in the molecule are those between the metal and the element(s) desired in the final product. Given that very few accurate organometallic bond dissociation energies have been reported in the literature [3-5], molecular design must be based, to some extent, on experience and necessities arising from ease (or lack thereof) of precursor synthesis. The following criteria are offered in this light.

Multiple Bonding--Multiple metal element (M-E) bonds offer the opportunity to insure that the preferred nonmetallic element remains bound to the metal during pyrolysis. The multiple bond can be of the form; $M=E$, $M\equiv E$ or it can involve bridging species $M-\mu-E-M$.

Polynuclear Compounds--Volatility in precursors of the second- and third-row metals can often be suppressed by synthesizing binuclear or trinuclear species. For some applications, a monomeric, but polymetallic precursor can serve as the precursor.

Nonoxygen, Nonhalogen--In general, metal-oxygen and metal-halogen bonds are quite strong. As such, the pyrolysis of precursors containing direct metal-oxygen or metal-halogen bonds should result in the formation of metal oxides and metal halides. Consequently, precursors should be designed to avoid this likely possibility. One alternative that may reduce this likelihood is to introduce species into the precursor that form even stronger bonds. Silicon represents one useful ligand. The Si-O bond strength is approximately 128 kcal/mole [11]. The formation of this bond may actually provide an effective way of removing extraneous ligands:

$$R_3SiM-OSiR_3 \xrightarrow{\Delta} M + R_3SiOSiR_3 \tag{1}$$

Anticipation of Polymer Linkages/Latent Reactivity--In many precursor systems, volatility or the need to introduce latent reactivity requires that the system contain moieties that permit polymerization/crosslinking to occur. Quite often, the only difference between these two objectives is the extent of polymerization. A general goal is to introduce simple moieties that can be caused to react, in a well defined fashion, so that careful control over tractability and viscoelastic properties can be maintained. Furthermore, useful moieties will be those that decompose to inocuous gases upon pyrolysis. Potential moieties include esters (CO_2 evolved as pyrolysis product), hydrazines (N_2 evolved as pyrolysis product) bridging alkyl or alkenyl groups (alkanes, alkenes, alkynes evolved as pyrolysis product), etc.

Given these design principles, how can they be used to choose potential precursors to refractory metals? The following section discusses their use for selecting precursors to tungsten carbide and molybdenum nitride from compounds reported in the literature.

Precursors to Tungsten and Molybdenum Nitride

Compounds 1-4 can be used to illustrate the application of these principles in choosing a precursor for tungsten carbide. Of the four compounds, all have multiple metal-carbon bonds. In compounds 1-3, the bonds are simple carbyne/alkylidyne bonds. In compound 4, the bonding arrangement involves two tungsten atoms and two carbon atoms in a tetrahedral arrangement such that each atom bridges two of the other atoms.

$\underline{1}^{12}$ (tBuO)$_3$W≡CCH$_3$

$\underline{2}^{13}$ CH$_3$C≡W(CO)$_4$Br

$\underline{3}^{14}$ CH$_3$C≡W(CO)$_2$Cp

$\underline{4}^{15}$ Cp(CO)$_2$W ⟷ W(CO)$_2$Cp

Based on the fact that compounds 1 and 2 contain direct metal-oxygen or metal-halogen bonds and are volatile, we can rule out these complexes as precursors to tungsten carbide. Compound 3 contains no direct metal-oxygen or metal-halogen bonds but is volatile, ruling it out (except for CVD applications). Compound 4, in addition to containing multiple carbon-metal bonds is relatively non-volatile and does not contain M-O or M-X bonds. Furthermore, this air-stable compound contains two ester groups of potential value in the preparation of oligomeric derivatives. Thus, compound 4 represents a potential precursor to WC.

Our efforts to develop a precursor to molybdenum nitride (MoN) illustrate the immature state of work in the area of precursor design. Theory predicts that MoN should be super-conducting at 29.4 K [16]; however, experimentally it was found to become super-conducting at 13 K [17]. It was suggested that the current preparative techniques were inadequate for the synthesis of high purity MoN[17]. Given that Sugiyama, et al. [18] have shown that vapor-phase pyrolysis of $Ti(NMe_2)_4$ generates good purity titanium nitride, we sought to examine the feasibility of preparing high purity MoN from a nonvolatile precursor. The relatively nonvolatile compound, $Mo_2(Me_2N)_6$, which is readily prepared according to the method of Chisholm [19], appeared to be a good candidate for the preparation of MoN.

The following section describes our efforts to convert compound 4 and $Mo_2(Me_2N)_6$ into WC and MoN, respectively.

Pyrolysis Studies

Unfortunately, the pyrolysis chemistry of organometallics has received little attention and, in general, the studies that have been reported were more concerned with the organometallic products that survived pyrolysis rather than with the decomposition products [20, 21]--the sole interest of the work presented here. As such, we were unable to find experimental examples to guide our planned studies.

Because our initial goals were to reduce reaction temperatures, we decided to limit pyrolysis temperatures to under 1000°C and preferably below 900°C. Sample size was kept at approximately 30-60 mg and all pyro-lyses were initially performed in 6-mm quartz tubes sealed or open and under vacuum or under Ar or N_2. As described earlier, samples pyrolyzed in a sealed tube for various times at 750°C formed mixtures of tungsten dioxide, WO_2, and tungsten oxycarbide $W_2(C,O)$, as shown in Figure 2. If these same pyrolyses are conducted in open tubes under an Ar or N_2 atmosphere, then the major product observed by X-ray powder diffraction (see Figure 3) is $W_2(C,O)$. It was concluded that WO_2 was formed by back reaction with the gases generated during pyrolysis.

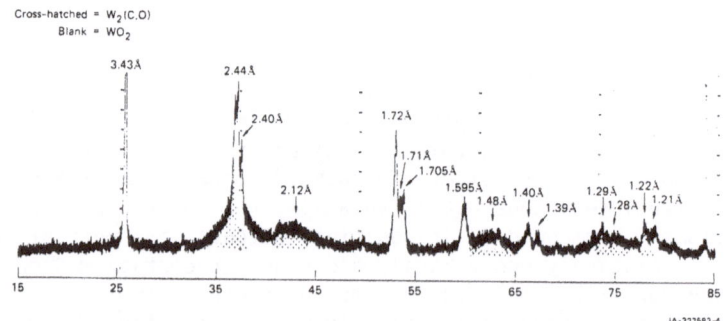

FIGURE 2 X-RAY POWDER DIFFRACTION PATTERN FOR $Cp_2W_2(CO)_4$ (DMAD) PYROLYZED AT 750°C FOR 30 MIN

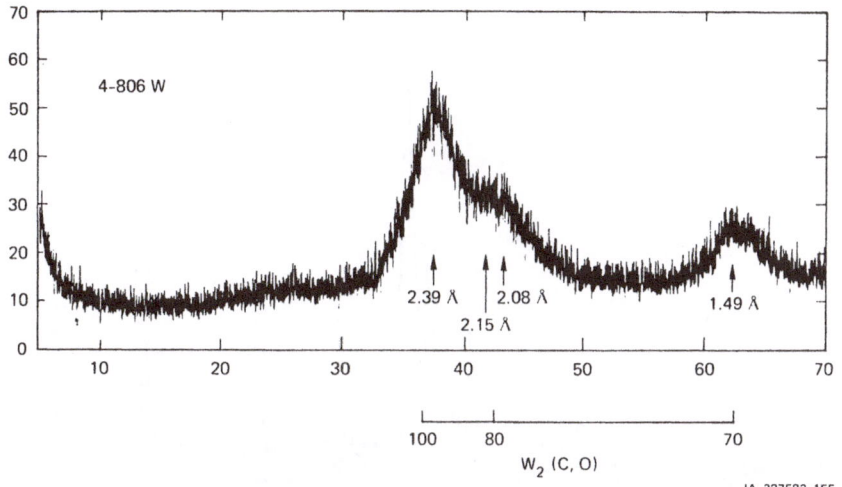

FIGURE 3 X-RAY POWDER DIFFRACTION PATTERN OF $Cp_2W_2(CO)_4$ DMAD PYROLYZED UNDER N_2 AT 750°C FOR 20 MIN.

Given the above monomer criteria, we were still surprised to observe the formation of $W_2(C,O)$ rather than WC, as one might (perhaps naively) expect from the starting material. However, reflecting on the fact that the pyrolysis process must strip the tungsten of stabilizing ligands, such as the 6CO groups per tungsten atom, we surmised that the resulting molecular species might be exceptionally reactive towards the SiO_2 oxygen in the quartz tubes. Indeed, the use of a nickel boat as carrier for 4 during pyrolysis led to a significant change in product selectivity. At 750°C, the pyrolysis of 4 in a standard nickel boat normally used for C, N, and H combustion analyses gives W_2C as the major product (identified by Auger, X-ray powder diffraction and chemical analysis, Figures 4 and 5). Traces of WO_2 were also observed in the powder diffraction pattern and chemical analysis revealed the presence of excess carbon.

Although the introduction of the nickel pyrolysis tube eliminates the formation of $W_2(C,O)$ and promotes the formation of W_2C, it does not provide an explanation for the formation of W_2C rather than WC. It has been argued that W_2C is thermodynamically more stable than WC below 800°C [22]. Again, because we are working in a temperature regime and under preparative conditions quite different from any known processes it is quite difficult to predict what products should form. The results of our studies on the pyrolysis of $Mo_2(Me_2N)_6$ only serve to illustrate this point.

Pyrolysis of $Mo_2(Me_2N)_6$ in a 6-mm stainless-steel tube at 800°C for $\sim$ 0.5-1h gives ceramic products with the X-ray powder pattern shown in Figure 6. Based on literature powder patterns, the ceramic product could be any of three products. Mo, Mo_2C, or Mo_6N_{17}, but not MoN. Surprisingly, chemical analysis of several samples, produced under similar, but not identical, conditions reveals that there are only traces of nitrogen in the

28

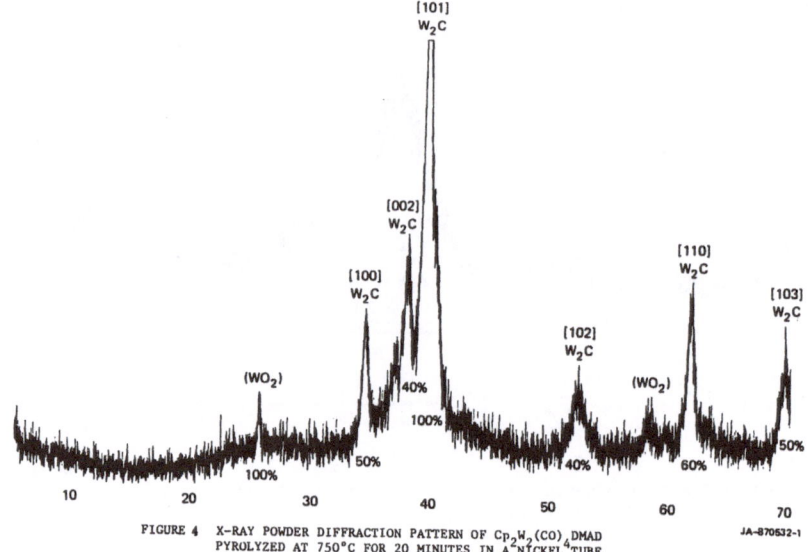

FIGURE 4 X-RAY POWDER DIFFRACTION PATTERN OF $Cp_2W_2(CO)_4$DMAD
PYROLYZED AT 750°C FOR 20 MINUTES IN A NICKEL TUBE

JA-870532-1

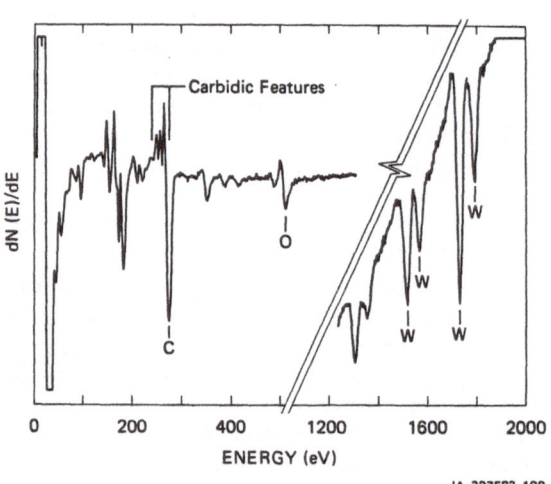

JA-327583-198

FIGURE 5 AUGER OF $Cp_2W_2(CO)_4$ DMAD PYROLYZED AT 750°C
FOR 20 MINUTES

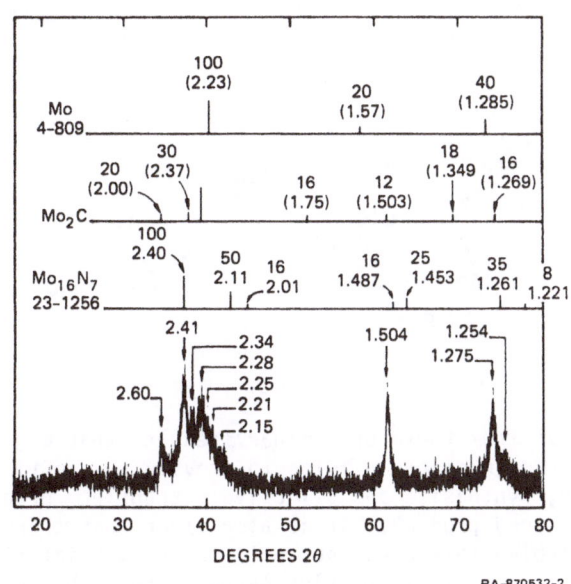

FIGURE 6   X-RAY POWDER DIFFRACTION PATTERN
OF $Mo_2(NMe_2)_6$ PYROLYZED AT 800°C
FOR 25 MINUTES

product.   More importantly, the analyses (e.g., found Mo, 83.14; C, 16.72, N, 0.39) suggest a product with an empirical formula, $MoC_{1.6 \pm 0.1}$.   This suggests that the major products are actually the Mo_2C compound together with excess carbon.   However, as we have previously noted [9], the products formed at 800°C are often only partially crystalline.   Because x-ray powder diffraction is only useful with crystalline material, the products identified by their powder patterns may not be representative of all (and in some instances, most) of the products formed at 800°C.   In addition, attempts to obtain crystalline materials by heating to higher temperatures may be counterproductive in that heating will transform the low temperature product to one commonly produced by a metallurgical process.   The value of the low temperature experiment will then be lost.

The potential problems associated with x-ray powder diffraction and heating point out an extremely difficult problem for researchers intent on establishing the value of organometallic routes to materials.   They emphasize the need for novel analytical techniques.

The fact that we obtain high yields of molybdenum carbide rather than the expected MoN points to high temperature organometallic reaction chemistry for $Mo_2(Me_2N)_6$ wherein one of the methyl groups must bind to one or both of the metals to form a strong M-C bond.   Our studies on C-N bond cleavage in metal catalyzed hydrodenitrogenation [23,24] and those of Adams et al. [25] support the formation of species such as 5 and their transformation to carbenoid derivatives such as 6, which could readily lead to the Mo_2C product.

5   6

It is clear from these preliminary results that a great deal of effort must be conducted in several fields to provide a scientific basis whereby one can design, synthesize and selectively transform organometallic precursors into a desired product. It is also clear that there is considerable value in developing this area, not only for commercial reasons, but also because it will most likely provide entree into hitherto unknown materials with unknown properties, an exciting prospect for scientists in all domains.

Acknowledgements

We would like to thank the Advanced Materials Program at SRI International and CommTech International for support of this work. We would like to thank Mr. Eldon Farley and Ms. Penni L. Lundquist for conducting the X-ray powder diffraction studies. We thank Dr. Joesph Leonelli for his generous gift of $Mo_2(Me_2N)_6$.

REFERENCES

1. K.J.A. Brookes, "World Directroy and Handbook of Hard Metals", 2nd Ed., Pub. by Eng. Dig., 1979, Hertfordshire, U.K.
2. P. Schwarzkopf, R. Kioffer, W. Leozynoki, and F. Benosowsky, "Refractory Hard Metals: Borides, Carbides, Nitrides and Silicides"; MacMillen and Co., New York, 1953.
3. G. P. Smith and R. M. Laine, J. Phys. Chem. 85, 1620 (1981).
4. K. E. Lewis and G. P. Smith, J. Am. Chem. Soc. 106, 4650 (1984).
5. G. P. Smith and R. Patrick, Int. J. Of Chem. Kinet. 15, 167-185 (1983).
6. D. McMillan and D. Golden, Ann. Rev. Phys. Chem. 33, 493-532 (1982).
7. Y. D. Blum, R. M. Laine, K. B Schwartz, D. J. Rowcliffe, R. C. Bening and D. B. Cotts in "Better Ceramics Through Chemistry II," Mater. Res. Soc. Symp. Proced., Vol 73, C. J. Brinker, D. E. Clark, D. R. Ulrich, Eds. (1986), p. 389-394.
8. Y. Blum and R. M. Laine, Organometallics 5, 2081-2086 (1986).

9. K. B Schwartz, D. J. Rowcliffe, Y. D. Blum, and R. M. Laine, in
    "Better Ceramics Through Chemistry II," Mater. Res. Soc. Symp.
    Proced., Vol 73, C. J. Brinker, D. E. Clark, D. R. Ulrich, Eds. (1986)
    p. 407-412.

10. R. M. Laine and A. S. Hirschon in "Better Ceramics Through Chemistry
    II," Mater. Res. Soc. Symp. Proced., Vol 73, C. J. Brinker, D. E.
    Clark, D. R. Ulrich, Eds. (1986) p. 373-383.

11. R. Walsh, Acc. Chem. Res. 14, 246 (1981).

12. R. R. Schrock, M. L. Listemann, and L. G. Sturgeoff, J. Am. Chem. Soc.
    104, 4291 (1982).

13. E. O. Fischer, U. Schubert, J. Organomet Chem., 100, 50-81 (1975).

14. E. O. Fischer, T. L. Linder, and R. R. Kreissl, J. Organomet Chem.
    112, C27-C30 (1975).

15. R. M. Laine and P. C. Ford, J. Orgmet. Chem. 124, 29 (1977).

16. W. E. Pickett et al., Physica B. 36, 667, (1981).

17. H. Ihara et al., Phys. Rev. B. 32, 1816-1817 (1985).

18. K. Sugiyama, S. Pac, Y. Takahashi, and S. Motojima, J. Electrochem.
    Soc. 122, 1546-1549.

19. (a)M. H. Chisholm, F. A. Cotton, B. Frenz, W. W. Reichert, and
    L. Shive, J. Am. Chem. Soc., 98, 4469 (1976);
    (b)M. H. Chisholm, F. A. Cotton, M. Extine, and B. R. Stults,
    J. Am. Chem. Soc. 98, 4477 (1976).

20. N. T. Allison, J. R. Fritch, P. C. Vollhardt, and E. C. Walborsky,
    J. Am. Chem. Soc. 105, 1384-1386 (1983).

21. R. D. Adams, P. Mathur, and B. E. Segmuller, Organomet., 3, 1258-1259
    (1983).

22. D. T. Hurd, H. R. McEntee, and P. H. Brisbin, Ind. and Eng. Chem.
    44, 2432-2435 (1952).

23. Robert B. Wilson, R. M. Laine, J. Am. Chem. Soc., 107, 361-368 (1985),
    and references therein.

24. A. Eisendtadt, C. Giandomenico, M. F. Fredericks, and R.M. Laine,
    Organometallics, 4, 2033 (1985).

25. R. D. Adams, H-K. Kim, and S. Wang, J. Am. Chem. Soc. 107, 6107
    (1985).

REFRACTORY HARDMETALS - INDUSTRIAL RELEVANCE

ERIC A ALMOND
Division of Materials Applications, National Physical Laboratory, Teddington, Middlesex TW11 0LW, UK

ABSTRACT

The origins of the microstructures of hardmetals are described in relation to the optimum mechanical properties now available and the possible improvements that could be obtained by replacing conventional manufacturing methods by new techniques using raw materials derived from organo-metallic precursors.

INTRODUCTION

Whereas the solid-state chemist can supply data on the equilibrium constitution of hardmetal systems, and the molecular physicist can predict relevant bond strengths, the materials scientist usually concentrates on the the dependence of properties on microstructure on both an atomic and macroscale. For hard materials especially, the final microstructure is decided at an early stage of manufacture since the solid state processes involved in consolidation of the constituents do not allow the changes in microstructure and composition that, for example, can be achieved in the casting and hot working of metals. Consequently, any new raw-material precursor or manufacturing route for a hard material will have a better chance of success if the inventor is aware of the microstructural and chemical effects on processing and final properties.

The most widely used hard materials (hardness above 1000HV) are WC/Co hardmetals which will be used in this paper to provide an illustration of some of the limitations that the industrial requirements place on the scope for new processes and replacement materials. The microstructure consists of hard WC particles in a Co based binder phase (Fig 1a). Typical carbide grain sizes lie in the range 1-4μm with binder phase contents of 6-20wt% to give a range of hardnesses from 1000-1800HV30 and fracture toughnesses of 10-20MNm$^{-3/2}$. Cubic carbides are added to confer crater wear resistance for cutting steels (Fig 1b), and Ni based binder phases are sometimes preferred for corrosive environments. The applications for hardmetals are mainly tools for metal cutting, rock and coal cutting, rock drilling and metal forming, and wear resistant components, but there is a very wide diversity of applications. Some examples are shown in Fig 2 [1]. The WC/Co hardmetals used in indexable inserts for machining steel are often coated with titanium nitride and titanium carbide,and sometimes with alumina.

Defining desirable properties for hard materials presents several interrelated problems: firstly, the most utilised characteristic of hard materials is their wear resistance, for which there are no truly relevant laboratory tests; whilst secondly, even the simplest properties such as strength and toughness present complications in measurement because of the difficulties in testing low ductility materials and uncertainties in the analyses for converting test results into meaningful data [2].

R. M. Laine (ed.), Transformation of Organometallics into Common and Exotic Materials: Design and Activation, 32–48.

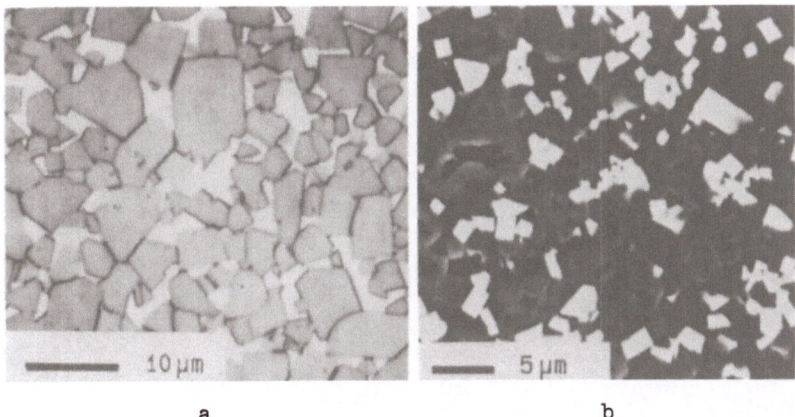

a   b

Fig 1   a) Typical angular microstructure of a 9%Co/WC hardmetal;
         b) a 58%WC-13%TaC-16%TiC-13%Co hardmetal

HARDMETALS AND CERMETS

    The term hardmetal when used in its broadest sense includes a wide range
of cermet compositions which are used for tools and wear resistant components
and consist of a high volume fraction of hard constituents in a metal binder
phase which confers toughness. The presence of the latter also facilitates
consolidation of powders of the hard constituent by enabling liquid phase
sintering to be performed at temperatures of about 1500°C whereas manufacture
of a binder-less product would require 2000°C and hot pressing, and usually
results in a coarse grained structure. The hard particles are borides,
carbides or nitrides of the transition groups IVB, VB, and VIB metals of the
periodic table, which are W, Ti, Ta, Hf, Nb, Zr, Mo, Cr, V; the basic
constituents of the binder phase are the transition group VIII metals Co, Ni,
Fe alloyed to various degrees with the hard particle's soluble constituents
and other deliberate additions. For example, the consolidation of the WC/Co
hardmetal occurs by a liquid phase sintering process which involves partial
dissolution of the WC particles into the Co. Thus single-phase Co-W-C solid
solutions can exist at room temperature within the composition limits
Co-0.05at%W-0.72at%C to Co-5.8at%W-0.5at%C [3]. These values would also
represent the solubility limits for the binder phase in the hardmetals except
that W does not diffuse fast enough to reach equilibrium and the W content
can be about 10at%, but will be lower for fine grain sizes. Outside the
solubility limits at the high and low C levels respectively, graphite and Co_3
W_3C are formed.
    The most effective crystallographic lattice for presenting resistance to
deformation is the diamond structure which has its analogue in compounds in
the zinc blende structure which is adopted by the next hardest materials,
cubic BN and SiC. Of the potential hardmetal constituents, the borides are
usually hardest, followed by the carbides, then nitrides but this distinction
applies only to metal-C, -N and -B compounds which form the tightly packed
Hagg-type interstitial structures [4] which are mainly simple hexagonal or
sodium chloride lattices with directional atomic bonding (Table I) [5].

34

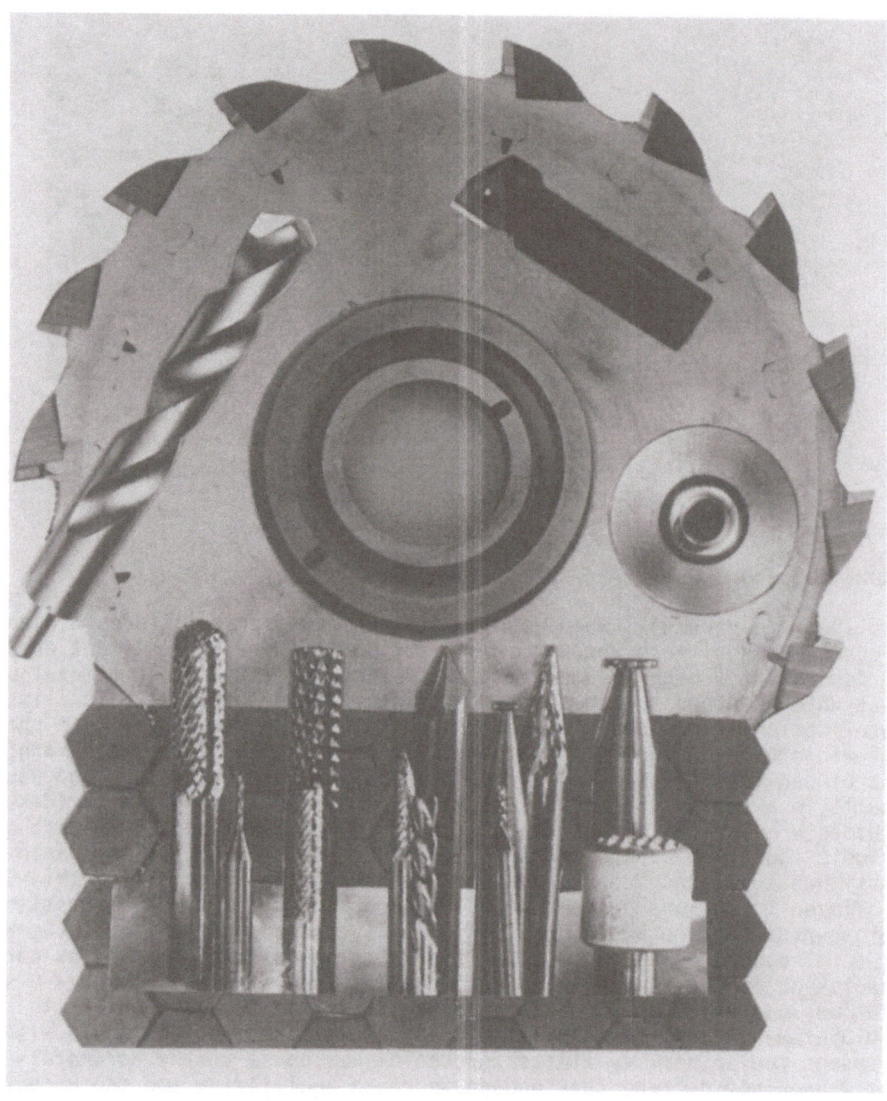

Fig 2 Examples of hardmetal tools and components: background - carbide
tipped circular saw, and wear resistant tiles (below) : lower foreground -
drills and burrs used for wood, composite boards and printed circuit boards :
centre - rotary seals : upper left - masonry drill : upper right - indexible
insert in holder, and wire drawing die

TABLE I   Pyramid Microhardnesses of Borides, Carbides and Nitrides (Quoted in
Cermets, Naukova Dumka, Kiev, USSR 1985)

	Boride	Carbide	Nitride
Cr	2100	1800	-
Hf	2900	2830	(1640)
Mo	2350	(1800)	-
Nb	2600	2600	1520
Ta	2500	2380	(1220)
Ti	3400	3170	2050
V	2800	2480	1310
W	2660	2100	-
Zr	2250	2950	1670

These values are presented solely to illustrate trends
since there are no absolute values for microhardness.

In this respect the compounds of the group VIII metals are excluded, and
oxides are relatively soft because of their non-directional ionic bonding.
These rules reflect trends in behaviour rather than scientific guidelines
since hardness is not an intrinsic property, and of equal concern are high
temperature properties, toughness and chemical inertness.

MECHANICAL PROPERTIES

Using this background information, the effect of some of the more
important microstructural parameters that are determined by processing
conditions can be examined. These will be referred to as first order effects
and arise from the defect content (porosity and inclusions), grain size and
binder phase composition.
Considering the resistance of hardmetals to compression, the hardness
and compressive strength increase as grain size decreases [6] (Fig 3).

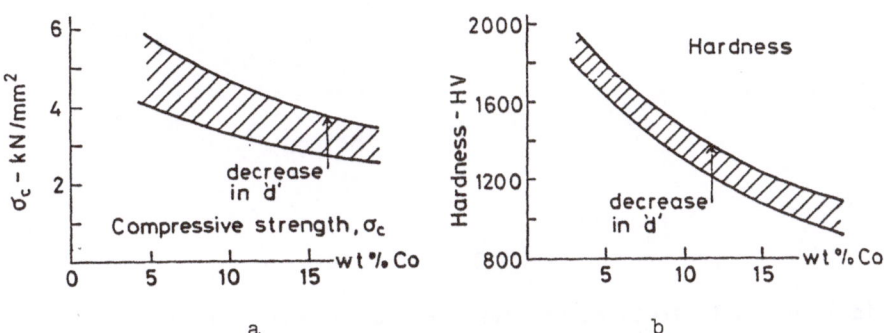

Fig 3   Effect of increasing WC grain size, d, on
a) compressive strength;
b) hardness versus Co content

The effect of grain size on tensile strength is less straightforward because of the limitations of the TRS (transverse rupture strength) test which is used to measure the tensile strength in bend. Thus the strength is calculated from the breaking load of the specimen on the assumption that failure initiates at the specimen surface at some stress level, characteristic of the material. Examination of the fracture surfaces of broken TRS specimens reveal that this is not true, since it is found that in some grades, failure in most of the specimens originates at pores and inclusions situated below the specimen surface [7]. Analysis of the results indicates that the the stress at the initiation site and the defect are related by a Griffith type dependence. If the results are depicted on a stress versus defect-area$^{-1/4}$ or defect radius$^{-1/2}$ (Fig 4), the Griffith relation is obeyed for specimens that fail from defect sizes greater than about 25µm but the remainder fail from near the surface at some maximum or limiting stress. When the specimen design is altered so that defect initiated fracture is suppressed it is found that the tensile strength increase with decrease in grain size but the dependence is much weaker than that for the compressive strength [8].

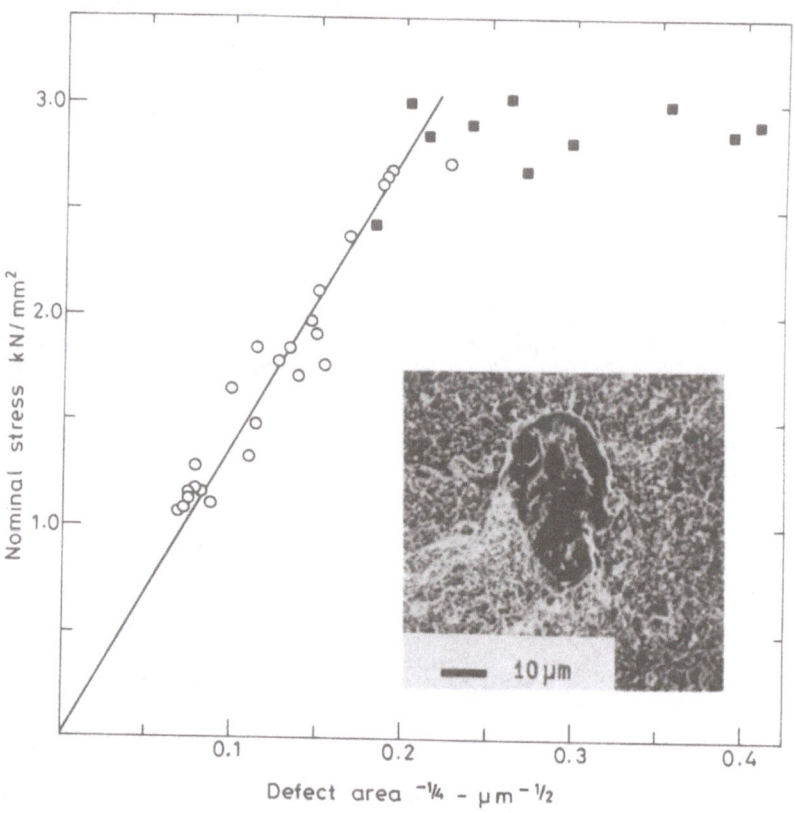

Fig 4   Nominal stress at fracture initiation site versus defect size
in transverse rupture tests on fine-grained 11wt%Co-WC hardmetal:
inset - pore at fracture initiation site

The occurrence of a Griffith dependence on defect size indicates that a fracture mechanics approach is applicable to determining the strength of hardmetals. The relevant parameter for this approach is the fracture toughness which is a measure of the difficulty of propagation of a crack or crack-like defect. As can be seen in Fig 5, the fracture toughness decreases with decrease in grain size [6] and thereby counterbalances the benefits of increased hardness that accompanies grain refinement. In fact, for both toughness and hardness, the binder phase spacing is probably the predominant factor controlling properties, and the dependence on WC grain size arises mainly from the scaling relation with binder phase spacing for a given composition. The third and less reported parameter affecting mechanical properties is the composition of the binder phase. Hardness and compressive strength increase with decrease in the C content of the hardmetal but there is a maximum in fracture toughness and in the compressive strain to fracture at a specific composition within the binder phase solubility limits [9].

To see how the more important microstructural parameters originate, it is pertinent to consider the manufacturing route, before examining other effects and developments in hardmetals.

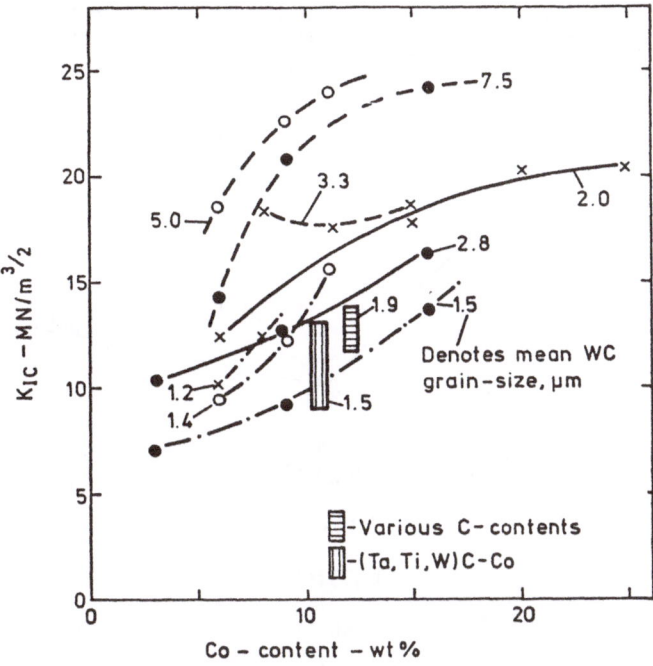

Fig 5 Fracture toughness versus Co content for various WC grain sizes

MANUFACTURE

In the conventional method for manufacturing WC/Co hardmetals (Fig 6), wolframite or scheelite ores are converted to a high purity ammonium paratungstate or tungstic acid, which in turn is converted to tungsten powder by way of tungstic oxide and controlled hydrogen reduction. The production of the tungsten monocarbide WC involves reacting the tungsten powder with carbon at 1400-2650°C. In the next stage, the Co and carbide powders are mixed by ball milling or attritor milling (Fig 7). The objective is to produce a very intimate and uniform distribution of the two constituents and it is achieved by extending the milling far beyond the time taken for mixing.

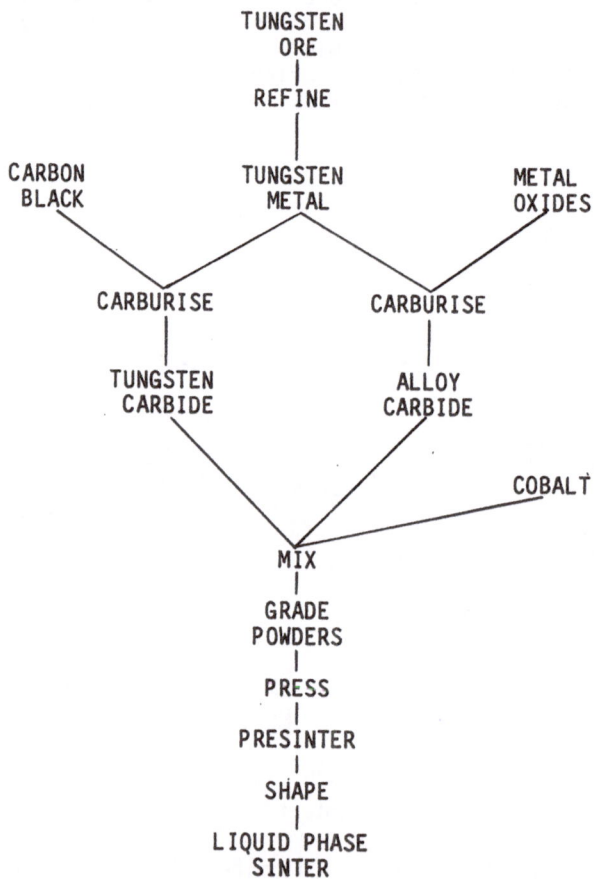

Fig 6   Flow chart of manufacturing process
for WC/Co based hardmetals

To facilitate the pressing of the powders into their final shape, they are mixed with a lubricant such as paraffin wax, and pelletised. The waxed powders are pressed into dies of the appropriate internal profile to give the final shape of the product. A dual operation of removal of the lubricant by volatilisation and of presintering is performed by heating slowly to 700-750°C, which provides a preform that can be machined either to remove defects or to provide the final shape.

The final sinter is performed at 1350-1650°C. A post sinter HIPping (hot isostatic pressing) treatment at 1500°C and 1kbar (100MPa) in argon is often used to close-up defects which may cause failure in large dies and drills, and for reducing porosity in tools such as dies and rolls which need to impart a high quality surface finish to a product.

POTENTIAL ROLES AND CHARACTERISTICS FOR NEW PRECURSOR POWDERS

There are many aspects of the manufacturing process which are relevant to the introduction of new forms of powders. Four will be examined in detail.

WC Grain-size and Grain-size Distribution

The WC grain size after carburising is determined by the tungsten grain size, and is little affected by milling which breaks down polycrystalline aggregates of WC grains but does not fragment individual grains significantly [10,11]. This milling characteristic of WC is one of the penalties resulting from the high toughness of WC which does not have a characteristic cleavage plane and displays a high resistance to transcrystalline fracture. In contrast, TiC comminutes readily and the mechanisms for particle size reduction include transgranular fracture. Nevertheless mechanical processes appear to be limited to a useful particle size diminution of about 1μm (Fig 8). Although lower values may be obtained, milling produces a wide particle size distribution and much of the lower size fraction is not suitable for processing. The particle size distribution and the WC grain size of the sintered product have lognormal distributions (Figs 9 and 10) [10,12,13].

Recently a demand has grown for WC powders and products with average grain sizes of about 0.3μm for applications in high precision machining of printed circuit boards, cutting of timber, soft metals and composites and in dot matrix printers. The considerable difficulties in producing appropriate powders by comminution, is stimulating research on their manufacture by plasma-torch [14], sol-gel and co-precipitation methods. The suitability of such powders depends not only on their average particle-size but also on their size-distribution since coarse particles may cause weakness in the sintered product while very small particles, known as fines, may produce sintering problems because of their high chemical reactivity.

Access to a source of pure WC powders with a very narrow grain size distribution would offer many possibilities for scientific studies and technological developments. If grain coarsening can be prevented during sintering by suitable additions, the product should have excellent strength and reproducible properties. The scientific interest would be in investigating the types of grain size distributions produced by sintering single size powders, and whether the contiguity of the structure is decreased for the angular WC structure, since a lowering of contiguity could produce increased toughness.

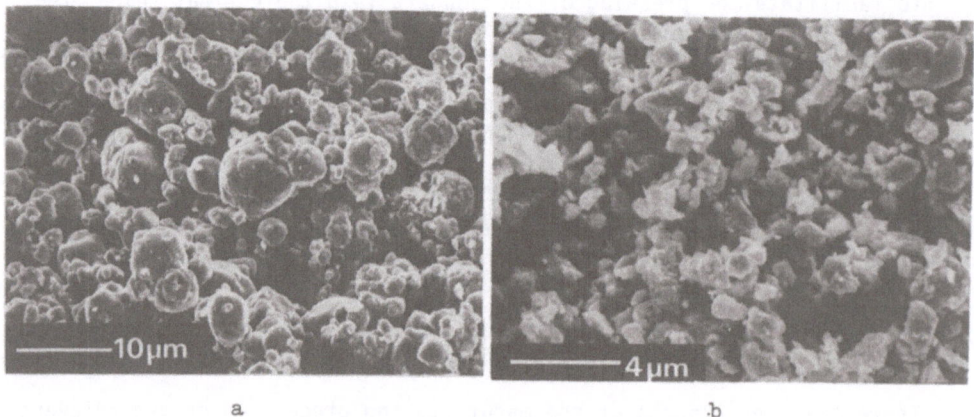

Fig 7 Coarse particle size WC powders:
a) in the as-received condition;
b) after milling for 1800 min

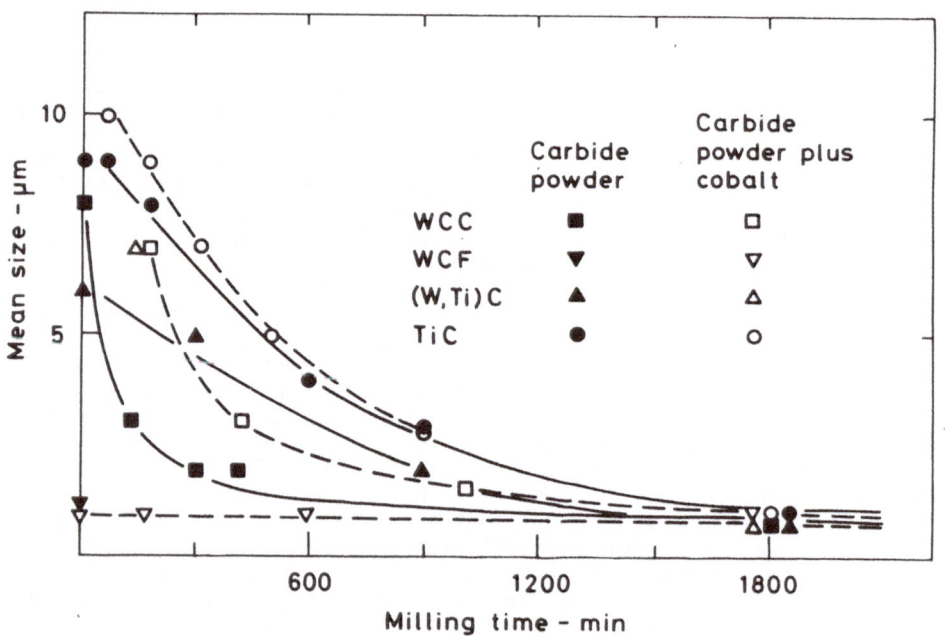

Fig 8 Mean size of particles in WC and TiC powders versus milling time

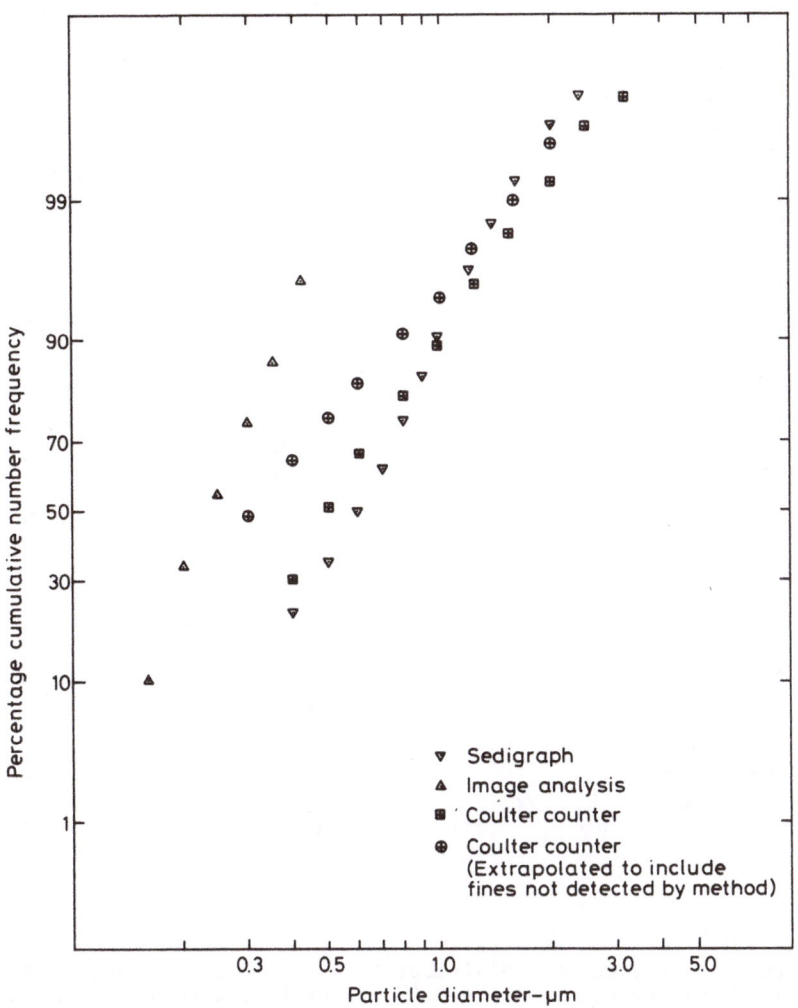

Fig 9 Cumulative frequency particle diameter distribution for WC powder: comparative measurements from image analysis, X-ray sedimentometry and Coulter counter

Consideration should also be given to current developments in new methods of processing hardmetal powders into the green compacts which are sintered into preforms. These include slip casting and injection moulding which could benefit from using powders with better flow characteristics and particle size uniformity than currently available powders.

42

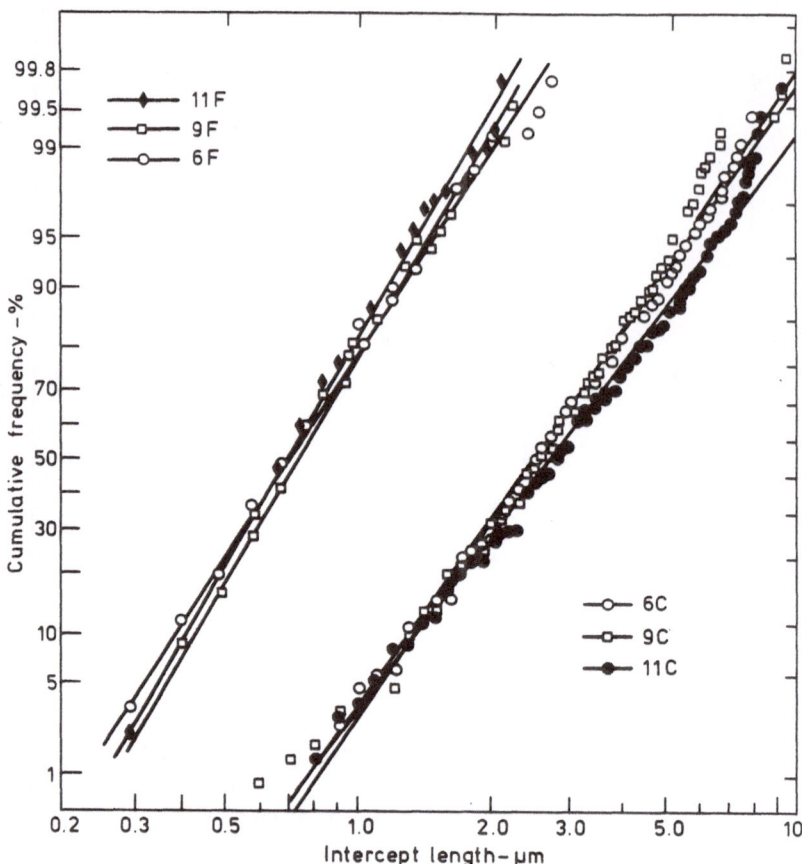

Fig 10   Cumulative frequency distribution of WC intercept lengths in
fine-grained (F) and coarse-grained (C) hardmetals with 6, 9 and 11%Co
contents

WC Substructure

Within individual particles of milled WC powder there is a dislocation
substructure which can survive sintering treatments (Figs 11 and 12) [10,11].

A similar substructure, found in TiC powders, appears to be introduced at some processing stage prior to milling. There is evidence that when the dislocations are uniformly distributed throughout the carbide grains, the mechanical properties are improved [15]. Consequently, in order to ensure that new precursor WC powders give the same or better properties than those obtained using existing powders it may be necessary to introduce some form of dislocation structure to provide some ductility in the WC structure.

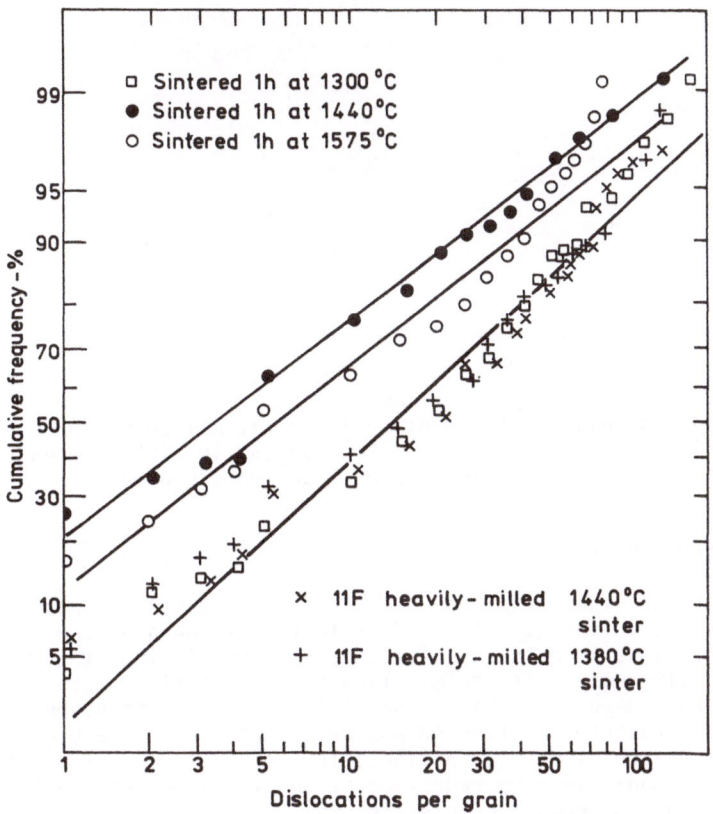

Fig 11 Cumulative frequency distribution of dislocations in WC grains in an 11%Co/WC hardmetal sintered for 1h at 1300°C, 1440°C and 1575°C

44

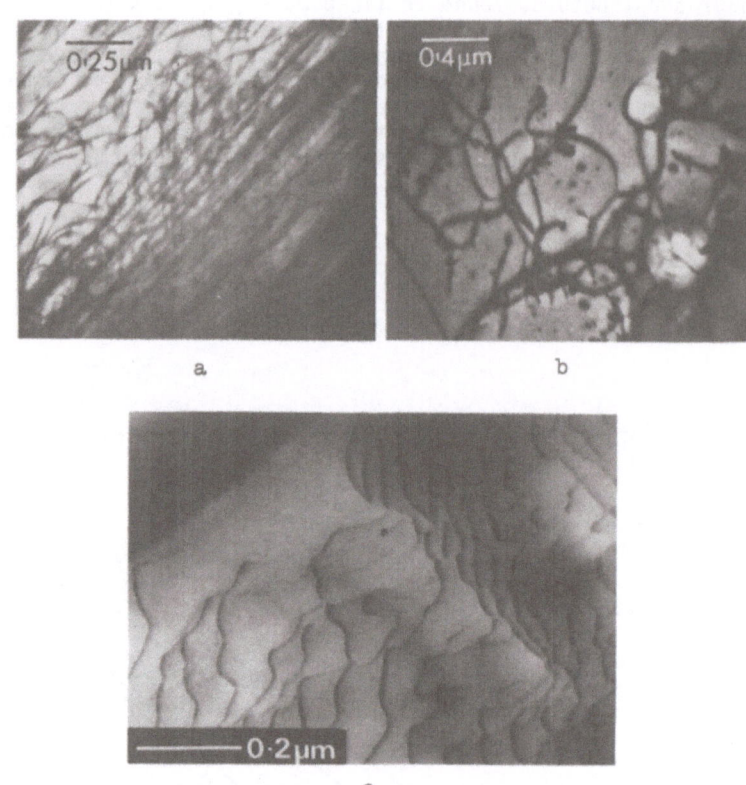

a   b

c

Fig 12   a) High dislocation density in spirit milled 11%Co/WC powder;
         b) dislocations in 11%Co/WC hardmetal in sintered condition;
         c) dislocation network in as-received TiC powder

Substitute Binder Phases

    A contributory factor to the success of Co as a binder phase in
hardmetals is the uniformity of its distribution throughout the
microstructure. This is mainly achieved during the milling stage when the Co
disperses readily because its deformation characteristics cause it to harden
rapidly and break down into smaller particles which can be dispersed
uniformly. The provision of fine Co powders is an area where organometallic
precursors have already made significant inroads in hardmetal manufacture,
and they are commonly made by reduction of cobalt oxide derived from organic
salts such as cobalt formate or oxalate [16].
    Recently there has developed a requirement for metal alloy powders for
use in Ni based binder phases to replace Co in WC hardmetals, but it is
difficult to obtain these in a small particle size and brittle form which
breaks down easily during milling and mixing.

Impurity Derived Defects

Impurities in the powders can persist or generate porosity which can lower the strength of the sintered product. According to Kny and Ortner [17], the origins of the various microstructural defects are as follows:
  i) Pores:
low carbon content, insufficient comminution, pressing defects, uneven powder density, entrapped gases, impurities and trace elements;
  ii) Large WC crystals:
trace elements, insufficient comminution, poor grain size distribution,
  iii) Mixed carbide clusters:
insufficient comminution, poor grain size distribution;
  iv) Weak interfaces:
segregated trace elements, alloying elements;
  v) Unwanted phases and inclusions:
low or high carbon contents, heterogeneous impurities, trace elements.

The elements Cr, Fe and Ni are introduced mainly during the processing as a result of wear of stainless steel vessels, attritor mill surfaces, etc; they are concentrated in the binder phase. The heterogeneous impurities, Ca, Al and Si, are initially present as ceramic inclusions. They are residues from ore processing or contaminants of hardmetal powders. The increased use of alumina-coatings frequently increases the Al-content of of recycled metal powders. The non-metals, O, N, H, S and P originate from varying sources either during processing (O, N, H) or from the raw materials (S, P). The source of P is often the W or WC powder, while S is usually an impurity in the carbon black, which is used for carburising. Typical impurity levels from these sources are listed in Table II for two hardmetals made from raw materials and one made from a mixture of recycled WC and raw materials.

Examination of the pores in fractured specimens of these hardmetals revealed that they usually contained Ca, Al and S but not Si which probably evaporated as SiS_2. It was found that large pores survived a HIPping treatment and were found to be lined with Ca, S and K. A proposed mechanism for pore formation involves conversion of Ca-containg silicates to CaS by a thermochemical reaction with S. The Si and S will be lost mainly during sintering. Some of the S will remain in the binder phase. Al from heterogeneous impurities is dissolved in the binder phase. The silicate oxygen forms CO and is partly emitted resulting in pore formation and a change of wetting behaviour of the binder phase in the vicinity of the CaS inclusion.

TABLE II   Trace Element Contents in wt% of 6%Co/WC Hardmetals manufactured from Raw Materials and from Recycled Materials

Element	Manufactured from raw materials		Manufactured from raw materials and recycled powders
Fe	0.13	0.09	0.156
Ni	0.04	0.04	0.055
Cr	0.18	0.0115	0.11
Al	0.02	0.005	0.014
Ca	0.01	0.002	0.007
S	0.0006	0.0005	0.0026
Si	0.056	0.019	0.0457
Mo	0.0055	0.016	0.0716

OTHER CONSIDERATIONS

Secondary Property Requirements

The most important practical consideration in devising a new material or a new route to manufacture is the expected benefit in terms of either cost or improvement in technical performance. Apart from the barrier of industry's need to recoup capital investment in existing practices, there are often unappreciated secondary properties that are responsible for a specific material's dominance in a particular role.

Thus when considering WC/Co hardmetals it is pertinent to refer to some of the major technological stumbling blocks which could be faced by a new raw materials, new manufacturing processes or new cermets. The good mechanical properties of WC hardmetals in laboratory tests have been referred to, and this is is evidenced by their dominance in applications needing impact toughness, such as percussion mining and hot forging, which represent the major proportion of the hardmetal market. Also, very few materials have equivalent abrasion resistance. Secondly, there are many useful ancillary properties. Amongst these are an amenability to ease of joining by brazing and diffusion bonding, and to net-shape forming with the minimum amount of finishing; a sufficiently high electrical conductivity to be machined by electro-discharge methods; a composition that is recyclable and reclaimable; physical properties that suit non-destructive quality control measurements such as predictable magnetic properties; an amenability to some degree of surface hardening and shape correction by heat treatment. When WC hardmetals show unpredictability or a poor performance in any of the above aspects the problem often originates from the presence of excessive amounts of the impurities listed in Table II and the defects they produce. This underlines the importance of keeping the impurity contents of precursor powders low, as was explained in describing the origin of defects.

Competing Materials

Although WC/Co hardmetals will probably be irreplaceable in some applications for the near future, they are facing increased competition from new cermets, ceramics and fibre-composites which could also provide possibilities for metallo-inorganic precursors [18].

Of primary interest are the superalloy compositions which fall into the two categories of Ni-Co-Cr-Mo and Ni-Cr-Mo-Al based. The former rely for strengthening on solid solution hardening while in the latter it occurs by the precipitation of the Ni_3Al, gamma prime phase. As expected the creep and probably the corrosion resistance are better than those of equivalent WC-Co grades whilst preliminary results indicate that their toughness may be superior also [19].

In Ti based hardmetals the binder phase has usually been Mo and Ni, but the recent developments have been directed towards achieving an increased toughness by using a Ti(C,N) hard phase and by increasing its cohesion to the matrix. In addition, as with WC-Co hardmetals the concept of using a superalloy binder phase has been adopted to enable improvements to be obtained in creep and corrosion resistance. Although the latter development is similar in the two classes of hardmetal the detailed chemistry of the reactions of the binder phase constituent with the Ti(C,N) is far more complicated than with WC. The hard constituent is usually a (Ti,Mo)(C,N) phase with a higher concentration of Mo and N in the outer rim of the particles, and although it appears to be tougher than TiC, the hardmetals still lack the impact toughness of the WC grades. In attempts to improve toughness further, a range of (Ti,W)(C,N) hardmetals are being developed [20].

The use of ceramics such as Al_2O_3 and Al_2O_3-TiC combinations for machining has recently been extended to include the Si_3N_4, Si-Al-O-N and ZrO_2 systems. However in their un-reinforced form, the simple ceramics are either adequately hard but too brittle, or reasonably tough but too soft to compete with hardmetals in the majority of applications. This weakness is also hindering the introduction of ceramics into high temperature structural applications.

The mechanisms for property improvements are based on particle or fibre reinforcement. The introduction of ZrO_2 into Al_2O_3 to produce toughening by transformation and microcracking has been studied extensively but other systems are now receiving attention such as TiC particles in Al_2O_3 and in Si_3N_4 [21].

Fibre research has concentrated on SiC because of its hardness and stiffness which it retains at higher temperatures than other ceramics. It is being tried mainly in Si_3N_4, Al_2O_3, ZrO_2 and Ti(C,N), and it has been used in at least one commercial cutting tool of hardness 2722 HVO.2 with an Al_2O_3 matrix.

CONCLUSIONS

Since the microstructure and properties of hardmetals are strongly dependent on the form and chemistry of the starting powders it is important to know whether new types of precursor powders can be processed into sintered products with uniform fine grain structures, and a freedom from embrittling inclusions and large pores. The products should also have the same favourable ancillary properties associated with conventional hardmetals such as amenability to joining, coating, recycling, electro-machining and non-destructive testing.

There is a requirement for: i) very fine WC powders with a narrow particle size distribution for the manufacture of high precision tooling; ii) fine metal alloy powders with similar good milling and mixing characteristics to those of Co, for the manufacture of WC hardmetals with substitute binder phases.

An advantage of new types of precursor powder is that they may be more suitable than existing powders for new manufacturing techniques involving slip casting and injection moulding.

WC hardmetals are facing increased competition from new cermets and fibre reinforced ceramics which may derive more benefit from the using new precursor powders and will probably replace WC hardmetals in some applications.

REFERENCES

1. E. A. Almond, to be published in Materials and Design

2. E. A. Almond, M.G.Gee and B. Roebuck, in "Science of hard materials" Inst. Physics CS75, Bristol, UK, 155 (1986)

3. B. Roebuck, A.M. Cottenden and E. A. Almond, Mater. Sci. Eng. **66**, 179 (1984)

4. P. Schwarzkopf and R. Kieffer, "Refractory hard metals" The Macmillan Company, New York, (1953)

5. P. S. Kisly, N. I. Bodnaruk, M. S. Borokiv "Cermets: Progress in Science and Technology Series, Ukrainian Academy of Sciences, Naukova Dumka, Kiev, USSR, (1985)

6. E. A. Almond, in "Science of hard materials" Plenum Press, London, 517 (1983)

7. E. A. Almond and B. Roebuck, Met. Sci. 11, 458 (1977)

8. B. Roebuck Mater. Sci. Eng. 44, 173 (1980)

9. B. Roebuck and E. A. Almond, to be published

10. E. A. Almond and B. Roebuck, High Temp. High Pressure 14, 143 (1982)

11. S. A. Horton, M. B. Waldron, B. Roebuck and E. A. Almond, Powder Metall. 27, 201 (1984)

12. B. Roebuck and E. A. Almond, Powder Metall., 29, 119 (1986)

13. H. E. Exner, Inter. Met. Rev. 24, 151 (1979)

14. P. Ronsheim, L. E. Toth, A. Mazza, E. Pfender and B. Mitrofonov, J. Mater. Sci., 12, 2665 (1981)

15. B. Roebuck and E. A. Almond in "Recent advances in hardmetal production" Loughborough University and Metal Powder Report, 31 (1979)

16. M. H. Tikannen, A. Taskinen and H. Taskinen, Powder Metall. 18, 259 (1975)

17. E. Kny and H. M. Ortner, in "Science of hard materials" Inst. Physics CS75, Bristol, UK, 1033 (1986)

18. R. W. Rice, Bull. Amer. Ceram. Soc. 62, 889 (1983)

19. B. Roebuck, E. G. Bennet and E. A. Almond, J. Mater. Sci. Letters 5, 473 (1986)

20. H. Doi, in "Science of hard materials" Inst. Physics CS75, Bristol, UK, 489 (1986)

21. J. G. Baldoni, S. T. Buljan and V. Sarin, Ibid, 427

USE OF AMMONOLYTIC INTERMEDIATES FOR THE SYNTHESIS OF NITRIDES AND CARBONITRIDES

LEON MAYA
Chemistry Division, Oak Ridge National Laboratory, P. O. Box X, Oak Ridge, TN 37831 USA

ABSTRACT

Liquid ammonia has been used as a reaction medium in which inorganic moieties such as metal halides or hydrides undergo total or partial ammonolysis to produce intermediates which can be converted into precursors of ceramic materials through reactions that utilize the residual halogen for substitution by additional amide ligands or acetylide. The precursors obtained in this manner have a relatively high metal content and readily convert into nitrides or carbonitrides at <800°C. Systems examined include aluminum hydride, titanium IV bromide, titanium III chloride and niobium V bromide.

INTRODUCTION

Advanced ceramic materials are chemically inert compounds with high thermal stability and mechanical strength. These characteristics make these materials attractive candidates for applications such as heat engines cutting tools and turbine blades which at present are made with expensive super alloys. Examples of these materials are SiC, Si_3N_4, TiC, TiN, VC, WC, BN. The chemical inertness of advanced ceramic materials makes fabrication of forms a difficult task and places stringent demands on the purity and morphology of starting materials. It is in this area that the contribution of chemists is required in order to design synthetic approaches that produce powders, coatings or fibers of high quality. One such synthetic approach is the use of precursors prepared at low temperatures which can be converted via pyrolytic reactions into the final ceramic materials. In a recent review [1] Rice gave a historical perspective of this approach. One of the potential advantages of using precursors is the possibility of obtaining finished forms after the pyrolysis, as it is done in the production of carbon objects. This requires that the precursor undergo a relatively small weight loss so that shrinking would be minimized.

Avoiding the introduction of impurities in the preparation of precursors is of great importance in order to realize the potential advantages of this approach. One such pervasive impurity is oxygen, in the form of oxides, thus an oxygen-free reaction medium is imperative in most cases. Liquid ammonia is a classic oxygen-free solvent which offers many advantages in preparative inorganic chemistry since it allows the synthesis of compounds that cannot be formed in other media. Liquid ammonia, similar to water, is a reactive medium towards many inorganic moieties, producing ammonolysis, a process parallel to hydrolysis in aqueous media. This characteristic can be turned into an advantage since the resulting ammonolytic products contain amide, imide and in some instances even nitride ligands. These ligands are strongly bound to the central element and persist even after pyrolysis to yield nitrides.

R. M. Laine (ed.), Transformation of Organometallics into Common and Exotic Materials: Design and Activation, 49–55.
© 1988 by Martinus Nijhoff Publishers.

The present study was undertaken to exploit the many advantages that liquid ammonia offers as a reaction medium and is illustrated by the preparation of nitrides of aluminum and niobium as well as carbonitrides of titanium and niobium.

RESULTS AND DISCUSSION

Aluminum Nitride

Simpler combinations of aluminum and nitrogen that could yield aluminum nitride upon thermal treatment are $Al(N_3)_3$ and $Al(NH_2)_3$. Earlier synthetic work on these materials was done by Wiberg et al [2,3]. In the present study the former compound was discarded because of the low anticipated yield of only 27% AlN and the potential hazards associated with azides. The amide was then chosen as a target compound. In their work, Wiberg and May [3] could prepare that compound only as a low temperature intermediate. Our study concentrated in finding alternate synthetic approaches, and the characterization of the room temperature species, which turned out to be an amide-imide $Al(NH_2)NH$. The conversion of this compound into aluminum nitride was also studied. Synthetic approaches leading to aluminum amide that were explored are described below.

Anodic dissolution of aluminum metal in liquid ammonia

A scheme based on the following reactions was tested:

$$Al^\circ + 4KNH_2 \xrightarrow[NH_3]{liq} 3K^\circ + KAl(NH_2)_4 \text{ (soluble)}$$

$$KAl(NH_2)_4 + NH_4Br \xrightarrow[NH_3]{liq} Al(NH_2)_3 + KBr \text{ (soluble)}$$

It was found that the process was feasible but the efficiency was lower than anticipated because of parasitic cathodic reactions involving the production of solvated electrons, which migrated into the anode compartment. A detailed description of these experiments is given in Ref [4].

Metathesis of aluminum bromide and potassium amide

A series of experiments were conducted based on the reaction

$$AlBr_3 + 3KNH_2 \longrightarrow Al(NH_2)_3 + 3KBr$$

The reaction proceeded as expected; however, the aluminum amide precipitate occluded some of the KBr by-product. Extensive washings with liquid ammonia yielded a product with a residual KBr content of 6.4%. The balance corresponded to $Al(NH_2)NH$. This product was characterized by physico-chemical means and also subjected to pyrolysis. Treatment to 800°C under dynamic vacuum produced the nitride according to

$$Al(NH_2)NH \xrightarrow{\Delta} AlN + NH_3$$

Ammonolysis of aluminum hydride

This is essentially the same scheme initially examined by Wiberg and May [3] and consists of the following reactions

$$3LiAlH_4 + AlCl_3 \xrightarrow{Et_2O} 3LiCl + 4AlH_3 \cdot Et_2O$$

$$AlH_3 \cdot Et_2O + 2NH_3 \xrightarrow{Et_2O} Al(NH_2)NH + 3H_2$$

The product of these scheme is essentially the same as that obtained by metathesis but in this case the impurities consist of LiCl (~1.7% by weight).

The product obtained in the ammonolytic scheme was pyrolyzed to produce AlN. In this case the pyrolysis was followed by IR spectroscopy. The N-H bending vibrations of either the NH or NH_2 moieties were used to monitor the process and revealed mechanistic pathways in the conversion of the initially polymeric material into the ionic lattice of aluminum nitride [5]. Crystalline aluminum nitride is initially detected at 600°C. Thermogravimetric analysis under He or dynamic vacuum showed that conversion, based on weight loss, is 90% at 800°C and complete at 1200°C.

Synthesis in the metal atom reactor

An additional synthetic scheme was explored using a metal atom reactor. This is a vacuum container in which ammonia vapors react with aluminum vapors evaporated from a resistively heated source. The reaction was conducted while holding the reactor in liquid nitrogen. The initial product of the reaction is a transient intermediate, stable only under an ammonia overpressure, which converts spontaneously at room temperature into very finely divided aluminum nitride. This product was initially amorphous but it crystallized upon a heat treatment.

The goal behind exploring different synthetic approaches, as described above, was to provide a choice of products with different morphologies and levels of impurities. It appears that the scheme that can be controlled best and scaled-up is the one involving the ammonolysis of aluminum hydride.

Titanium Carbonitrides

A series of ammonolytic intermediates was obtained by bringing in contact ammonia and either $TiBr_4$ or $TiCl_3$. Some of the reactions were performed in the presence of KBH_4 with the expectation of obtaining borohydride derivatives which could serve as precursors for borides. Borohydride derivatives could indeed be obtained but these were produced in low yields and required larger excesses of KBH_4. Evidently the borohydride ion cannot compete for sites in the coordination sphere of the metal in the presence of the more nucleophilic amide ions provided by the reaction medium. On the other hand and as an unanticipated effect the borohydride ion promotes the ammonolysis by reacting with the ammonium halide by-product thus preventing a back reaction that tends to solubilize the ammo-nolytic product. An additional effect of the borohydride ion, that is evident in the case of $TiBr_4$, is its ability to act as a reducing agent producing a mixed valence $Ti(IV)-Ti(III)$. The mixed-valence product of the ammonolysis of $TiBr_4$, carried out in the presence of KBH_4, is an ionic tetramer having a composition described by $[NH_4^+ \cdot NH_3]_2[Ti_4Br_4(NH_2)_{12}]^{2-}$. This assignment is based on chemical and spectroscopic evidence [6], which include the identification of ammonium ions by FTIR. The oxidation state was established both by ESCA and potentiometric titrations. The charge of the ion was established by treatment of the solid with sodium acetylide in liquid ammonia which releases acetylene in a 1:1 mole relation to the amount of ammonium ions present. This observation was validated, in an independent experiment, by treating a known amount of ammonium bromide with sodium acetylide.

The product of the acetylide treatment of the mixed valence compound produced a material with an idealized composition corresponding to $Ti_4(C\equiv C)(NH)_6$. This derivative is one of the precursors, among others to be described below, that produced titanium carbonitride upon thermal treatment to 800°C.

Titanium III chloride, similar to the other halides of $Ti(III)$, does not ammonolyze giving instead a series of solvates of which $TiX_3 \cdot 4NH_3$ is the room temperature species. On the other hand contact of $TiCl_3$ with liquid ammonia in the presence of borohydride produces ammonolysis leading to a compound having a formula corresponding to $[NH_4^+ \cdot NH_3]_4[Ti_4Cl_8(NH_2)_8]^{4-}$ which was characterized in a similar fashion as the mixed valence compound. In this case the acetylide derivative obtained had a composition corresponding to $Ti_4(C\equiv C)_2(NH)_4$. It is interesting to note that this acetylide derivative has a C/Ti mole ratio twice of that derived from the mixed valence compound. This relation parallels the halogen to titanium ratio in the corresponding reactants.

Titanium IV bromide ammonolyzes, in the absence of borohydride, producing a material with an empirical formula corresponding to $TiBr(NH_2)_3 \cdot 2NH_3$ which upon continued pumping at room temperature yields $TiBr(NH_2)_3$ [7,8]. The latter was examined in the course of the present study and was found to be also an ammonium salt in spite of the fact that in the sequence of the $Ti(III)$ derivative $Ti_4X_8(NH_2)_8^{4-}$ and the $Ti(IV)-Ti(III)$ derivative $Ti_4Br_4(NH_2)_{12}^{2-}$ it would be expected that the $Ti(IV)$ derivative would be the neutral molecular $Ti_4Br_4(NH_2)_{12}$. Evidently elimination of the ammonia of solvation leads to condensation and formation of imide bridges to yield $[NH_4^+]_2[Ti_4Br_4(NH)_4(NH_2)_6^{2-}]$. The sodium acetylide treatment of this compound produced a derivative with a composition corresponding to $Ti_4(C\equiv C)(NH)_7$.

Pyrolysis of titanium acetylide derivatives

The pyrolysis of representative acetylide derivatives was conducted under dynamic vacuum up to 800°C. The condensible volatile products were trapped in liquid nitrogen and examined by FTIR finding ammonia, acetylene and traces of HCN. Weight losses were in the range of 20-30% and the residues contained carbon in the range of 3-9%. The carbon content of the residue was proportional to the initial amount of carbon in the derivative. Micrographs of the residue showed dispersed powders with an average particle-size of about 1 μm. The x-ray diffraction patterns of these materials corresponded to that of titanium carbonitride which is very similar to that of titanium nitride. Apparently the system forms solid solutions. Raman spectroscopy gave similar results. This technique does not distinguish between the nitride and the carbide showing only very small shifts of the bands for different compositions.

Niobium Nitride and Carbonitrides

The chemistry of niobium V chloride in liquid ammonia was initially explored by Fowles and Pollard [9] and more recently by Grebtsova et al [10]. The ammonolysis proceeds to the extent of replacing two halides to yield the equivalent amount of ammonium halide and a soluble species. This was established on the basis of tensimetric measurements. Evaporation of the ammonium from the ammonolytic mixture leaves a residue with an empirical formula corresponding to $NbCl_5 \cdot 8NH_3$ which is actually $NbCl_3(NH_2)_2 \cdot 2NH_4Cl \cdot 4NH_3$. The amount of ammonia varies between 7 and 8 moles/mole Nb. Long term ammonolysis produces a precipitate of a compound having an empirical formula corresponding to $NbCl(NH_2)_2NH$.

The ammonolytic intermediates described above were of interest in the context of our efforts to obtain derivatives. In the present study $NbBr_5$ was subjected to ammonolysis both at low temperature and on a long term basis finding essentially the same results as those described above.

The ammonolytic mixture obtained after evaporation of the ammonia was subjected to thermal treatment up to 230°C under dynamic vacuum in order to separate the ammonium bromide. This treatment produced additional ammonolysis leaving a brownish-black intermediate, compound A, showing mole ratios of Nb:Br:N of 1:1.5:1.17 in agreement with Fowles and Pollard observations [9]. On the other hand Grebtsova et al [10] did not observe this intermediate apparently because of a too rapid heating rate in their thermogravimetric determination.

Compound A was characterized by a variety of techniques, including FTIR, ESCA and magnetic susceptibility measurements. It was found that the 1:1.17 Nb:N relation was significant leading to a formulation of $Nb_6Br_9N_7$ for compound A. Pyrolysis of this material liberated N_2 and $NbBr_5$ leaving a residue of Nb_4N_3.

Sodium acetylide treatment in liquid ammonia of compound A produced a derivative having a stoichiometry corresponding to $Na_5Nb_6(C \equiv C)_5(N)_5(NH)_5$. This compound converts, upon thermal treatment at 800°C into a carbonitride containing 13.7% C and 3.5% N. The calculated carbon content for a carbonitride containing 3.5% N is 7.5% thus, evidently there is an excess of carbon, to the extent of 6.3%, which is present in an amorphous elemental form and was detected by Raman spectroscopy. The sodium originally present was sublimed during the pyrolysis.

Compound A undergoes additional ammonolysis at -50°C to yield a product with an empirical formula corresponding to $Nb_6Br_5(NH)_{10}(NH_2)_5$. On the other hand, ammonolysis at room temperature produces a material with a composition corresponding to $Nb_6Br_4(NH)_{11}(NH_2)_4$.

Long term ammonolysis of $NbBr_5$ produces compound B $NbBr(NH)(NH_2)_2$. This material can also be prepared with KBH_4 as promoter of the ammonolysis in a shorter period (overnight instead of two weeks). Sodium acetylide treatment of compound B produces a derivative having a composition corresponding to $Na_4Nb_6(C\equiv C)_4(N)_8(NH)$. This material yields a carbonitride containing 6.2% C and 4.4% N upon thermal treatment at 800°C. In this case there is no free carbon in the product.

It is seen that the ammonolysis of $NbBr_5$ can provide intermediates having Br/Nb ratios of 1.5, 1.0, 0.83 and 0.66 depending on the conditions of the reaction. The residual bromide in the ammonolytic intermediates constitutes a handle to introduce carbon in the form of acetylide or some other carbanion and thus obtaining derivatives of varying carbon content.

CONCLUSIONS

Liquid ammonia constitutes a useful reaction medium as well as a reactant to convert inorganic moieties such as metal halides or hydrides, that can be obtained in very pure form, into amides or imides having residual halide. The remaining halide serves as a synthetic handle to introduce additional amide by reaction with alkali amide or to introduce carbon in the form of acetylide by reaction with sodium acetylide. The amides or acetylide-amide derivatives have a relatively high metal content and convert readily into the corresponding nitrides or carbonitrides.

ACKNOWLEDGMENT

Research sponsored by the Division of Materials Sciences, Office of Basic Energy Sciences, U. S. Department of Energy under contract DE-AC05-84OR21400 with Martin Marietta Energy Systems, Inc.

REFERENCES

1. R. W. Rice, Bull. Ceram. Soc. $\underline{62}$, 889 (1983).

2. E. Wiberg and H. Michand, Z. Naturforsch. B $\underline{9}$, 495 (1954).

3. E. Wiberg and A. May, ibid $\underline{10}$, 229 (1955).

4. L. Maya, "Aluminum Electrochemistry in Liquid Ammonia," Oak Ridge National Laboratory Report ORNL/TM-9762 (Sept. 85).

5. L. Maya, Advan. Ceram. Mat. $\underline{1}$, 150 (1986).

6. L. Maya, Inorg. Chem. (in press).

7. G. W. A. Fowles and D. Nicholls, J. Chem. Soc. 990 (1959).

8. T. Kottarathil and G. Lepoutre, J. Chim. Phys. (France) $\underline{73}$, 849 (1976).

9. G. W. A. Fowles and F. H. Pollard, J. Chem. Soc. 4938 (1952).

10. O. M. Grebtsova, V. N. Troitskii and Yu. M. Shulga, Zh. Prikl. Khim $\underline{58}$, 2181 (1985).

CLUSTER DERIVED METAL PARTICLES IN CATALYSIS. HYDROGENATION, ISOMERI-
ZATION AND DEHYDRATION REACTIONS ON (η^5-C_5H_5)NiM$_3$(μ-H)$_3$(CO)$_9$ (M =
Ru, Os).
EFFECT OF THE METAL STOICHIOMETRY IN HOMOGENEOUS AND HETEROGENEOUS
REACTIONS AND SYNERGISTIC BEHAVIOUR OF THE SUPPORT.

Mario CASTIGLIONI[*], Roberto GIORDANO[*], Enrico SAPPA[*], Eugenio GUGLIEL-
MINOTTI[**], Giancarlo ALBANESI[§], Pietro MOGGI[§] and Giovanni PREDIERI[§§].

[*] Istituto di Chimica Generale ed Inorganica, Università di Torino,
    Corso Massimo d'Azeglio 48, 10125 Torino, Italy
[**] Istituto di Chimica Fisica, Università di Torino, Corso Massimo
    d'Azeglio 48, 10125 Torino, Italy
[§] Istituto di Chimica Organica, Università di Parma, Via Massimo
    d'Azeglio 85, 43100 Parma, Italy
[§§] Istituto di Chimica Generale ed Inorganica, Università di Parma,
    Via Massimo d'Azeglio 85, 43100 Parma, Italy

Summary
     The title complexes catalyze the hydrogenation of unsaturated
hydrocarbons under homogeneous conditions showing considerable
selectivity; the homometallic cluster fragments catalyze isomerization reactions on the same substrates.
     These clusters are precursors of heterogeneous catalysts, which
show better activity and selectivity with respect to systems formed
by homometallic clusters or by metal salts.
     The support material used in some experiments has a synergistic
effect on the reactions by acting as a dehydration catalyst.

     Homo- and hetero-metallic carbonyl clusters have met only
limited success as homogeneous catalysts [1]; this is due to (i) the
limited stability of most clusters especially under the severe
reaction conditions required, and (ii) to the difficulty of identifying the catalytically active intermediates and proving unambi-
guously that "intact clusters" act as catalyst. Thus, the better
selectivity expected for cluster catalysis cannot be fully exploited
[2]. There is however, some evidence for cluster catalysis [3] and
also for synergistic effect of different metals [4] under homogeneous conditions.
     Attempts to heterogeneize clusters, by impregnation or by
anchoring on supports, have been performed with the aim of improving
control of heat exchange, separation of the catalyst from the products [5], as well as selectivity. However, in most cases, partial
or total decarbonylation occurs during the catalytic conditions;
polynuclear sub-carbonyl species or cluster derived metal particles
(CDMP) are formed [6].

R. M. Laine (ed.), Transformation of Organometallics into Common and Exotic Materials: Design and Activation, 56–75.
© 1988 by Martinus Nijhoff Publishers.

CDMP frequently behave as heterogeneous catalysts and show remarkable activity and selectivity when compared with systems obtained from metal salts. Indeed, carbonyl clusters offer two major advantages as catalyst precursors; (i) the metals do not require reductive activation and (ii) the extremely small particles formed display activities and selectivities different from those associated with larger crystallites. Moreover, the heterometallic clusters give metal particles of uniform composition, not easily obtainable with conventional methods [7] and different metals may exert synergistic effects [8][9].

Here we report examples of synergistic effects of different metals either under homogeneous or heterogeneous conditions; we also discuss a synergistic effect of the support and of the CDMP in hydrogenation-dehydration reactions. The heterometallic clusters $(\eta^5 - C_5H_5)NiM_3(\mu - H)_3(CO)_9$ (M = Ru complex $\underline{1}$ [10]; M = Os complex $\underline{2}$ [11]) have been considered.

We have already reported that complex $\underline{2}$ is homogeneous "intact cluster" catalyst for the selective hydrogenation of terminal double bonds of dienes and for the isomerization of pent-1-ene and non-conjugated dienes [12]; we hypothesized that the cluster acts as a hydrogenation catalyst, and that $HOs_3(CO)_9(MeC.CH.CMe)$, found in small amounts as a side-product in the reaction solutions, acts as an isomerization catalyst or precursor [13].

A short account of the catalytic activity of $\underline{2}$ when supported on Al_2O_3 and activated thermally has also been given; we observed methanation of CO and CO_2 under relatively mild conditions, with considerable activity and selectivity [14]. Here we report on attempts to identify the nature of the CDMP formed upon thermal treatment of $\underline{2}$; these met with limited success. However we could compare the activity and selectivity of the heterogeneous catalyst formed from $\underline{2}$ with those of other systems obtained from clusters containing different Ni/Os stoichiometries and activated under the same conditions. We supported the following clusters on Al_2O_3 $(Cp)_2Ni_2(CO)_2$ (complex $\underline{3}$), $(Cp)_3Ni_3Os_3(CO)_9$ (complex $\underline{4}$) [15]; $H_2Os_3(CO)_{10}$ (complex $\underline{5}$) and $Os_3(CO)_{12}$ (complex $\underline{6}$) and we also impregnated Al_2O_3 with $Ni(NO_3)_2.6H_2O$ and $OsCl_3$. We found that complex $\underline{2}$ forms the best catalyst, and that, especially in the hydrogenation of CO_2, the behaviour of the CDMP from homo- or hetero-metallic clusters is considerably different.

A detailed examination of the behaviour of acetone in the presence of supported CDMP derived from complex $\underline{2}$ is reported here; a synergistic effect of the CDMP, which are active hydrogenation catalysts, and of the g.l.c. support material, which acts as a dehydration catalyst, is discussed [16][17].

Preliminary results on the behaviour of cluster $\underline{1}$ supported onto chromatographic materials are also given. The cluster forms an active catalyst for the hydrogenation of cyclic monoenes and dienes, and for the demethylation and C-C cleavage of benzene and toluene.

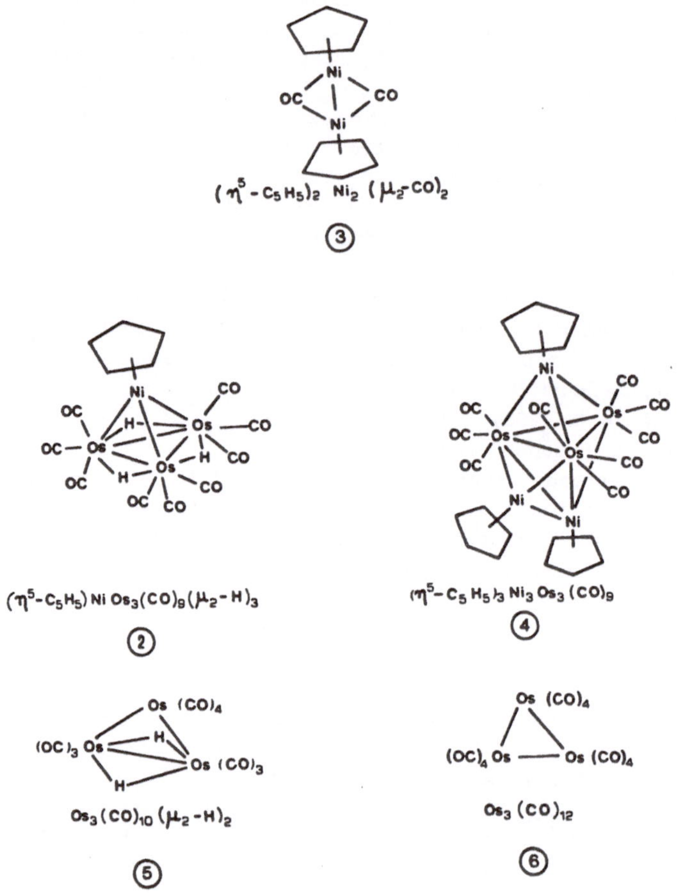

$(\eta^5\text{-}C_5H_5)_2\ Ni_2\ (\mu_2\text{-}CO)_2$

③

$(\eta^5\text{-}C_5H_5)\ Ni\ Os_3(CO)_9(\mu_2\text{-}H)_3$

②

$(\eta^5\text{-}C_5H_5)_3\ Ni_3\ Os_3\ (CO)_9$

④

$Os_3(CO)_{10}(\mu_2\text{-}H)_2$

⑤

$Os_3(CO)_{12}$

⑥

Figure 1. Structures of complexes 2-6.

Experimental section.

    Materials. Complex 1 was prepared using modification of establi-
shed procedures [10] [18], consisting of treating $Ru_3(CO)_{12}$ and
$(Cp)_2Ni_2(CO)_2$ under hydrogen in cellosolve at 95°C for 3 hours
(Yields of 75%); t.l.c. purification, dissolution of the product in
hexane-chloroform and cooling to -10°C under nitrogen afforded a
microcrystalline powder of 1. Complex 2 was obtained and purified as
previously reported [11a] , the clusters 4 15 , 5, 6 [19] were also
obtained by established procedures.

    The organic substrates, dienes, monoenes, benzene, toluene,
acetone, isopropanol, di-isopropyl-ether and other authentic samples
for blank experiments were commercial products and were used as
received after checking their purity using [1]H n.m.r.

    Gaseous reactants were purchased from Matheson and used as

received; ultrapure N_2, H_2, CO, CO_2 or deuterium were used as reactants or carrier gases.

Homogeneous hydrogenations in the presence of cluster 1. These reactions were performed in the same solvent, under the same conditions and temperature, and with the same procedures as previously described for cluster 2[12] ; also, deuteriated 1 was obtained and purified following the same procedure as for 2 (D_3), relevant differences being a shorter reaction time (40 min) and a lower yield (85%) for 1 (D_3) with respect to the osmium homologue. The [1]H n.m.r. spectrum of 1 (D_3) indicated a deuteration yield of ca. 99.8% [20].

The organic products in the solutions after the isomerization and hydrogenation experiments were analyzed with a Carlo Erba 4200 FID gas-liquid chromatograph.

The nature of the deuteriated organic products and of the unknown liquid or gaseous reaction products in the homogeneous and heterogeneous experiments was detected by means of a Carlo Erba 4200 FID and Kratos MS-50 gas-chromatographic mass-spectrometric system and, in part, with a JEOL JNM GX 270 FT n.m.r. instrument.

The solutions were also checked both for decomposition of the cluster and for formation of new complexes. Indeed, complex 1 decomposes extensively after only 20–40 min reaction time (at 120°C), giving mainly insoluble (inorganic of polymeric) materials and very small amounts of organometallic derivatives (intermediates or side-products); their identification was based mainly on the comparison of i.r. spectra with those of known products and on mass spectrometric analyses.

Heterogeneous hydrogenations in the presence of CDMP derived from clusters 2-6 and metal salts. The support material used in all the experiments was obtained from a lot of microspheroidal (6-8 mesh) γ-Al_2O_3 (Snam Progetti); this was crushed and sieved and the 40-80 mesh fraction was collected and used after drying at 120°C in a thermostated oven.

Chloroform, dichloromethane or THF solutions of the clusters, depending on their solubility, were added to accurately weighed portions of alumina in order to obtain a final catalyst with less than 2% of metals (Os+Ni) of the total weight. The solvent was then evaporated at room temperature in vacuo and the dry material further sieved; the above operations were performed under dinitrogen. The supported clusters were then heated at temperatures in the 100-300°C range, in a dihydrogen current for several hours.

The catalyst from metal salts was prepared by dissolving $Ni(NO_3)_2.6H_2O$ and $OsCl_3$ in distilled water. After addition of alumina, the mixture was dried in vacuo and further in an oven at 70°C for 4 h; after sieving, the catalyst was heated at 230°C for 2.5 h and further at 400°C for 12.5 h as described in the literature [21].

Analysis of the content of metals after thermal activation. The Electron Prode analyses on golden plated samples (and on standards containing 2-10% total weight of metals) were performed with a Cambridge Stereoscan 250 instrument.

The AA analyses were performed with a Perkin Elmer 303 instrument (air-acetylene flame) calibrated with standard solutions containing 1-10 ppm of metal. The catalytic material, accurately weighed was treated with 37% HNO_3; the solution was filtered and diluted with distilled water, in order to bring the metal concentration to about 5 ppm.

The Ion Coupled Plasma spectroscopy analyses were performed with a Leeman Lab ICP 2.5 Plasma Spec on samples and standards with concentrations in the 1-15 ppm range, by using the 231.60 nm line of Ni and the 225.58 nm line of Os.

The analytical results for the different catalysts are collected in Table 1.

Table 1

Metal content of the samples

Starting cluster Treatment	% Ni	%Os	% (Ni+Os)	Os/Ni molar ratio
2 room T	0.085	0.83	0.92	3.0
180°C, 7h	0.050	0.63	0.68	3.9
200°C, 3h	0.045	0.51	0.55	3.5
300°C, 4h	0.041	0.51	0.55	3.8
3 room T	0.111	-	0.111	-
4 room T	0.190	0.61	0.80	1.0
300°C, 4h	0.015	0.17	0.19	3.5
5 room T	-	0.96	0.96	-
6 room T	-	1.04	1.04	-
Ni/Os salts[a]	0.049	0.74	0.79	4.7

[a]Activated as described in the text.

Both samples containing 2 and 4 undergo loss of metal and modifications of the Os/Ni ratio at increasing temperatures. These modifications can be explained by the preferential volatilization of nickel in the dihydrogen current. Similar behaviour has been discussed elsewhere for the synthesis of nickel containing heterometallic clusters [11a],[11b].

Hydrogenation experiments in a pulse microreactor. The catalytic activity tests were performed with a pulse microreactor, charged with about 0.3-0.4 g of the catalyst, operating in a current of dihydrogen (1.9-2.4 atm, 40 ml/min, space velocity about 2400 h^{-1}).

Injections of 0.1–1.0 μl of the gaseous substrates or of 1.0–10.0 μl of the liquid substrates were made and the products obtained were detected in the effluent gas with a DANI 3400 gas-chromatograph equipped with a 2 m x 5 mm i.d. Porapak T (150–200 mesh) column, a thermal conductivity detector and a Hewlett Packard 3373B integrator. The catalytic activity (C.A.) was expressed as moles of reagent converted per gram atom of metals (Ni+Os).

Attempts at identifying the thermal decomposition products of cluster 2. (a) Unsupported cluster. Thermogravimetric high resolution gas-chromatographic experiments in a stream of nitrogen showed gradual loss of carbonyls; however we could not obtain quantitative results because part of the complex sublimes unaltered in the N_2 stream and can be recovered in the g.c. columns. Also, traces of C_5H_6 could be detected (g.l.c.-mass spectrometry); cyclopentadiene dimers were not observed, probably because these are retained by the g.l.c. columns. A DTA analysis under nitrogen (Mettler TA 3000) showed a reversible endothermic modification at 166.6°C followed by a less intense change at 172°C. Infrared analyses on the original cluster and on the samples heated under nitrogen in the DTA experiments and rapidly quenched showed only slight modifications in the CO stretching patterns (KBr disc, Nicolet FT-IR instrument). The samples heated and allowed to cool slowly gave i.r. spectra identical to those of the original cluster.

(b) In the presence of alumina. As discussed in the section above, (Table 1) heating of 2 after impregnation on Al_2O_3 in the presence of a hydrogen stream results in partial loss of nickel. Attempts were made to detect the size and composition of the particles by using Scansion Electronic Mycroscopy methods (Cambridge Stereoscan 250 instrument). A uniform distribution of the metal content was observed on samples of the catalyst after reaction. Unfortunately, the size of the particles is beyond the resolution of the system and too small to allow their detection.

Attempts to "mimic" the cluster-alumina interactions were also made: treatment of 2 in sealed vials with small amounts of water (under vacuum) for 96 hours at 120°C led to partial decomposition to metals. Some intact cluster could also be recovered. The same reaction in the presence of alumina gives decomposition to metals and trace amounts of $H(OH)Os_3(CO)_{10}$ (comparison with original samples) [22]. I.r. studies in a vacuum cell (Perkin Elmer 580 B) were also performed on Halon C disc impregnated with 2 (40 mg of 2 on 2 g of Halon C) [23]; treatment with 1 atm of ultrapure hydrogen from room temperature up to 200°C resulted only in minor modifications. At 250°C however, considerable amounts of water and carbonates were formed. Similar runs, in the presence of CO/H_2 (1:1) at temperatures below 150°C, gave minor modifications; above 150°C however, spectroscopic manifestations typical of $Os_3(CO)_{12}$ or of $H_2Os_3(CO)_{10}$ chemisorbed and heated onto alumina were observed [24] [25] and at 300°C Os(0) and Os(II) superficial species could be observed. The above samples, after pumping off the gas, readily readsorbed CO; however, a band attributable to CO on nickel [24] disappeared irreversibly after pumping off the gases.

Heterogeneous hydrogenation-dehydration of acetone in the presence of CDMP derived from 2. These studies were performed by using the gas-chromatographic catalytic-analytic device previously described [16],[17]. The catalytic column used was formed by a 1 m x 0.6 cm i.d. glass coil filled with 12 g of silanized Chromosorb P (J. Manville) 60/80 mesh, previously wetted with a light petroleum solution containing 35 mg of 2, and dried under reduced pressure at room temperature. The column was then heated at 155°C for 25 hours in a dihydrogen current (25 ml/min), while carefully controlling the possible sublimation of part of the cluster; previous experience 16 indicates that, under these conditions, 2 is totally decomposed to metal particles. Extraction of this material with $CHCl_3$ shows the absence of any soluble products; very limited hydrogenating activity was observed using pure silanized Chromosorb P at temperatures up to 230°C.

In order to evaluate the effect of the chromatographic support material on the dehydration reactions, in some experiments simple Chromosorb P columns were substituted for the catalytic one; these were operated under N_2 or H_2 under the same reaction and analysis conditions used for the catalytic column, as described below.

Gas-chromatographic catalytic experiments and analysis of the products. The catalytic column was thermostatted in an oven and connected with the analytic gas-chromatographic apparatus via a thermostatted junction; the analytic instrument, acting also as a flow regulator, was a Carlo Erba 4200 FID and HWD operated with a 2 m x 0.6 cm i.d., n-octane/Porasil C 80/100 mesh column, with a H_2 flow of 25 ml/min.

Under the conditions described the retention times on the catalytic column were of about 70 sec. Unless otherwise specified the substrates were injected as vapours; glass gas containers equipped with a rubber septum were evacuated and accurately freed from moisture and the liquid substrates were then introduced in amounts sufficient to form saturated vapour phases in equilibrium with the residual liquid. Variable amounts of the vapours were transferred with gas syringes while checking accurately the temperature of the vapour before each injection.

Results and discussion

Homogeneous hydrogenation and isomerisation in the presence of 1. A comparison with the behaviour of cluster 2. For an evaluation of the hydrogenating properties of cluster 1 we chose as substrate cis-1,3-pentadiene, which had been extensively studied in the presence of 2 [12]; the same solvent, temperature (120°C) and procedures were used. The results of the hydrogenation experiments with 2 are shown in Table 2.

Complex 1 shows a higher reaction rate (a turnover of 177 after

Table 2

Hydrogenation reactions for cis-1,3-pentadiene in the presence of 1

Experiment	Reaction time (min)	Turnover[a]	Per-cent composition of the reaction mixtures (organic products)					Extent of cluster decomposition[b]
			pentane	1-pentene	t-2-pentene	c-2-pentene	substrate	
1[c]	10	1	–	tr[d]	–	–	99.9	80%
	20	2	–	0.3	–	–	99.6	
	30	4	–	0.7	–	–	99.3	
	40	7	–	1.1	–	–	98.8	
2[e]	10	15	0.2	0.2	1.4	0.4	97.8	70%
	20	48	0.2	1.2	5.2	1.4	91.9	
	40	117	0.2	2.1	14.6	3.6	79.4	
3[f]	20	44	0.2	0.6	5.0	1.7	92.5	50%
	20	20	0.2	0.2	1.9	0.8	96.9	
	20	15	0.2	0.2	1.1	0.7	97.7	
	20	10	0.2	0.2	0.5	0.4	98.6	
4[g]	20	105	0.3	1.9	12.7	3.6	81.5	40%
	20	160	0.3	3.8	18.6	5.6	71.6	
	20	221	0.6	5.3	23.4	10.1	60.6	
	20	226	1.4	7.3	18.4	13.1	59.8	
5[h]	10	8	tr	0.1	0.5	0.15	99.2	100%
	20	23	–	0.4	2.3	0.8	96.5	
	30	19	–	0.3	2.0	0.6	97.1	
	40	16	–	0.2	1.6	0.5	97.7	

(a) moles of substrate reacted per mole of 1: (b) the degree of decomposition is the maximum observed for long reaction times: (c) in the absence of hydrogen: (d) tr = traces: (e) under 0.9 atm of dihydrogen: (f) hydrogen pressure = 1.00, 0.75, 0.50, 0.25 atm respectively for the four entries: (g) diene concentration = 1.70, 0.85, 0.42, 0.21 mol l^{-1} respectively for the four entries: (h) in the presence of hydrogen and CO (1:1 v/v 0.9) added.

40 min. is obtained, compared with 21 after 1 hour for 2); however, the overall catalytic efficiency is greater for complex 2. In our opinion this is due to the better stability of 2, which can be recovered nearly unaltered after hours, whereas 1 decomposes by about 80% after 40 min; this behaviour, as well as the correlation of slower reaction rates with the decomposition of the clusters, point to "intact cluster catalysis".

Cyclohexene, 1,3-cyclohexadiene, 1,4-cyclohexadiene and benzene had not been previously investigated; these were reacted in the presence of 1 under the same conditions as the linear substrates. The results are collected in Table 3. 1,3-cyclohexadiene, a conjugated molecule, is hydrogenated to cyclohexene with a reaction rate comparable to that found for the linear cis-1,3-pentadiene; only one double bond is hydrogenated. Moreover, the hydrogenation of cyclohexene to cyclohexane occurs very slowly. Complex 1 can be considered, hence, as a selective hydrogenation catalyst for one double bond both in linear and in cyclic dienes.

1,4-cyclohexadiene shows a behaviour intermediate between that of the 1,3-isomer and of cyclohexene; in the early stages of the reaction considerable amounts of 1,3-cyclohexadiene are formed, that disappear after longer reaction times. Thus, apparently, 1,4-cyclohexadiene is first isomerized to 1,3 which is hydrogenated to cyclohexene. Benzene is not hydrogenated; after 4 hours at 120°C only traces of cyclohexane and 1,3-cyclohexadiene are observed.

Isomerization experiments on 1-pentene (terminal monoene), 1,4-pentadiene (non conjugated diene) and cis-1,3-pentadiene (for cis-trans isomerization) were also attempted. The results are shown in Table 4 where some experiments involving 2 (previously not published) are also collected. The solutions containing 1 easily isomerize 1-pentene to trans- and cis-2-pentene with a turnover of 479 after 40 min. and trans to cis ratios of 3:1, compared with a turnover of 142 after 6 hours and a 2:1 ratio for complex 2 [12]. These solutions are also fairly active in the isomerization of 1,4-pentadiene to cis-1,3-pentadiene in the absence of H_2; in the presence of this gas however, the isomerization-hydrogenation occurs more slowly. By contrast complex 2 shows a low isomerizing ability in the absence of H_2 and is a good isomerization-hydrogenation catalyst in the presence of it; however a change in the reaction rate and type of products is observed after 2 hours. This could indicate the presence of a different catalytic species in solution.

Evidence for this hypothesis has been gained by reacting 1,4-pentadiene with $Ru_3(CO)_{12}$, $Os_3(CO)_{12}$ and $(Cp)_2Ni_2(CO)_2$ under the same conditions; the results are shown in Table 5. It is evident that the Ru_3 or Os_3 "fragments" show isomerizing ability (presumably by forming trimetallic, acetylide, allenyl or allyl hydridic clusters [13]), whereas the nickel "fragment" gives mainly hydrogenation and decomposes totally to metal powder.

Table 3

Hydrogenation (and isomerization) of cyclic monoenes and dienes in the presence of 1

Experiment	Reaction time (min)	Turnover[a]	cracking products hexenes[b]	c-hexane	c-hexene	1,3-c-hexadiene	substrate	Extent of cluster decomposition[c]
Cyclohexene								
1[d]	10	1	tr	0.1	99.8[e]	--	--	15%
	20	3	tr	0.4	99.5	--	--	
	30	11	tr	1.8	98.0	--	--	
2[f]	10	1	--	0.1	99.8	--	--	10%
	30	2	--	0.2	99.7	--	--	
1,4-cyclohexadiene								
1[d]	10	15	tr	--	0.7	1.8	97.4	20%
	20	33	tr	--	2.8	2.6	94.4	
	30	38	tr	--	6.1	0.3	93.4	
2[f]	10	3	--	--	tr	0.4	99.5	20%
	30	8	--	--	0.3	0.8	98.7	
1,3-cyclohexadiene								
1[d]	10	17	0.1	tr	2.8	97.0[e]	--	90%
	20	70	0.1	tr	12.1	87.7	--	
	30	85	0.2	tr	14.4	85.2	--	
2[f]	10	7	tr	--	1.0	98.8	--	80%
	30	7	tr	--	1.2	98.7	--	

(a) moles of substrate reacted per mole of 1; (b) C_1 and C_2; (c) the extent of decomposition is the maximum observed after long reaction times; (d) in the presence of 0.9 atm. of dihydrogen; (e) when underlined, the same as the substrate; (f) in the absence of hydrogen.

Table 4

Isomerization of mono-enes and dienes in the presence of clusters 1 and 2

Experiment	Reaction time	Turnover[a]	Per-cent composition of the reaction mixtures (organic products)					substrate
			pentane	1-pentene	t-2-pentene	c-2-pentene	c-1,3-pentadiene	
1-pentene								
1[b]	40'	-	-	-	-	-	-	100
2[c]	10'	17	tr	-	2.3	1.0	-	96.6
	20'	235	tr	-	34.0	13.1	-	52.9
	30'	473	tr	-	70.0	25.0	-	5.0
	40'	479	tr	-	71.6	24.5	-	3.9
cis-1,3-pentadiene								
1[b]	40'	-	-	-	-	-	100[d]	100
2[c]	10'	5	-	tr	0.11	0.07	99.81	99.81
	20'	9	-	tr	0.60	0.34	99.05	99.05
	30'	19	-	tr	1.70	0.92	97.38	97.38
	40'	27	-	tr	2.81	1.29	95.90	95.90
1,4-pentadiene								
1[b]	40'	-	-	-	-	-	-	100
2[c]	10'	5	-	0.1	-	-	0.7	99.1
	20'	34	-	0.7	-	-	5.7	93.6
	30'	67	-	1.6	-	-	11.0	87.4
	40'	79	-	2.1	-	-	12.8	85.1
3[e]	1 h	1	-	-	-	-	tr	99.9
	2 h	2	-	-	-	-	0.2	99.7
	4 h	11	-	-	-	-	1.9	98.0
4[f]	10'	12	tr	0.2	-	-	1.9	97.8
	20'	27	1.1	1.2	-	-	2.8	94.9
	30'	42	1.4	1.7	-	-	4.7	91.3
5[g]	1 h	77	0.2	5.50	15.10	-	-	79.20
	2 h	119	4.7	20.80	6.70	-	-	67.80
	3 h	106	3.2	22.40	3.30	-	-	71.10
	4 h	265	4.1	11.20	38.20	18.40	-	28.10

(a) moles of substrate reacted per mole of cluster; (b) experiment 1, in the absence of the cluster; (c) experiment 2, in the presence of 1 and in the absence of hydrogen (extent of decomposition of 1 = 70%); (d) when underlined it is the substrate; (e) experiment 3, in the presence of 2 and in the absence of hydrogen; (f) experiment 4, in the presence of 1 under 0.9 atm of hydrogen; (g) experiment 5, in the presence of 2 under 0.9 atm of hydrogen (see ref. 12).

Hydrogenation-isomerization experiments for 1,4-pentadiene in the presence of $M_3(CO)_{12}$ (M=Ru,Os) and $(Cp)_2Ni_2(CO)_2$ (120°C, 40 min., 0.9 atm. H_2)

Cluster	Turnover	Per-cent composition of the reaction mixtures (organic products)				
		Cracking	pentane	1-pentene	1,3-pentadiene	substrate
$Ru_3(CO)_{12}$	113	tr	-	6.5	38.9	54.5
$Os_3(CO)_{12}$	104	tr	-	7.4	22.0	70.5
$(Cp)_2Ni_2(CO)_2$	17	tr	0.3	12.9	0.8	85.9

The observed lower hydrogenating and greater isomerizing ability of 1 with respect to 2 can be explained easily when considering the stability of the intact clusters under the reaction conditions; apparently the hydrogenation occurs on the "intact cluster" and the effect of (Cp)Ni is probably that of stabilizing the cluster frame by "capping" the M_3 triangle [26]. Another effect could be the labilization of the CO's trans to it, the axial ones; indeed, all of the phosphine substitution products of 2 show axial substitution [27]. By contrast, isomerization occurs on M_3 "fragments" formed upon loss of nickel; the use of more than one coordination site on different metal atoms is apparently required. Based on the above results, we propose an overall reaction pattern similar to that hypothesised for 2 and shown in Figure 2.

Heterogeneous hydrogenation in the presence of cluster 2. A comparison with other heterogeneous systems having different Ni/Os stoichiometries. Clusters 2-6 are chemically and structurally related; indeed the synthesis of complexes 2 and 4 is accomplished by reacting 3 with 5 or 6 [11a]. Also, clusters 2 and 4 are formed by triangular arrangements of osmium atoms "capped" and stabilized by nickel cyclopentadienyl groups.

Methanation of CO. The results obtained for CO hydrogenation over the CDMP derived from metal salts and from clusters 2-6 are collected in Table 6.

Methane is the main product in all of these experiments, together with some C_2 products, especially at the highest reaction temperatures. The catalyst obtained from metal salts gives good selectivity but lower conversions with respect to 2 and 5 at comparable reaction temperatures. At activation temperatures below 300°C complex 2 is by far the best catalyst giving good conversions and high selectivity (although 100% methane is obtained only in the reaction temperature range of 210-240°C against 190-255°C for 5).

The other complexes give low conversions (but high selectivity). After activation at 300°C a general increase of activity is observed for all the systems and the efficiency of 5 approches that of 2.

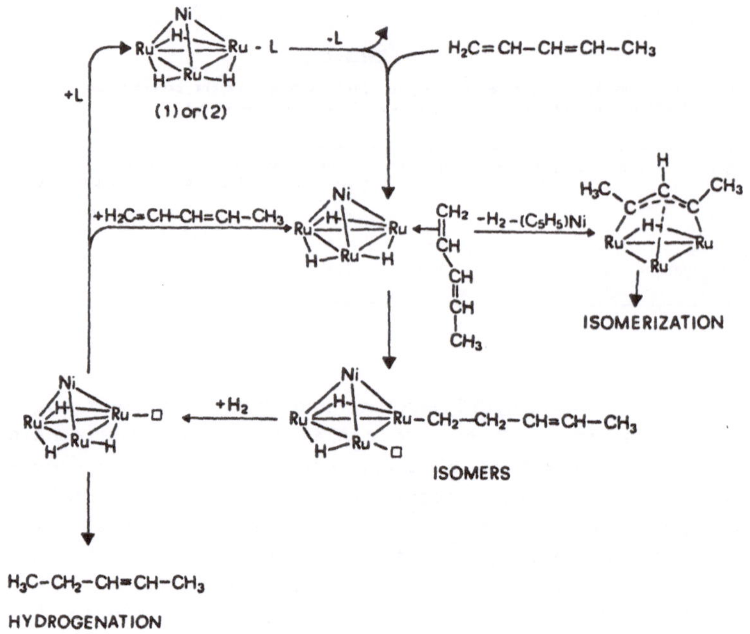

H₂C=CH−CH=CH−CH₃

(1) or (2)

+L

+H₂C=CH−CH=CH−CH₃

CH₂
‖
CH
‖
CH
‖
CH₂
|
CH₃

−H₂−(C₅H₅)Ni

H₃C−C=C−CH₃ (ISOMERIZATION)

+H₂

Ru−CH₂−CH₂−CH=CH−CH₃

ISOMERS

H₃C−CH₂−CH=CH−CH₃

HYDROGENATION

Figure 2.

Table 6

Methanation of CO on catalysts derived from clusters 2-6 and from metal salts.

Catalyst precursor	Activation temperature (°C) and time	Reaction temperature (°C)	Activity[a]	Selectivity[b] CH₄	C₂
Salts	460 (12 h)	192	0.02	100.0	-
		275	0.25	97.1	2.9
		328	0.53	92.0	8.0
3	180 (4 h)	183	0.01	100.0	-
	250 (4 h)	210	0.04	100.0	-
2	250 (4 h)	198	0.23	92.5	-
		236	0.33	100.0	-
		250	0.44	97.2	1.4
	300 (4 h)	215	0.47	90.1	1.3
		244	0.57	97.1	2.9
		278	0.79	96.1	3.9
		295	0.83	96.0	4.0
4	250 (4 h)	254	0.03	100.0	-
	300 (4 h)	303	0.04	100.0	-
5	250 (4 h)	196	0.01	100.0	-
		255	0.16	100.0	-
	300 (4 h)	206	0.13	98.4	-
		307	0.80	87.4	2.4
6	250 (4 h)	250	0.03	100.0	-
	300 (4 h)	245	0.08	100.0	-

(a) Moles of CO converted per gram atom of metals; (b) Per cent of product on the total of products; balance is represented by unidentified C₂₊ hydrocarbons.

Methanation of CO_2. Carbon dioxide is converted to CH_4 or CO by all of the above cluster-derived catalysts and by metal salts. The results are shown in Table 7.

Table 7

Methanation of CO_2 over catalysts derived from clusters 2, 3, 5 and 6 and from metal salts as precursors.

Catalyst precursor	Activation temperature (°C) and time	Reaction Temperature (°C)	Activity[a]	Selectivity[b] CH_4	CO
Salts	460 (12 h)	290	0.17	97.7	2.3
		338	0.51	96.4	3.6
3	250 (4 h)	230	0.06	8.9	91.1
		270	0.40	30.9	69.1
2	250 (4 h)	220	0.20	89.0	11.0
		236	0.37	91.5	8.5
	300 (4 h)	207	0.25	94.7	5.3
		249	0.43	96.8	3.2
		289	0.58	96.7	3.3
		298	0.59	96.5	3.5
5	250 (4 h)	252	0.13	91.6	8.4
	300 (4 h)	245	0.30	96.8	3.2
		300	0.53	91.8	8.2
6	250 (4 h)	270	0.09	44.8	55.2
	300 (4 h)	290	0.26	61.5	38.5
		320	0.31	66.6	33.4

(a) Moles of CO_2 converted per gram atom of metals: (b) Per cent of product on the total of products.

We observed however, significant differences in the catalytic activity of these clusters, which can be reasonably attributed to the nature and stoichiometry of the metals in the supported CDMP.

High conversion mainly to CO are observed for 3 and 6 after activation at 250°C; by contrast 2 and 5 give methane as the main product (in terms of conversion and selectivity 2 is the most effective catalyst). After activation at 300°C the activity increases for all these systems; 3 and 6 give higher amounts of methane and the activity and selectivity of 2 and 5 become close to each other.

The CH_4/CO ratio would hence depend on the WGSR catalytic ability of the CDMP versus their hydrogenating ability; it is known that hydridic and heterometallic cluster-based catalysts are good WGSR catalysts [28]; thus a possible explanation is that 2 and 5 are good catalysts both for WGSR and for hydrogenation, whereas 3 and 6 cannot afford good yields of WGSR products to be hydrogenated further in the short reaction time allowed.

Heterogeneous hydrogenation-dehydration of acetone in the presence of CDMP derived from 2. Synergistic effect of the CDMP and of the support material. The gas chromatographic catalytic-analytic system easily hydrogenates acetone to isopropanol and propane (and trace amounts of propylene). The hydrogenation results for acetone are in Table 8.

The presence of considerable amounts of isopropanol led to the hypothesis that the hydrogenation of acetone occurs in a multi-step process which requires first the formation of the alcohol, and then its dehydration to ether. Evidence for the intermediacy of isopropanol has been gained by reacting this latter with hydrogen under the same conditions as acetone; the results are in Table 9.

The products are essentially the same as for acetone; C_1+C_2 cracking products [25], propane traces of propylene, hexane, hexenes and considerable amounts of di-isopropyl-ether. The results of the hydrogenation of diisopropylether, which can hence be considered another intermediate of this reaction, are collected in Table 10.

Considerable amounts of propane and propylene are observed; once again, the presence of propylene would indicate that the ether is further cleaved; however, dehydration experiments (see below) indicate that simple elimination of water is not favoured and hence the most likely processes could be the hydrogenolysis to isopropanol and propane or the cleavage to isopropanol and propylene.

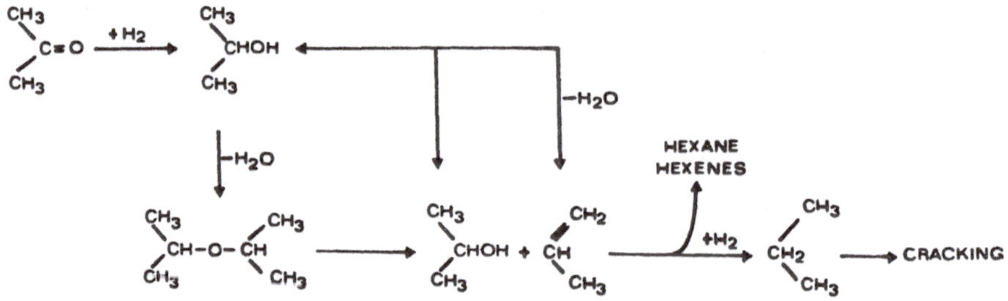

We found evidence that, in the above reactions, the hydrogenating ability of the CDMP's is enhanced by the dehydrating properties of the gas chromatographic support material. For this purpose, we used simple chromosorb columns without supported 2 instead of the catalytic one. The results are collected in Table 11. Under N_2, the isopropanol gives considerable amounts of diisopropyl ether and propylene. Under hydrogen propane and lower yields of ether are observed; thus apparently the chromosorb is the dehydrating agent, and in the presence of hydrogen, the ether is further cleaved.

In the same conditions, di-isopropyl-ether gives only small amounts of dehydration products; under N_2 small amounts of propylene are observed and this supports the hypothesis of a cleavage to propylene and isopropanol (and further dehydration of the latter).

Table 8

Hydrogenation of acetone. Per-cent composition of the effluent gases

Reaction temperature (°C)	ml gas[a]	Cracking products[b]	Propane[c]	isopropanol[c]	unreacted acetone
140	0.5	1.3	1.6	6.9	90.2
	1.0	0.6	0.7	3.8	94.9
	2.0	0.3	0.4	2.9	96.4
170	0.5	2.3	3.7	14.6	79.4
	1.0	1.7	2.7	14.1	81.5
	2.0	1.2	1.8	4.1	92.5
200	0.5	2.6	5.0	15.7	76.7
	1.0	2.0	3.3	8.0	86.7
	2.0	1.1	1.8	7.3	89.8
230	0.5	5.8	6.9	8.9	78.4
	1.0	2.7	3.5	5.1	88.7
	2.0	2.1	2.6	4.1	91.1

(a) ml of gaseous acetone injected (vapour pressure at 25°C), see experimental.
(b) Methane and ethane or ethylene: not separated in the analytical conditions adopted. (c) Traces of proylene and of C_6 derivatives were also observed.

Table 9

Hydrogenation of isopropanol. Per-cent composition of the effluent gases

Reaction temperature (°C)	ml gas[a]	Cracking products[a]	Propane[b]	di-isopropyl-ether	unreacted substrate
140	0.5	2.8	5.0	26.0	66.2
	1.0	2.0	4.1	30.8	63.1
	2.0	1.5	2.7	38.3	57.5
170	0.5	8.7	9.3	27.8	54.2
	1.0	5.0	8.2	34.9	51.9
	2.0	3.7	7.1	46.8	42.4
200	0.5	12.3	19.8	34.6	33.3
	1.0	10.3	18.3	39.1	32.2
	2.0	8.1	16.1	47.5	28.3
230	0.5	16.2	31.1	38.3	14.4
	1.0	12.8	29.6	43.3	14.3
	2.0	11.6	27.8	45.7	14.8

(a) See Table 8; (b) Traces of proylene and C_6 products were also observed.

Table 10

Hydrogenation of di-isopropyl-ether. Per-cent composition of the effluent gases

Reaction temperature (°C)	ml gas[a]	Cracking products[a]	Propane	Propylene	Hexane	Hexenes	Unreacted substrate
140	0.5	tr	3.0	1.1	0.6	tr	95.2
	1.0	tr	2.5	0.8	0.7	tr	96.0
	2.0	tr	2.1	0.7	0.5	tr	96.6
170	0.5	tr	5.6	3.0	0.4	0.1	90.8
	1.0	tr	3.4	2.3	0.3	0.1	93.9
	2.0	tr	2.3	0.8	0.6	tr	96.3
200	0.5	0.3	10.4	5.8	0.4	0.1	83.0
	1.0	0.3	9.9	5.0	0.5	0.1	84.3
	2.0	0.3	9.6	4.6	0.4	0.2	84.9
230	0.5	1.0	16.7	6.9	0.2	tr	75.1
	1.0	1.0	16.3	5.1	0.3	0.2	76.9
	2.0	1.0	15.5	4.7	0.2	0.3	78.2

(a) See Table 8.

Table 11

Dehydration reactions in the absence of 1. Per-cent composition of the gaseous effluents.

Reaction temperature (°C)	ml gas[a]	Cracking products[a]	Propane	Propylene	Hexane	Hexenes	di-isopropyl-ether	Unreacted substrate
Acetone								
140[b]	0.5	1.9	0.7	0.2	-	-	-	97.1
	2.0	0.3	0.1	tr	-	-	-	99.5
230[b]	0.5	8.4	2.4	1.3	-	-	-	87.9
	2.0	7.7	2.1	0.9	-	-	-	89.2
Isopropanol								
140[b]	0.5	3.2	3.2	1.5	-	-	23.2	68.9
	2.0	0.7	0.7	0.4	-	-	38.6	59.6
140[c]	1.0	-	-	6.7	-	-	-	93.3
140[b,d]	-	3.7	5.9	7.2	3.8	-	71.1	8.3
140[c,d]	-	-	-	2.3	-	2.5	84.6	10.6
230[b]	0.5	7.1	5.6	2.5	-	-	39.1	45.7
	2.0	6.3	4.9	2.3	-	-	50.3	36.9
230[c]	1.0	-	-	22.0	-	-	-	78.0
230[b,d]	-	11.5	14.8	20.1	3.0	-	45.8	4.8
230[c,d]	-	-	-	30.9	-	2.2	61.2	5.7
Di-isopropyl-ether.								
140[b]	0.5	tr	0.5	0.3	0.2	0.1	98.9[e]	98.9
	2.0	tr	0.2	0.2	0.1	tr	99.4	99.4
140[c]	1.0	-	-	3.3	-	1.3	95.4	95.4
140[b,d]	-	-	0.3	0.2	0.3	0.2	98.9	98.9
140[c,d]	-	-	-	0.3	-	0.6	99.1	99.1
230[b]	0.5	tr	1.0	0.7	0.1	tr	98.1	98.1
	2.0	tr	0.7	0.5	tr	tr	98.7	98.7
230[c]	0.1	-	-	6.0	-	1.3	92.7	92.7
230[b,d]	-	1.7	7.5	9.5	0.8	2.2	78.2	78.2
230[c,d]	-	-	-	7.7	-	1.3	91.0	91.0

(a) See Table 8; (b) under hydrogen stream; (c) under nitrogen stream; (d) liquid sample injected; (e) when underlined it is the substrate.

Behaviour of cluster 1 when supported onto gas-chromatographic
material. Preliminary results. A catalytic gas-chromatographic column
analogous to that described for 2 [17] was prepared by dissolving in
light petroleum 35 mg of 1 and adding the solution to 12 g of
silanized Chromosorb P, drying under reduced pressure and then treating the material packed into the column in a current of dihydrogen
(25 ml/min) for 7 h at 230°C.

The column was used for comparing the hydrogenating ability of
1 when supported, with that displayed in homogeneous conditions; only
preliminary results are available at present. The CDMP derived from 1
form an efficient heterogeneous catalyst that hydrogenates completely
or only 80° C cyclohexene 1,3-cyclohexadiene and 1,4-cyclohexadiene
to cyclohexane; as expected the selectivity shown by the homogeneous
catalysts in these reactions is lost for the heterogeneous system.

Benzene is hydrogenated mainly to cyclohexane in the temperature range 80-150°C; however, some hexane is also formed, which indicates a tendency for the hydrogenolysis of C-C bonds. By contrast,
complex 2 supported on alumina hydrogenates benzene to cyclohexane
and only traces of hexane in the temperature range of 100-150°C; when
supported onto chromosorb at 200°C, complex 2 gives only 10% of
cyclohexane, thus showing to be less active than 1. At 230°C the
supported 1 gives mainly cracking products.

When toluene is reacted in the temperature range 80-150°C
methylcyclohexane is the main product (at 200°C under the same
conditions 2 gives only 1% of this product); however a demethylation
to cyclohexane is also observed. The intermediacy of methylcyclohexane in this process has been established by blank experiments; methyl
cyclohexane is demethylated to cyclohexane under the same conditions. Ring opening of this latter also occurs to give n-hexane in
small amounts. At 230°C, once again, the cracking to C_1 and C_2
products becomes the main process.

Concluding remarks. The examples discussed in this work represent attempts of evaluating the catalytic reactivity of a heterometallic cluster as an "intact" species and as supported metal particles.
It is noteworthy that similar trends, although associated with very
different reaction rates, can be found for the homogeneous and
heterogeneous species.

References and notes

1. (a) Zwart J., Snel R., J. Molec. Catal., (1985), 30, 305–352.
   (b) Whyman R., in B.F.G. Johnson Ed. "Transition Metal Clusters",
   J. Wiley, New York (1980) p. 545.
2. At present only the ethylene-glycol synthesis based on rhodium
   clusters is exploited: Pruett R.L., Walker W.E., U.S. Pat.
   3.833.634 (1974): 3.957.857 (1976). Kaplan L., U.S. Pat.
   3.944.588 (1976): 4.162.261 (1979). Dougherty J.S., Wells R.C.,
   U.S. Pat. 4.001.289 (1977).
   However, the complex solution equilibrium between the clusters is
   not fully understood: Whyman R., Phyl. Trans. R. Soc. London,
   Ser. A, (1982), 308, 131.
3. See for example: Pittman C.U. Jr, Ryan R.C., Chemtech., (1978,
   march) 170.
4. Venalainen T., Iiskola E., Pursiainen J., Pakkanen T.A.,
   Pakkanen T.T., J. Molec. Catal., (1986) 34, 293.
5. (a) Bailey D.C., Langer S.H., Chem. Rev., (1981) 81, 109.
   (b) Evans J., Chem. Soc. Rev., (1981) 10, 159.
6. See for example: (a) Psaro R., Ugo R., Zanderighi G.M., Besson
   B., Smith A.K., Bassett J.M., J. Organomet. Chem., (1981) 213,
   215. (b) Budge J.R., Scott J.P., Gates B.C., J. Chem. Soc. Chem.
   Commun., (1983) 342.
7. (a) Basset J.M., Ugo R., in R. Ugo Ed., "Aspects of Homogeneous
   Catalysis" vol. 3, p. 137, Reidel, Dordrecht (1977).
   (b) Anderson J.R., "Structure of metallic Catalysts", Academic
   Press, New York, (1975) p. 244.
8. Recent examples are: (a) Ungermann C., Landis V., Moya S.A.,
   Cohen H., Walker H., Pearson R.G., Ford P.C., J. Am. Chem. Soc.,
   (1979) 101, 5922.
   (b) Evans J., Jingxing G., J. Chem. Soc. Chem. Commun., (1985) 39.
   (c) Marrakchi H., Haimeur M., Escalant P., Lieto J., Aune J.P.,
   Nouv. J. Chim., (1986), 10, 159.
9. (a) Pierantozzi R., Valagene E.G., Nordquist A.F., Dyer P.N., J.
   Molec. Catal. (1983) 21, 189.
   (b) Bender R., Braunstein P. J. Chem. Soc. Chem. Commun., (1983)
   334.
10. For the original synthesis see: Sappa E., Manotti Lanfredi A.M.,
    Tiripicchio A., J. Organomet. Chem., (1981) 221, 93.
    The complex has been fully characterized and the structure has
    been proposed on spectroscopic data, see ref. 11a.
11. Synthesis and reactivity: (a) Castiglioni M., Sappa E., Valle M.,
    Lanfranchi M., Tiripicchio A., J. Organomet. Chem., (1983), 241,
    99.
    X-ray structural characterization: (b) Churchill M.R., Bueno C.,
    Inorg. Chem., (1983) 22, 510.
    (c) Lavigne G., Papageorgiou F., Bergounhou C., Bonnet J.J.,
    Inorg. Chem., (1983) 22, 2485.
12. Castiglioni M., Giordano R., Sappa E., Tiripicchio A., Tiripic-
    chio Camellini M., J. Chem. Soc. Dalton Trans., (1986) 23.

13. (a) Vaglio G.A., Valle M., Inorg. Chim. Acta, (1979) 30, 161.
    (b) Vaglio G.A., Osella D., Valle M., Transition Met. Chem., (1977) 2, 94.

14. Moggi P., Albanesi G., Predieri G., Sappa E., J. Organomet. Chem., (1983) 252, C 89.

15. Sappa E. Lanfranchi M., Tiripicchio A., Tiripicchio Camellini M., J. Chem. Soc. Chem. Commun., (1981) 995.

16. Castiglioni M., Giordano R., Sappa E., Predieri G., Tiripicchio A., J. Organomet. Chem., (1984) 270, 7.

17. Castiglioni M., Giordano R., Sappa E., Volpe P., J. Chromatog., (1985) 349, 179.

18. Lavigne G., Papageorgiou F., Bergounhou C., Bonnet J.J., Inorg. Synth., XXV, in press.

19. Sappa E., Valle M., Shore S.G., Siriwardane U., Inorg. Synth., XXV, in press.

20. Registered with a long relaxation time; plus delay = 30 sec.

21. Wheaterbee G.D., Bartholomew C.H., J. Catal., (1981) 68, 67.

22. Johnson B.F.G., Lewis J., Kilty P.A., J. Chem. Soc. (A), (1968) 2859.

23. The Al_2O_3 used as support in the catalytic experiments was not transparent enough for allowing the registration of the i.r. spectra of the supported cluster.

24. (a) Choplin A., Leconte M., Basset J.M., Shore S.G., Hsu W.L., J. Molec. Catal., (1983) 21, 389.
    (b) Basset J.M., Choplin A., . J. Molec. Catal., (1983) 21, 95-108, and references therein.

25. Studies on comparable ruthenium systems are in; (a) Kuznetsov V.L., Bell A., Yermakov Y., J. Catal., (1980) 65, 374-389.
    (b) Evans J., McNulty G., J. Chem. Soc. Dalton Trans., (1984) 1123.
    (c) Zecchina A., Guglielminotti E., Bossi A., Camia M., J. Catal., (1982) 74, 225; Idem, J. Catal., (1982) 74, 240; Idem, J. Catal., (1982) 74, 252.

26. See for example: (a) Masters C., Adv. Organomet. Chem., (1979) 17, 61.
    (b) Vahrenkamp H., Adv. Organomet. Chem., (1984) 22 169.

27. (a) Sappa E., Valle M., Predieri G., Tiripicchio A., Inorg. Chim. Acta, (1984) 88, L 23.
    (b) Sappa E., Predieri G., Tiripicchio A., Tiripicchio Camellini M., J. Organomet. Chem., (1985) 297, 103.
    (c) Sappa E., Nanni Marchino M.L., Predieri G., Tiripicchio A., Tiripicchio Camellini M., J. Organomet. Chem., (1986) 307, 97.
    (d) Predieri G., Tiripicchio A., Vignali C., Sappa E., Braunstein P., J. Chem. Soc. Dalton Trans., (1986) 1135.

28. See for example: (a) Ford P.C., Rinker R.G., Ungermann C., Laine R.M., Landis V., Moya S.A., J. Am. Chem. Soc., (1978) 100, 4595.
    (b) Laine R.M., J. Am. Chem. Soc., (1978) 100, 6451.
    (c) Laine R.M., J. Molec. Catal., (1982) 14, 137, and references therein.

29. Methane and ethane and/or. ethylene; the latter two could not be separated in the analysis conditions adopted.

Organometallic Polymers from Metal 2,3-Dihydro-1,3-diborol Complexes

W. Siebert, Anorganisch-Chemisches Institut der Universität,
Im Neuenheimer Feld 270, 6900 Heidelberg (FRG)

Organometallic dinuclear complexes may serve as building blocks for the construction of conducting polymers and as starting material for organometallic pyrolysis chemistry. This report describes the preparation and characterisation of the first polydecker sandwich complexes which exhibit, depending on the number of valence electrons (VE) per stack (metal + ligand) different electrical conductivities.

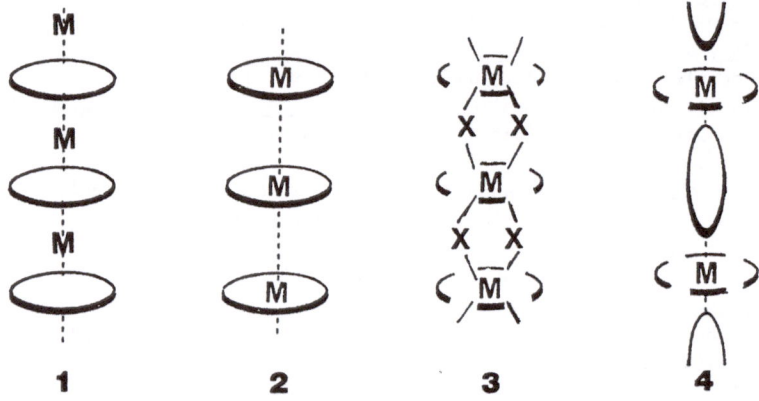

In the polydecker sandwich $\underline{1}$ the metals are coupled via the planar cyclic ligands, whereas in type $\underline{3}$ the metal metal interaction may be realized via chalcogen or halogen bridges. Type $\underline{2}$ represents the many complexes of d^8-metals (Ni^{24}, Pt^{24}) with direct metal-metal bonding. $\underline{4}$ is another type of a polymer in which the metals are linked via heterocycles.

Ligands for Oligodecker and Polydecker Complexes

In the early days of sandwich chemistry polydecker complexes with bridging cyclopentadienyl rings were envisioned, but no compound is known today. The bonding in polydeckers with different transition metal centers M (M=Mn, Fe, Co, Ni, Cu, Zn) and a variety of five-membered π ligands from cyclopentadienyl to the boron ring B_5H_5 has been studied recently [1].

R. M. Laine (ed.), Transformation of Organometallics into Common and Exotic Materials: Design and Activation, 76–85.
© 1988 by Martinus Nijhoff Publishers.

It is, however, possible to construct the triple-decker complexes <u>5</u> - <u>8</u> with C_5H_5 [2], C_6H_6 [3a,b], P_5 [4], and P_6 [5] rings in bridging position:

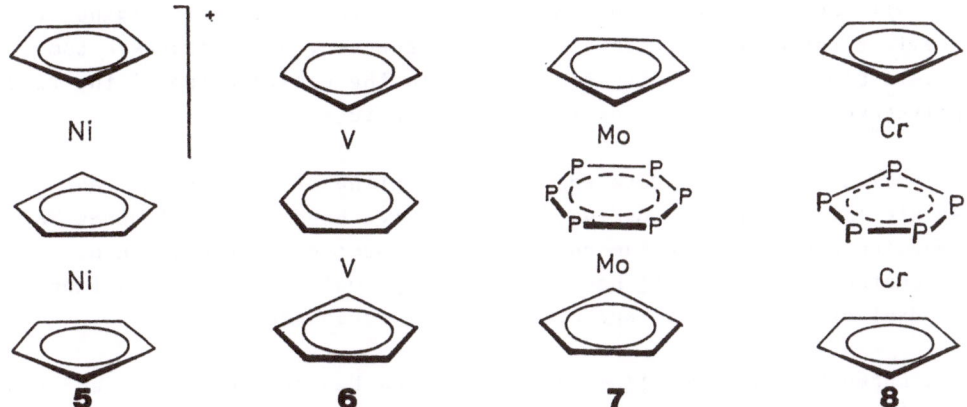

These homonuclear ligands serve as five (C_5H_5, P_5) or six electron donors (C_6H_6, P_6), their acceptor properties are weak. For this reason it is unlikely that they are able to act as ligands in stacking units.

For the construction of polydecker sandwich complexes a combination of good acceptor and donor capabilities of the ligands is required. This may be realized by formal replacement of CH groups in C_5H_5 and C_6H_6 by BH groups [6]:

Among the five-membered heterocycles 9 - 15 the ligands 9 [7], 10 [8], 12 [9], and 13 [9] have been used to prepare triple-decker complexes. The 2,3-dihydro-1,3-diborolyl heterocycle 10 acts as 3e donor and formally as 3e acceptor to complete a π^6 system. With this ligand we realized tetradecker , pentadecker , hexadecker, and octadecker complexes [10]. These results finally led to the construction of the first polydecker sandwich complexes with 10 as bridging ligand [11].

9 is also a good candidate for the preparation of polydeckers, and so are 11, 12, and 13. Derivatives of 9 have been used as ligands in sandwich and triple-decker complexes [7]. Numerous attempts in our laboratory to prepare 11 have failed. 12 as well as 13 have no independent existence, they are obtained from carbaboranes [9].

The boron heterocycles 16 - 18 derived from benzene act as one and two electron acceptors, respectively [12]. 17 and 18 [13] should also be candidates for the construction of polydecker, since they allow the synthesis of sandwich and triple-decker complexes [12].

Number of Valence Electrons (VE)

In the series of oligodecker sandwich complexes the magic numbers of VE are 30 [14], 42 [6], 54, 66 etc. Starting from ferrocene (18 VE) the first insertion of the 12 VE stack ($\eta\Delta$-2,3-dihydro-1,3-diborolyl)-cobalt leads to the closed-shell triple-decker sandwich 19 and the second insertion yields the high-spin tetradecker 20 [10]:

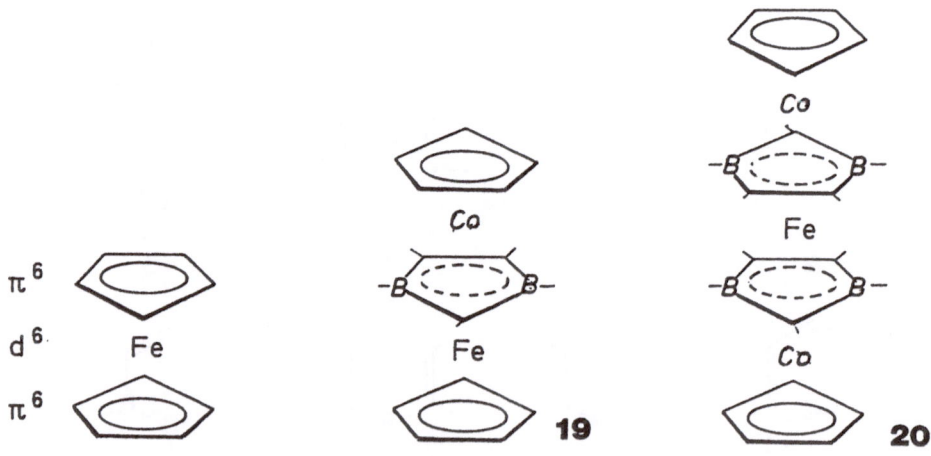

For a given number of stacks (n) the magic number of VE calculates

$$12n + 6 \ VE$$

The maximum number of VE may be obtained by $14n + 6$ VE. A series of oligodecker complexes with $13n' + 6$ VE has been realized, too.

For polydecker complexes the number n should be a large one; the additional 6 VE of the terminal ligand do not influence the total number of VE much. In order to construct polydecker complexes we shall look for ligand/metal combinations which yield stacks with 12, 13 or 14 VE. In our case the ligand 10 supplies 3 VE, in combination with the d^9 and d^{10} metals cobalt and nickel the stacks have a total of 12 and 13 VE, respectively.

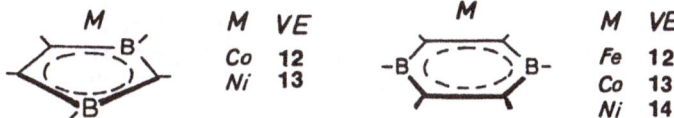

M	VE		M	VE
Co	12		Fe	12
Ni	13		Co	13
			Ni	14

Using the 1,4-diboracyclohexadiene ligand 17 the number of VE could reach 14, provided that the one-dimensional stack is stable enough and does not decompose in bis(1,4-diboracyclohexadiene)nickel (18VE) and elemental nickel (d^{10}). The requirements for stacking are donor proper- ties of the metal and acceptor properties of the monofacial coordinated ligand in the unit, thus allowing head-to-tail stacking reactions.

Preparation of Polydecker Sandwich Complexes

The problem of stacking ligands and metals can be tackled from diffe- rent directions. Obviously, a reaction of metal atom vapour with ligands could lead to the desired products; however such experiments are difficult to control with respect of the reagent concentrations.

Another approach may start from a sandwich dianion and its reaction with the dication M^{24} should lead to a polymer, provided no redox reaction occurs. A possible disadvantage is the formation of alkaline metal halide, which most likely precipitates with the polymer.

Alternatively, a neutral sandwich may react with a complex ML., which readily dissociates and allows a growing of the stack.

This route has not yet been tested with derivatives of bis(2,3-dihydro-1,3-diborole)nickel 21 and Ni(C₂H₄)₃. We have, however, studied the reaction between derivatives of 10 and Ni(C₃H₅)₂ to give yellow-brown 21 or the red diamagnetic tetradecker 22 (42 VE) depending on the ratio of the components. Heating of 22 at temperatures > 150°C in vacuo results in the formation of a black material, most likely having a polydecker arrangement.

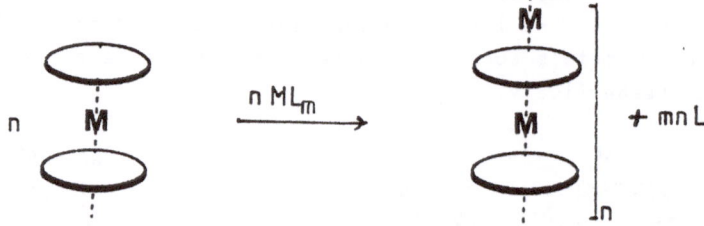

Tetraalkyl derivatives of 21 undergo an unusual reaction which leads to a new 2,3,5-tricarbahexaboranyl(6)-derivative by capping with CH_3-B groups. At the same time a stacking reaction produces a whole variety of oligodecker complexes.

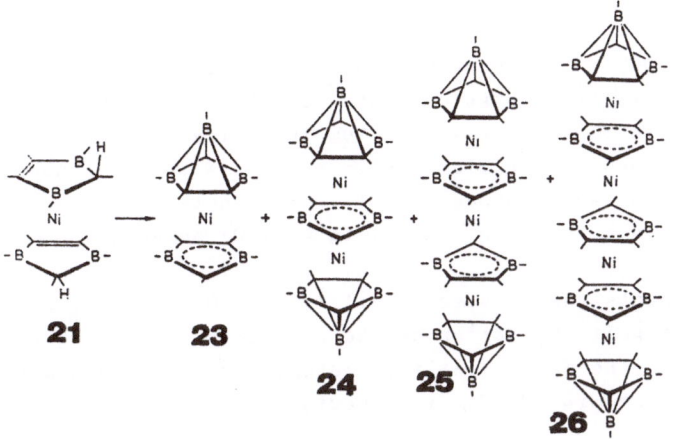

21   **23**   **24**   **25**   **26**

From the deep green reaction mixture the complexes <u>23</u> – <u>26</u> could be isolated and furthermore bis(carbaboranyl)oligodeckers up to ten decks were identified by ms studies.

To avoid the carbaboranyl formation the dinuclear complex <u>27</u> was prepared in which two nickel atoms prevent the capping to carbaboranyl compounds. When a thin film of <u>27</u> is heated in vacuo, bis(allyl)nickel and hexadiene are pumped off, leaving behind a black material. On cooling to room temperature it cracks into small shiny platelets [11].

10 $\xrightarrow[-1/2\ H_2]{2\ Ni(C_3H_5)_2}$ **27** $\xrightarrow[-(C_3H_5)_2Ni]{\Delta}$ **28**

Properties of the Polydecker 28

On heating up to 500°C the product does not change.. It is extremely
air-sensitive and ignites in pure oxigen. In the mass spectrum the
only fragment up to 350°C is allyl in low intensity. [11]B-and ^{13}C-MAS-
NMR yield broad lines in the diamagnetic region. X-band esr spectra in
the range 2 - 290 K show a broad signal, which further broadens on
cooling. Below 50 K a new sharp signal (g = 2.10) appears.

Magnetic measurements show a susceptibility of $2.05 \pm 0.15\mu_B$ per
stacking unit. Stepwise oxidation of the product by air leads to a
linear increase in the susceptibility. Saturation is reached with
$2.88\mu_B$ per deck, for Ni^{24} $2.87\mu_B$ is expected.

Structure of the Polydecker 28

The electron photomicrographs of the product clearly prove the amorphous nature of the material. EXASF investigations at 120, 300, and
340 K confirm the polydecker sandwich structure. The observed ratio Ni
to ligand is 1 : 1. Long Ni-B,C and shorter Ni-B,C distances are
found: 2,56(2) and 2,15(2) at 120 K. This result requires two Ni-Ni
distances which are found at 3.35(2) and 3.80 - 3.90 A. The rationali-
zation of the findings is a Peierls distortion of the structure, in
which two nickel atoms form a pair connected by a bridging ligand.

Conductivity and Electrochemistry

At 2500 bar the two-contact powder conductivity is near 1S/cm; it
drops considerably when oxigen or iodine attacks the polymer. The
four-point measurements give results which are characteristic for a
variable range hopping [11].

The determination of solid state cyclovoltammetry in $LiClO_4$/propylene-
carbonate yields a capacity for reversible electron exchange of only
0.1 electron per stack (15 Ah/kg).

2,3-Dihydro-1,3-diborolyl-Rhodium Polydecker

The formation of the nickel polydecker 28 is achieved via elimination
of bis(allyl)nickel. Another possibility to obtain polymers is the
elimination of the 1,3-diborolyl ligand. We have observed [15] that
the red Rhodium sandwich 29 (16 VE) easily forms the green triple-
decker sandwich 30 (28VE) by heating above its melting point (102° C).
Further heating of 30 in vacuo yields a black high-melting product,
which is insoluable in organic aprotic solvents. The analytical data
of 31 are in agreement with a rhodium-2,3-dihydro-1,3-diborolyl
stack. Preliminary conductivity measurements indicate the product to
be an insulator ($\sim 10^{-10}$ S/cm) [16].

The difference between the two polymers 28 and 31 stems from the metals.
The number of valence elctrons in 28 is 13 VE per stack, whereas 31
has 12 VE. The conductivity in 28 must originate from the extra
electron with respect to the 12 VE of an "ideal" stack in oligo and
polydeckers with 12n + 6 VE.

Possible Use of Polydecker Complexes

The first two polydecker compounds exhibit a high thermal stability.
The nickel polydecker 28 shows characteristics of a semiconductor,
whereas the rhodium complex 31 behaves as an insulator. Despite its
high sensitivity toward oxigen a potential use of 28 as a semiconducting material seems possible. We are convinced that chemical altera-
tion (e.g. substitution of the substituents of the boron heterocycle

by sterically or electronically active groups) could change its behavior toward oxidizing agents. The use of other ligands e.g. 17 and
metals (Fe) could also have a stabilizition effect.

In particular we hope that 11 VE stacks like Fe(η-2,3-dihydro-1,3-
diborolyl) are less prone to electrophilic attack.

The other potential area of the polydecker material and its organometallic precursors are transformation into materials of high strengths
and high thermal stability. Compounds with different element combinations can be designed, which upon pyrolysis result in the formation of
refractory metals. For instance oligodecker complexes composed of the
elements Ti, C, B, H may form unusual pyrolysis products.

Acknowledgement: This work was generously supported by the Deutsche
                 Forschungsgemeinschaft, by the Fonds der Chemischen
                 Industrie, by the state Baden-Württemberg (Forschungs-
                 schwerpunkt Komplexchemie), by the BASF AG and by
                 Degussa.

References

[1] M.C. Böhm, Z. Naturforsch. A39 (1984) 223

[2] H. Werner, A. Salzer, Synth.React.Inorg.Met.Org.Chem.2 (1972) 239;
A. Salzer, H. Werner, ibid. 2 (1972) 249; Angew.Chem. 84 (1982) 949;
Angew.Chem.Int.Ed.Engl. 11 (1972) 930

[3] a) A.W. Duff, K. Jonas, R. Goddard, H.-J. Kraus, C. Krüger,
J.Am.Chem.Soc. 105 (1983) 5479;
b) The triple-decker $(Me_3C_6H_3)_3Cr_2$ has been prepared:
W.M. Lamanna, J.Am.Chem.Soc. 108 (1986) 2096

[4] O.J. Scherer, J. Schwalb, G. Wolmershäuser, W. Kaim, R. Groß,
Angew.Chem. 98 (1986) 34

[5] O.J. Scherer, H. Sitzmann, G. Wolmershäuser, Angew.Chem. 97
(1985) 358; Angew.Chem.Int.Ed.Engl.24 (1985) 351

[6] W. Siebert in A. Müller, E. Diemann (Eds.): Transition Metal
Chemistry, Verlag Chemie, Weinheim 1981, p. 161; Adv.Organomet.
Chem. 18 (1980) 301

[7] G.H. Herberich, B. Heßner, W. Boveleth, H. Lüthe, R. Saive,
L. Zelenka, Angew.Chem. 95 (1983) 1025; Angew.Chem.Int.Ed.Engl.
22 (1983) 996

[8] J. Edwin, M. Bochmann, M.C. Böhm, D.E. Brennan, W.E. Geiger,
C. Krüger, J. Pebler, H. Pritzkow, W. Siebert, W. Swiridoff,
H. Wadepohl, J. Weiss, U. Zenneck, J.Am.Chem.Soc. 105 (1983) 2582

[9] D.C. Beer, V.R. Miller, L.G. Sneddon, R.N. Grimes, M. Mathew,
G.J. Palenik, J.Am.Chem.Soc. 95 (1973) 3046; R.N. Grimes, D.C.
Beer, L.G. Sneddon, V.R. Miller, R. Weiss, Inorg.Chem.13 (1974) 1138

[10] W. Siebert, Angew.Chem. 97 (1985) 924; Angew.Chem.Int.Ed.Engl.24
(1985) 943

[11] T. Kuhlmann, S. Roth, J. Rozière, W.Siebert, Angew.Chem. 98
(1986) 87; Angew.Chem.Int.Ed.Engl. 25 (1986) 105

[12] G.E. Herberich in G. Wilkinson, F.G.A. Stone, E.W. Abel (Eds.):
Comprehensive Organometallic Chemistry, Vol. 1, Pergamon,
Oxford 1982, p.381;

[13] G.E. Herberich, B. Heßner, M. Hostalek, Angew.Chem. 98 (1986) 637

[14] J.W. Lauher, M. Elian, R.H. Summerville, R. Hoffmann, J.Am.Chem.
Soc. 98 (1976) 3219

[15] K. Geilich, W. Siebert, Z.Naturforsch. 41b (1986) 671

[16] K. Geilich, W. Siebert, to be published

SECTION B

PRECERAMIC POLYMERS

CERAMICS VIA POLYMER PYROLYSIS: (1) GENERAL PRINCIPLES AND (2)
THE CRYSTAL AND MOLECULAR STRUCTURE OF μ-IMIDO-BIS-
[BIS(TRIMETHYLSILYLAMINO)(TRIMETHYLSILYLAMINO)]BORANE

KENNETH J. WYNNE

Chemistry Division, Office of Naval Research, Arlington, VA
22217-5000

ABSTRACT

Preceramic polymer chemistry is reviewed in terms of the
intrinsic limitations of this approach and with respect to how
polymer structure affects ceramic yield. Boron nitride precursor
polymer chemistry is discussed together with the synthesis of the
title compound (I), which is a model for the product of a boron
nitrogen (=B-Cl + H-N=) condensation polymerization. The crystal
and molecular structure of (I) has been determined and the
presence of a central N_2BNBN_2 skeleton is confirmed.

POLYMER PRECURSOR CHEMISTRY

Introduction. The activity in synthesis and characterization
of polymers containing elements such as Si and N or Si and C
stems from the desire to extend the well-known route by which
polymers such as polyacrylonitrile or pitch are pyrolyzed to
carbon materials. In contrast to preceramic carbon polymers where
emphasis is on improved processing and elucidation of pyrolysis
chemistry of known materials, the synthesis of new materials
plays a key role in non-carbon preceramic polymers, as no
commercial products exist. Indeed, in some areas where a great
deal of molecular chemistry is known (e. g. for B-N) virtually
nothing is known of macromolecular chemistry. In other areas (e.
g., Si-C, Si-N) known polymer chemistry has been expanded and new
sythetic routes have been developed to incorporate required
properties. Presently there still exists a significant chemical
synthesis challenge for the organometallic chemist in
constructing suitable polymeric materials, as rather stringent
requirements must be met for success. Nevertheless, the basic
limitations of other methods of preparing and processing high
temperature ceramic materials (e.g., sintering, chemical vapor
deposition (CVD)) have focussed increased attention on pyrolysis
of novel polymer precursors.

Before going farther, one intrinsic limitation of the polymer
pyrolysis route to ceramic materials should be noted. This is the
significant disparity between polymer and ceramic densities. For
carbon polymers densities of 1 g/cc are typical, while the
density of graphite is ca. 2 g/cc. Thus, substantial shrinkage and
porosity are anticipated for a carbon product produced by
polymer pyrolysis. The situation is exacerbated for other
ceramic materials, as polymer densities increase only slightly

R. M. Laine (ed.), Transformation of Organometallics into Common and Exotic Materials: Design and Activation, 89–96.
© *1988 by Martinus Nijhoff Publishers.*

with incorporation of heavier atoms such as Si, while ceramic densities are higher (e. g., SiC, 3.2 g/cc). Even with higher yields than those typical for carbon (40-60%) shrinkage of 50% or more must occur on passing from a Si-C or Si-N polymer to a ceramic material. This intrinsic shrinkage with attendant incorporation of porosity, cracks and flaws limits the applicability of preceramic polymers. It is unlikely that such materials could be used alone in the "injection molding" of ceramic parts requiring even moderate control of dimensional tolerances. Nevertheless, there are important application areas where the polymer precursor approach promises improved materials. As binders for ceramic bodies, such materials would be considerably better than currently utilized organic materials which are completely burned off during firing and often leave undesirable open porosity. In the pyrolysis of polymer to ceramic fiber, radial shrinkage can occur without exessive sacrifice of final strength. The potential for high temperature high modulus fibers has already been realized through the appearance of a commercial material, predominately SiC, offered by Nippon Carbon and resulting from the work of Yajima [1].

Structure Yield Correlations. Successful polymer pyrolysis requires identifying a suitable polymer, devising an efficient preparative route, effecting polymer characterization and developing pyrolysis conditions which will yield a practical quantity and quality of desired ceramic. The strength requirements for a polymer as a ceramic fiber precursor are particularly demanding. From carbon polymer science it is known that polymer fiber formation requires a molecular weight high enough to effect chain entanglement (generally ca. 20,000 amu for hydrocarbons). Alternatively, the polymer must contain polar groups which bring about intermolecular interactions (e. g. hydrogen bonding as found in nylons). The intrinsic polarity of many preceramic polymers is fortunate, as few exhibit molecular weights higher than a 2-3 thousand. Thus the Yajima polycarbosilane, subsequently utilized by Nippon carbon for the manufacture of a SiC containing fiber, has Mw typically in the range of 1000-2000 [1,2].

Polymer pyrolysis is preferably carried out at ambient pressure, although the atmosphere surrounding the fiber may be controlled. Despite a lack of detailed chemical knowledge, some trends with respect to structure/yield relationships on unconfined pyrolysis may be summarized [3].

1. Linear polymers give negligible ceramic yields because of reversion reactions, i. e., the generation of large, volatile molecules and cyclics. Thus Wesson and Williams [4] reported that poly(dimethylsilylene) when heated in an open system under argon at one atmosphere gives a negligable yield of either carbosilane preceramic polymer or SiC. Rather, a reversion reaction occurs whereby cyclic hexamer is the primary product. Here, the relatively small size of the methyl group facilitates chain mobility and provides minimum steric hindrance to unzipping and other reactions leading to cyclics (Figure 1).

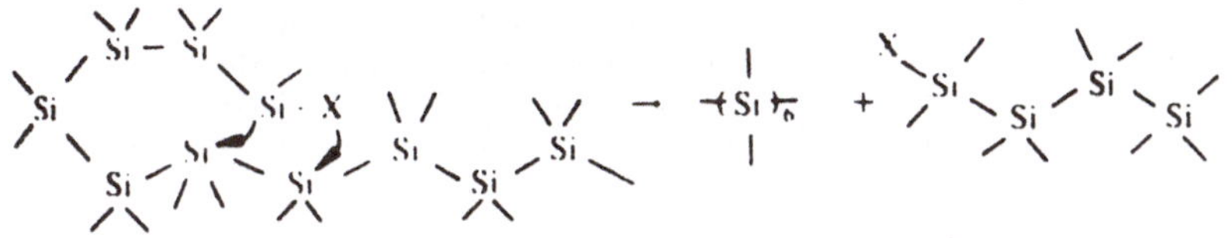

Figure 1. A schematic representation of
unzipping to cyclics in a polysilylene.

2. Polymer structures containing rings (or cages) give good
ceramic yields. Ring or cage structures act to slow the rate of
reversion reactions, as the liberation of relatively large
volatile molecular fragments requiring multiple chemical bond
ruptures (e. g., cyclics) is sterically hindered. However, solid
state cross linking necessary for preceramic structure buildup,
which involves localized reactions and the evolution of smaller
molecules (H_2, HCl, CH_4)) proceeds unabated. As an example,
poly(arylenesiloxanes) give a high yield of ceramic materials (C,
SiO_2) on pyrolysis in nitrogen at 600-900°C, while related linear
poly(siloxanes) not containing rings in the chain structure
undergo depolymerization to cyclics under similar conditions
[5,6].
3. Branched polymer structures give high ceramic yields. The
existence of branched structures has a pronounced effect on
ceramic yields. Such structures slow the rate of reversion
reactions and provide a higher fraction of more reactive end
group fuctionality. It seems likely that this higher fraction of
reactive end groups is important in providing more facile cross-
linking of the early stages of thermolysis to the "ultimate"
cross-linking found in the ceramic structure.
Shilling et al [7,8] focussed on the criticality of branched
structures in poly(carbosilane) preceramic polymers with the
preparation of tractable poly(carbosilanes) via the potassium
dechlorination of mixtures of vinylmethylchlorosilanes or
methyltrichlorosilane with other silane monomers. Percent yields
for branched poly(carbosilanes) were at least two orders of
magnitude higher than for similar linear poly(carbosilanes). As
an example, the potassium dechlorination of $ClCH_2SiMe_2Cl$ yields a
linear polymer, which upon pyrolysis gives a negligible yield of
SiC:

$$ClCH_2SiMe_2Cl \longrightarrow (CH_2SiMe_2)_n \longrightarrow trace\ SiC. \quad (1)$$

In contrast, when $ClCH_2SiMe_2Cl$ is copolymerized with $MeSiCl_3$
(mole ratio 2:1.3), the resulting branched poly(carbosilane) is
an effective SiC precursor, yielding 31% ceramic product (mostly
SiC) on unconfined pyrolysis:

$$ClCH_2SiMe_2Cl + MeSiCl_3 \longrightarrow (CH_2SiMe_2)_m(MeSi)_n$$

$$\longrightarrow \beta\text{-}SiC(31\%) \quad (2)$$

Many polymers with branched structures form fibers which are too brittle to survive spinning and rollup. Verbeek described the synthesis of polysilazanes from alkylchlorosilanes and alkylamines [9]. Equation 3 shows the synthetic route to one of these polymers "Resin A" and a formula based on an analysis [3] of the work of Penn [10]:

$$MeSiCl_3 + MeNH_2 \xrightarrow[-MeNH_3Cl]{} MeSi(NHMe)_3 \xrightarrow{520^{\circ}C}$$

$$[Me(MeNH)Si(MeN)]_xMeSi(MeN)1.5]_y$$
"Resin A"

(3)

To obtain spinnability, high molecular weight polyethylene oxide was added to the spinning solvent to create a novel polymer blend.

4. Polymers with latent reactivity enabling the formation of branched or ring structures give good or improved yields of ceramic products. A classic example of latent reactivity from carbon chemistry is the formation of branched-ring structures during thermolysis of solution-spinable polyacrylonitrile [11]. Such latent chemical reactivity provides a particular advantage for preparation of ceramic fibers, as the precursor polymer fibers can easily be melt or solution spun with subsequent formation of stabilizing ring/branched structures being carried out just before thermolysis and subsequent pyrolysis.

Cross linking by thermolysis is the most common method of "stabilizing" a polymeric material for shape retention. Thermolysis may be in an inert atmosphere, taking advantage of inter- and intrachain crosslinking via the presence of latent functionality (vide infra), or may be carried out in the presence of a reactive gas such as oxygen. Alternatively, radiation (e.g., UV) may be used to effect stabilization. Regardless of the approach, the key to success is the establishment of a properly distributed, crosslinked structure in sufficient quantity such that the material will survive subsequent pyrolysis with nominal shape retention.

Latent chemical reactivity requires carefully selected chemistry, as the reactive site must survive the normal conditions for polymer sythesis and processing, and cause branch/ring structures below the temperature at which reversion occurs. Latent reactivity generally requires the presence of reactive (unsaturated) organic functions such as vinyl or cyano, or reactive bonds such as Si-H or N-H. Incorporating the latter may be particularly important vis-a-vis the former if it is desired to minimize the formation of carbon.

An interesting example of latent reactivity is the incorporation of pendant vinyl functionality in polydiorganosilylenes [7,8]. Potassium reacts with vinylic halosilanes causing polymerization at carbon as well as at Si. In contrast, sodium in hydrocarbon/ether solvent only reacts with Si-Cl bonds, and the result is the preparation of a poly(vinyl alkyl)silylenes. Si-H functionality is also mostly unreactive, facilitates crosslinking, and increases ceramic yield:

$$0.85 \ Me_3SiCl$$
$$0.3 \ MeSiHCl_3 \qquad \xrightarrow{\text{Na,solvent}} Me_3Si(SiMe)_m(SiMe)_nSiMe_3$$
$$1.0 \ CH_2=CHSiMeCl_2 \qquad\qquad\qquad\quad \underset{H}{|} \qquad \underset{CH=CH_2}{|}$$

$$\xrightarrow{1200^\circ C} SiC \ (57\%)$$

$$(4)$$

The residual vinyl and Si-H functionality allows thermal crossliking through vinyl polymerisation and hydrosilation. A major advantage with these polymers is that curing occurs without loss or gain of weight and without the use of oxygen which could subsequently contaminate the resulting ceramic.

A number of other examples of this growing class of polymeric materials with residual functionality have been noted [3].

While the above considerations provide a framework for correlating polymer microstructure with the quantity and nature of the resulting ceramic, structurally extrinsic features involving control of processing conditions can play an important role in determining the nature of the ceramic product. Pyrolysis in a reactive atmosphere is illustrated by the work of Wiseman and Seyferth [12] where SiC-free Si_3N_4 was formed by pyrolysis of a silazane polymer in nitrogen, which converted the Si formed during pyrolysis to Si_3N_4, in addition to that produced directly from the polymer.

BORON NITROGEN POLYMERS

Introduction. The B-N bond strength is high (100 kcal) and boron nitrogen chemistry has been intensively studied but few tractable B-N polymers are known [13]. A Japanese patent [14] describes a resin derived from the thermolysis of B, B',B"-triaminotriphenylborazole, but these results have not proved reproducable [15]. In contrast to carbon which is a conductor, BN is a dielectric and a challenging target material for preceramic polymer synthesis. Success would offer a complementary route to the known method of converting boric oxide supported fibers to BN [16].

Equation 4 shows an example where the presence of fuctionality for condensation polymerization (either to rings or chains) was insufficient:

$$\overset{\overset{\textstyle Cl}{|}}{(Me_3Si)_2NBNHSiMe_3} \ \xrightarrow{\qquad} \ \underset{Me_3SiN-BN(SiMe_3)_2}{(Me_3Si)_2NB-NSiMe_3}$$

$$+$$

$$[Me_3SiN--BN(SiMe_3)_2]_n \quad (5)$$

Either in the presence of base or under thermolysis, the "monomer", containing a -B(Cl)-N(H)-group, proved surprisingly robust and no oligomeric material was isolated before thermal decomposition. A related reaction shows that the condensation of N-H and B-Cl moieties can occur intermolecularly and can lead to an isolable compound which is a model "dimer" (Equation 6).

$$
\begin{array}{ccc}
\overset{\displaystyle Cl}{\underset{\displaystyle |}{}} & & \overset{\displaystyle NH_2}{\underset{\displaystyle |}{}} \\
(Me_3Si)_2NBNHSiMe_3 & + & (Me_3Si)_2NBNHSiMe_3 \quad \text{-----} \rightarrow
\end{array}
$$

$$
\begin{array}{c}
(Me_3Si)_2NBNHSiMe_3 \\
| \\
NH \qquad\qquad + \ TEA.HCl \\
| \\
(Me_3Si)_2NBNHSiMe_3 \\
I
\end{array}
\qquad (6)
$$

The yield of the product, μ-imido-bis[bis(trimethylsilylamino)-(trimethylsilylamino)]borane, was 57%. A yield greater than 98% is required for a condensation reaction to give high molecular weight polymer. One encouraging feature, however, is the observation that I can be remelted in air without change in melting point (169-171°C). This attests to a significant degree of oxidative stability for this compound. In order to obtain insight into the nature of bonding in catenated BN species, the crystal and molecular structure of I was obtained [17].

Colorless I crystallizes in the monoclinic system, space group P21/n. The lattice constants are: a = 12.876(2) Å; b = 14.828(3) Å; c = 18.628(4) Å; α = 90.00°; β = 97.07°; γ = 90.00°; V = 3529(1) Å^2; Z = 4. The calculated molecular weight based on a calculated

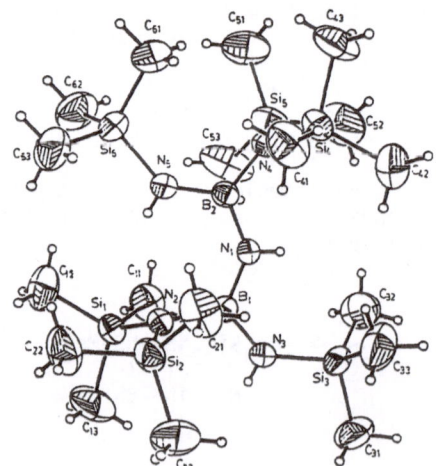

Figure 2. The molecular structure of I.

density of 1.01 g cm^{-1} was 534.8, in excelent agreement with the parent ion observed in the mass spectrum (533 amu).

The structure of I shown in Figure 2 confirms the presence of a central N_2BNBN_2 skeleton. The B-N and N-H bond distances are shown in Table I. The N2BNBN2 grouping is essentially planar with a B(1)-N(1)-B(2) angle of 135.9(2)^{o}. The remaining B-N-B angles are close to 120° and are not listed.

The B-N bond distances range from 1.405(3) to 1.487(3) Å. These distances reflect a competition for pi-electron density. Thus, the shortest bonds (B(2)-N(5) and B((1)-N(3) are to nitrogens bonded to one silicon, while the longest B-N bonds B(2)-N(4) and B(1)-N(2) are to nitrogens bonded to two silicons. The central B-N bonds are of intermediate length due to mutual competition of B(1) and B(2) for pi-electron density on N(1). The shortest distance for I (1.405(3) Å) is not as short as the shortest distance (1.394(3) Å) for the recently reported 1,3,5,7-tetra-tert-butyl-2,4,6,8-tetramethyl-1,3,5,7,2,4,6,8-tetrazatetraborocin [II,18], while the longest distance for I (1.487(3) Å) is not as long as the longest distance reported for II (1.518(4) Å).

Table I. Selected bond distances for I (Å).

B(1)-N(1)	1.428(3)
B(2)-N(1)	1.446(3)
B(1)-N(2)	1.487(3)
B(1)-N(3)	1.423(3)
B(2)-N(4)	1.485(3)
B(2)-N(5)	1.405(3)
N(1)-H(1)	0.77(2)
N(3)-H(3)	0.71(2)
N(3)-H(3)	0.71(2)

ACKNOWLEDGEMENT

The author thanks the Office of Naval Research for support of this research.

REFERENCES

1. Yajima, S., Hayashi, J., Okamura, K., Nature 266, 521, (1977).

2. Hasegowa, Y., Okamura, K., J. Mater. Sci., 18, 3633, (1983).

3. Wynne, K. J., Rice, R. W., Ann. Rev. Mater. Sci., 15, 297, (1984).

4. Wesson, J. P., Williams, T. C. J. Polym. Sci. Polym. Chem., 17, 2833 (1979).

5. Dvornic, P., Lenz, R., Polymer, 24, 763 (1983).

6. Ballistreri, A., Montaudo, G., Lenz, R., Macromolecules, 17, 1848, (1984).

7. Schilling, C. L. Jr,. Wesson, J. P., Williams, T. C1, Am. Ceram. Soc. Bull., 62, 912 (1983).

8. Schilling, C. L., Jr., Brit. Polym. J., Submitted for publication.

9. Verbeek, W., US Patent 3,853,567, (1974).

10. Penn, B. G., Ledbetter, F. E. III, Clemons, J. M., Daniels, J. G., J. Appl. Polym. Sci, 27, 3751 (1982).

11. Jenkins, G. M., Kawamura, K., Polymer Carbons, Carbon Fiber, Glass and Char, London: Cambridge Univ. Press, 1976.

12. Seyferth, D., Wiseman, G. H., Prud'homme, C., J. Am. Ceram. Soc., C13, 66 (1983).

13. Wagner, R. I., Bradford, J. L., Inorg. Chem., 1, 99 (1962).

14. Taniguchi, I., Harada, K., Maeda, T., Jpn. Kokai 76 53,000 (1976); Chem. Abstr., 85, 96582v.

15. Paciorek, K., Kratzer, R., unpublished results.

16. Economy, J., Sampe J., 5 (1976).

17. Wynne, K. J., Paciorek, K., Kratzer, R., Day, C., details on the synthesis and crystal structure will be published separately.

18. Franz, T., Hanecker, E., Noth, H., Stocker, W., Storch, W., Winter, G., Chem. Ber., 119, 900 (1986).

ORGANOSILICON PRECURSORS TO SILICON CARBIDE FOR ELECTRONIC AND CERAMIC APPLICATIONS

JAN G. NOLTES

Division of Technology for Society,

Netherlands Organization for Applied Scientific Research

3502 JA Utrecht, The Netherlands

INTRODUCTION

SiC is a compound with exceptional physical and mechanical properties. It is one of the hardest materials known to man, second only to diamond and cubic boron nitride. SiC has low density, high inherent strength, high thermal conductivity, is stable at high temperatures and is highly resistant to thermal shock. These properties make it very useful in demanding mechanical engineering applications, both as a reinforcing material in composite materials and as shaped bodies of substantially pure SiC. It is a major commercial commodity, produced in amounts up to a million tons annually.

SiC is the only group IV-group IV semiconductor. It has high electron mobility and high temperature stability, is easily doped either p- or n-type, and p-n junctions are readily produced. Therefore, SiC is already being used or is a candidate material for employment in high-temperature, high-frequency and high-power electronic and opto-electronic devices.

In view of the tremendous actual and potential importance of SiC it is not surprising that the search for alternative routes to this material is intensifying.

Currently the possibility to use designed organometallics for generating semiconductor coatings, refractory metal coatings, fibers and powders, or ceramic materials is receiving wide attention [1,2]. Traditional purification methods of the organometallic precursors in principle allow for high purity products. Variable atomic ratio's in the precursors allow for novel compositions not easily obtainable otherwise. Polymeric precursors allow the production of ceramic fibers and other 3-D shapes.

TNO is in the process of setting up a multi-disciplinary research program in this area. In the present paper the role of organosilicon precursors to SiC for electronic and ceramic applications is addressed.

R. M. Laine (ed.), Transformation of Organometallics into Common and Exotic Materials: Design and Activation, 97–102.
© *1988 by Martinus Nijhoff Publishers.*

ORGANOSILICON PRECURSORS TO SiC FOR ELECTRONIC APPLICATIONS

Most modern methods available for depositing thin layers have been applied to the deposition of SiC. CVD techniques have been widely used for depositing thin layers of SiC in the form of one of its crystallographic polytypes (α-SiC[3], β-SiC[4], 6H-SiC[5]). The input source of Si and C used in this work were the combinations $SiCl_4/C_3H_8$, SiH_4/C_3H_8 or $SiCl_4/CCL_4$, with H_2 being used as the carrier gas. Organosilicon precursors heve not been used for this purpose. It is, however, feasible to prepare pure or doped SiC via the organosilicon route. For example, poly-crystalline β-SiC has been prepared by the classical Van Arkel process starting from $MeSiCl_3$, Me_2SiCl_2 or Me_3SiCl [6]. Doping was accomplished by mixing in Me_3Al or B_2H_6. Best results were obtained in the 1400-1600°C temperature range. This explains why CVD techniques are not really suitable for depositing SiC on Si substrates. Si melts at 1410°C and higher temperatures are required for epitaxial growth of SiC (in the range of 1500-1700°C). Techniques that require lower temperatures such as photo-CVD or RF-plasma-CVD do not seem to have been applied to organosilicon precursors.

It would seem that the organometallic route to SiC for electronic and opto-electronic uses has been little explored. Further research into the possible role of organosilanes in low-temperature processes seems worthwhile.

ORGANOSILICON ROUTES TO CERAMICS

Monomeric organosilicon precursors

Thin organosilicon polymeric films with a thickness of ~0,5 μm have been deposited on steel and glass surfaces via RF plasma polymerisation of simple organosilanes such as Me_4Si and $(Me_3Si)_2NH$.

The polymeric films which displayed high crosslink density and excellent surface adherence were pyrolyzed into ceramic-like coatings which likewise showed excellent surface adhesion [7].

Uniform SiC powders have been produced by applying CVD techniques to simple organosilanes [7]. Particle size could be varied from submicron (~0,2 μm) to supramicron (up to 10 μm) dimension by controlling CVD process parameters. Sintered ceramic shapes were obtained from the submicron powders by compacting and sintering at 1500°C and atmospheric pressure.

These routes to SiC coatings and ceramic shapes seem to have been little explored and are worthy of further study.

Polymeric organosilicon precursors

The thermal degradation of linear organometallic polymers and their conversion into ceramic materials has been much less extensively studied than the conversion of organic polymers into high-surface area carbons or carbon fibers.

The feasibility of this approach in the case of SiC has been demonstrated by the late Professor Yajima [8] who has pioneered the commercial production of high-strength SiC fibers in Japan. Yajima used linear permethylpolysilanes $[(CH_3)_2Si]_n$ or the cyclo-hexasilane $[(CH_3)_2Si]_6$ as precursors to SiC. Thermolysis at 400-450°C results in the formation of

$$[(CH_3)_2Si]_n \xrightarrow[\text{Ar}]{450°C} \begin{bmatrix} \overset{\text{H}}{\underset{\text{CH}_3}{-\text{Si}-\text{CH}_2-}} \end{bmatrix}_n \xrightarrow[\text{2. N}_2/1300°C]{\text{1. Air}/350°C} \beta\text{-SiC} + CH_4 + H_2$$

a polycarbosilane. The high-molecular weight fraction (M.W. ~8000) obtained by fractionation of the polycarbosilane is spun into fibers which after surface oxidation in air are pyrolyzed to amorphous SiC at 800°C. Crystals of β-SiC form in the fibers upon raising the temperature to 1300°C. The SiC fibers made in this way are among the strongest substances known.

Baney et al. at Dow Corning starting from poly-methyl silane precursors obtained by Bu_4PCl-catalyzed rearrangement of commercially available methylchlorodisilanes applied a final pyrolysis step at 1473°C to obtain small grain β-SiC similar to the results obtained by Yajima [9].

Organosilicon precursor polymers should meet a number of requirements. If a shaping operation such as spinning is to be performed the polymer should be tractable. However, some degree of crosslinking is required in order to retain shape up till the temperatures where the pyrolysis process sets in. The ceramic yield should be high at the lowest possible pyrolysis temperature. The formation of large amounts of volatile low-molecular weight fragments during pyrolysis should be minimal.

From a ceramic yield point of view (94%) the monomethylpolysilane $[HSiCH_3]_n$ or the corresponding polycarbosilane $[-Si(H)_2CH_2-]_n$ look attractive, but these polymers have not yet been prepared and might be too volatile anyway.

So far, the only published alternative to the Yajima process has been the poly-silastyrene precursor polymer proposed by Robert West at the University of Wisconsin [10]. This polymer is obtained by cocondensation of Me_2SiCl_2 and $PhMeSiCl_2$ with sodium. The proportion between Me_2Si and $PhMeSi$ can vary, but the average composition is 1:1. Polymer fractions with mole-

$$
\begin{array}{l}
Me_2SiCl_2 \\
PhMeSiCl_2
\end{array}
\xrightarrow[100-600°C]{Na,\ hydrocarbon}
[(Me_2Si)_x(PhMeSi)_y]_n
\xrightarrow{>800°C}
\begin{array}{l}
2SiC + 2H_2 \\
+ CH_4 + C_6H_6
\end{array}
$$

cular weights in excess of 500,000 have been isolated, corresponding to chains of more than 5000 Si atoms. Polysilastyrene can be formed into films or fibers. The solid polymer becomes crosslinked upon exposure to u.v. light (350 nm) and thereby becomes insoluble and infusible. The cross-linked fibers can be directly converted into β-SiC fibers. A transformation into intermediate carbosilane polymer, fractionation of the latter and surface air oxidation of the fibers as required in the Yajima process are not necessary which would seem to be a considerable advantage. In recent work by Sinclair at the 3M Company the theoretical ceramic yield of 45% for the production of SiC from photo-crosslinked poly-silastyrene has been approached. Moreover, an identifiable β-SiC X-ray diffraction pattern was observed upon heating the SiC fibers at 1100°C [11, 12].

West and Mazdiyasni have used linear polysilane precursors with $Me_2Si/$ $PhMeSi$ ratio's from 3:1 to 20:1 for strengthening silicon nitride ceramics. The Si_3N_4 body was soaked in polysilane and refired. SiC was formed in the pore spaces resulting in increased strength [13].

Research at TNO on precursor polymers has not yet started. However, it seems relevant to recall some results from earlier research at TNO dealing with poly-p-phenylenesilanes. Such polymers which at that time were studied because of their potentially interesting heat stability were obtained by Wurtz-type condensation of diorganodichlorosilanes with diorganobis(p-chlorophenyl)silanes or by homocondensation of diorgano(p-chlorophenyl)chlorosilanes [14]:

$$R_2SiCl_2 + R_2Si(C_6H_4Cl)_2$$
$$ClR_2SiC_6H_4Cl$$
$$\xrightarrow{\text{Na}}$$
$$\left[-\underset{\underset{R}{|}}{\overset{\overset{R}{|}}{Si}} - \bigcirc - \right]_n$$

The molecular weights of the polymers were low (up to 4000; n ~30) and were not optimized. Thermal behaviour was studied by thermogravimetry (2.5°C/min. to 900°C under N_2). Three samples of $[-(CH_3)_2SiC_6H_4-]_n$ for which n was 5, 13 and 28 started to decompose at 350, 390 and 400°C, respectively. No appreciable further weight loss took place after the temperature had reached 600°C. The weight retained upon heating from 600-900°C for the samples with n=5, 13 and 28 was 15, 32 and 58%, respectively. The following thermolysis route leads to the formation of SiC with simple stoechiometry:

$$\left[-\underset{\underset{Me}{|}}{\overset{\overset{Me}{|}}{Si}} - C_6H_4 - \right]_n \xrightarrow{400\text{-}600°C} SiC + CH_4 + C_6H_6$$

The theoretical char yield for this reaction is 30% which is in nice agreement with the value of 32% observed for the n=13 sample. This interpretation of the scant experimental data is speculative at best. It might be of interest to prepare high-molecular weight poly-dimethyl-p-phenylenesilanes and study their crosslinking and pyrolysis behaviour to see whether such polymers have potential as precursors to SiC.

As regards the rational selection of polymeric precursors to SiC fibers and 3-D shapes suitable candidate organosilicon polymers must meet a number of requirements. Some of these are not easily met such as high ceramic yield at low pyrolysis temperatures and others seem conflicting such as low crosslink density for tractability and ease of shaping and high crosslinking density for high ceramic yield and for retention of shape upon pyrolysis. Although the feasibility of the approach has been beautifully demonstrated it seems to be no simple task to realize drastic improvements for the organometallic route to SiC.

ACKNOWLEDGEMENTS

I thank Carel I.M.A. Spee (MT-TNO) who provided me with details on the preparation of SiC for electronic applications.

REFERENCES

[1]   "Ultrastructure Processing of Ceramics, Glasses and Composites", L.L. Hench and D.R. Ulrich, Eds. John Wiley, New York, 1984.

[2]   "Emergent Process Methods for High-Technology Ceramics", R.F. Davis, H. Palmour III and R.L. Porter, Eds. Plenum Press, New York, 1984.

[3]   J.M. Harris, H.C. Gatos and A.F. Witt, J. Electrochem. Soc., 118 (1971) 335; R.B. Campbell and T.L. Chu, ibid., 113 (1966) 825.

[4]   S. Nishino, Y. Hazuki, H. Matsunami and T. Tanaka, J. Electrochem. Soc., 127 (1980) 2674.

[5]   S. Nishino, H. Matsunami and T. Tanaka, J. Cryst. Growth, 45 (1978) 144.

[6]   W. von Muench and E. Pettenpaul, J. Electrochem. Soc., (1978) 294.

[7]   C.L. Beatty, p. 272 in ref. [1].

[8]   S. Yajima, J. Hayashi and M. Omori, Nature, 261 (1976) 683; Y. Hasegawa, M. Jimura and S. Yajima, J. Mater. Sci., 15 (1980) 720.

[9]   L.H. Baney, J.H. Gaul Jr. and T.K. Hilty, p. 253 in ref. [2].

[10]   R. West, p. 235 in ref. [1].

[11]   R. West, L.D. David, P.I. Djurovich, H. Yu and R.A. Sinclair, Am. Ceram. Soc. Bull., 62 (1983) 2899.

[12]   R.A. Sinclair, p. 256 in ref. [1].

[13]   K.S. Mazdiyasni, R. West and L.D. David, J. Am. Ceram. Soc., 61 (1978) 504.

[14]   J.G. Noltes and G.J.M. van der Kerk, Rec. Trav. Chim., 81 (1962) 565.

CATALYZED DEHYDROGENATIVE COUPLING AS A ROUTE TO NOVEL ORGANOMETALLIC POLYMERS.

J.F.Harrod. Chemistry Department, McGill University, Montreal.

ABSTRACT

Dimethyltitanocene and dimethylzirconocene are very active catalysts for the dehdrogenative coupling of primary organosilanes to polysilanes and of primary and secondary organo-germanes to polygermanes. Characterization of the polymers produced by these reactions will be described. The reasons for the unique catalytic activity of the Ti and Zr compounds and the failure of analogous metallocene derivatives of the other early transition groups to catalyze silane polymerization is discussed. The generality of dehydrogenative coupling as a route to new families of organometallic polymers is reviewed.

INTRODUCTION

Silyl complexes of most transition metal complexes are known and show reasonable chemical stability. An exception to this generalization is group 4, especially Ti. To our knowledge only two silyltitanium complexes have been fully characterized,[1,2], and in view of the present results one of those should probably be re-evaluated. In the course of a search for a synthetic route to silyltitanocene complexes we investigated the reactions of silanes with dimethyltitanocene and, rather than the expected silyltitanocene(IV), obtained a novel mixed valence titanium hydride complex[3]. The subsequent screening of a variety of organosilicon hydrides in an effort to improve the purity of this complex lead to the adventitious discovery that dimethyltitanocene can catalyze the dehydrogenative coupling of primary organosilanes to poly(organosilanes)[4]. The rapid catalytic coupling is specific for primary silanes and early studies lead us to the conclusion that the polymers were linear and of the order ten silicons per average chain. Although secondary organosilanes react readily with dimethyltitanocene to give reduced Ti species they do not couple at a

R. M. Laine (ed.), Transformation of Organometallics into Common and Exotic Materials: Design and Activation, 103–115.
© *1988 by Martinus Nijhoff Publishers.*

detectable rate. Tertiary organosilanes do not react with dimethyltitanocene at a reasonable rate.

CHARACTERIZATION OF THE POLYMERS

Polymers produced from primary silanes have the following structure:

$$H-[-\underset{H}{\overset{R}{Si}}-]_{\overline{n}}H$$

In the absence of any stereoregulating influence the polymers are atactic. Because of this the NMR spectra of the polymers are of limited use for establishing the structure since the Si-H region appears as a broad envelope with little structure (see fig. 1). The spectra clearly show the presence of Si-H in the correct ratio to phenyl protons. However, use of a DEPT sequence for the natural abundance ^{29}Si-NMR spectrum establishes the presence of SiH_2 units in the polymers[4]. The presence of SiH_2 units is also manifest by an intense band at ca. $900cm^{-1}$ in the IR spectra of the polymers.(see fig.2)[5].

Preliminary studies of molecular weights by vapor pressure osmometry led to the conclusion that polymers derived from phenylsilane and from n-hexylsilane both had molecular weights in the range of 1000 to 1500 Dalton[4]. We have attempted to refine our knowledge of molecular weight by the application of exclusion chromatography (GPC)[5]. Using a column that resolves standard polystyrenes of M_n = 500, 1000 and 2000 ($M_w/M_n < 1.04$), all polymers examined exhibited extremely narrow dispersions, and apparently consisted of one or two molecular species (fig.3). The retention times corresponded to molecular dimensions equivalent to polystyrenes of M_n = 1000 to 1500.

Each sample characterized by GPC was also examined by MS using a direct inlet probe[5]. The mass spectra of polymers were in accord with the assigned structure. In particular, the linear as opposed to cyclic nature of the polymers is clear from a comparison of the spectrum of $Ph_5Si_5H_7$ prepared by catalytic coupling with that of the cyclic $Ph_6Si_6H_6$ reported by Hengge et al.[6](table I). The sample of the pentamer was prepared by running the catalytic reaction at $5^{o}C$ and although it is unusual in giving a detectable parent molecular ion, it otherwise behaves very similarly to molecules of higher molecular weight. Of particular note is the fact that the linear polymers give

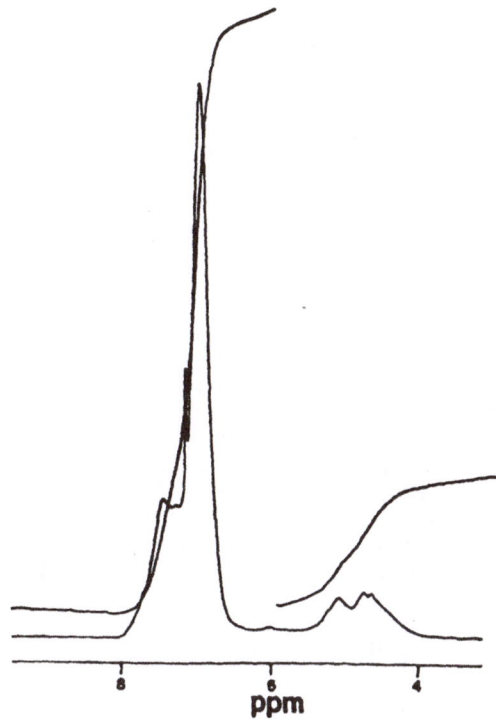

Figure 1. Proton NMR spectrum of poly(phenylsilylene).

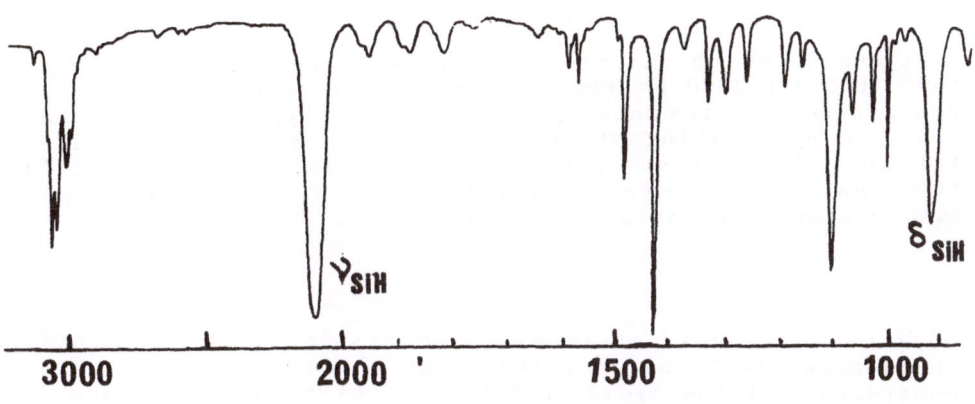

Figure 2. IR spectrum of poly(phenylsilylene).

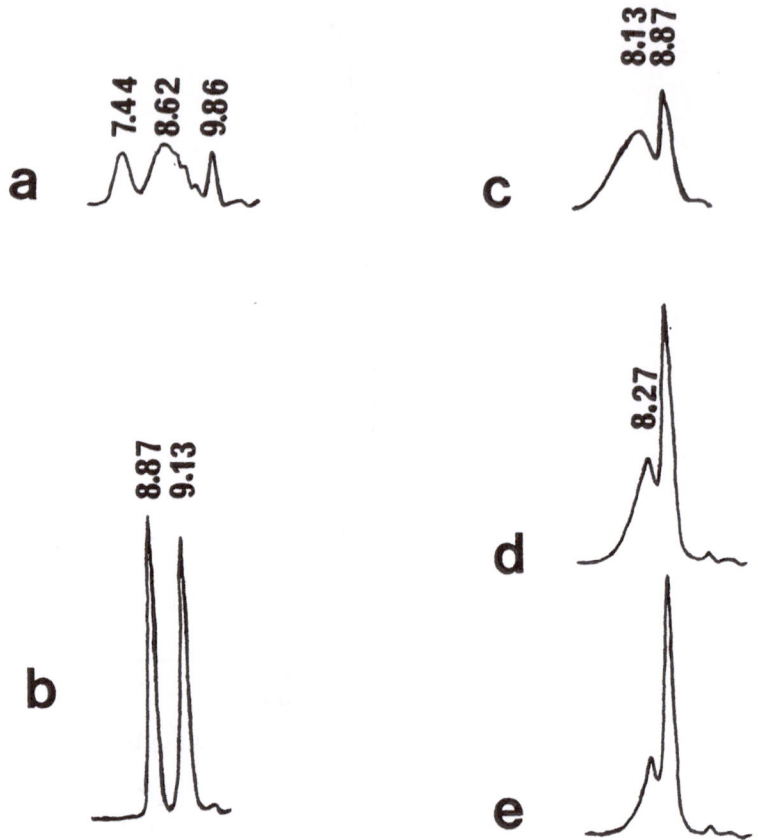

Figure 3. GP chromatograms of a) polystyrene standards of 500, 1000 and 2000 Dalton b) a mixture of phenylmethylcyclo-tri and tetrasiloxanes c), d) and e) polyphenylsilylenes produced by coupling of phenylsilane in toluene (ratio 1/0, 1/1 and 1/2 v/v respectively) with dimethyltitanocene (1 %) for a period of seven days. Conditions for chromatography are constant. Retention times are in min.

only small amounts of heavy fragments since even a single chain scission leads to a reduction in molecular size. This contrasts with the cyclic which gives even the parent ion in l·arge abundance with apparent stepwise loss of silicon . The linear molecules all give spectra characterized by very weak or absent parent ion, facile loss of alkyl or aryl groups and facile stripping of hydrogen atoms.

Table I. Mass spectra of polysilanes.

$Ph_5Si_5H_7$ [1]		$Ph_6Si_6H_6$ [2]	
Ion	Mass(abundance)	Ion	Mass(abundance)
$Ph_5Si_5H_7$	532(0.45)	$Ph_6Si_6H_6$	636(30)
$Ph_5Si_4H_3$	499(0.35)	$Ph_5Si_5H_5$	530(20)
$Ph_4Si_5H_4$	452(1.13)	$Ph_4Si_5H_4$	452(50)
$Ph_4Si_4H_2$	422(2.95)	Ph_4Si_4H	421(40)
$Ph_3Si_4H_3$	346(7.18)	$Ph_3Si_5H_5$	376(16)
Ph_3Si_3H	316(12.4)	Ph_3Si_4H	344(75)
Ph_3Si_2	287(27.0)	Ph_3Si_3	315(20)
Ph_3Si	259(24.0)	Ph_3Si	259(60)
$Ph_2Si_3H_2$	240(25.0)	Ph_2Si_3	238(12)
Ph_2Si_2H	211(42)	Ph_2Si_2	209(16)
Ph_2Si H	183(47)	Ph_2Si H	183(80)
PhSi	105(100)	PhSi	105(100)

1. Polymer prepared from $PhSiH_3$(1 ml) and Cp_2TiMe_2(5 mg.) at
−2C for 7 days.
2. Data from ref.6.

From the above studies we conclude that the polymers
are linear and hydrogen terminated. The molecular sizes are
of much too narrow a distribution and much too independent
of reaction conditions and silane structure to be accounted
for by a conventional propagation process. It seems more
likely that some kind of size selective process is involved.
Since achievement of molecular weight control is one of the
outstanding goals of this work, an understanding of this
process will be essential to progress.

THE GENERALITY OF THE REACTION.

Other catalyst systems. A number of cyclopentadienyl
derivatives of early transition metals and actinides have
been screened for catalytic activity for the dehydrogenative
coupling of silanes. The results of this screening are
summarized in table II. Dimethylzirconocene and dihydrido-
zirconocene dimer were both highly active and gave oligomers
of the same type as obtained with titanium catalysts. No
significant catalytic activity was obtained for the metallo-
cenes and metallocene hydrides of groups 5 and 6[7].

Attempts to activate Cp_2MoH_2 and Cp_2TaH_3 photochemically and by the addition of proton and hydride sponges were all unsuccessful.

Table II   Reactions of Cp_2MR_n with phenylsilane.

M	R	n	PRODUCT	CATALYSIS
Cr	–	0	NONE	NO
Mo	H	2	$Cp_2MoH(SiH_2Ph)$	NO
W	H	2	NONE	NO
V	Me	2	NONE	NO
V	–	0	NONE	NO
Nb	H	3	$Cp_2NbH_2(SiH_2Ph)$	NO
Ta	H	3	$Cp_2TaH_2(SiH_2Ph)$	NO
Th	Me	2	?	DIMERIZATION
U	Me	2	?	OLIGOMERS

$Cp_2^*ThMe_2$ was active for the catalytic dimerization of phenylsilane in ether solvent but was deactivated by aromatic hydrocarbon solvents [8]. The uranium analogue catalyzed the oligomerization of phenylsilane in ether, but in poor yield[8]. Given the difficulty in using these compounds and the lack of any advantage over the titanium group catalysts they have not been studied in detail.
Other substrates. Phenylgermane is rapidly coupled to a three dimensional gel in the presence of catalytic amounts of dimethyltitanocene [9]. Diphenylgermane was found to undergo a slow dimerization in the presence of catalytic amounts of pure dimethyltitanocene [7]. On exhaustion of the substrate the solution suddenly changes colour from pale yellow to dark purple with a surge of gas evolution. The resulting purple solution is highly active for the oligo-merization of diphenylgermane. The polygermane chain builds

up in a stepwise fashion and the presence of dimer, trimer and higher oligomers is evident from 1H-nmr spectra of reacting solutions. The purple product can also be prepared from a stoichiometric reaction of diphenylgermane with dimethyltitanocene. The catalytic dimerization cannot be sustained with impure dimethyltitanocene due to premature transformation to the purple species.

Unlike triphenylsilane and triphenylstannane, triphenyl-germane reacts with dimethyltitanocene to give an isolable methyltriphenylgermyltitanocene according to equation 1:[7]

$$Cp_2TiMe_2 \ + \ Ph_3GeH \ \longrightarrow \ Cp_2TiMe(GePh_3) \ + \ CH_4 \quad (1$$

Borane adducts of the kind LBH_3 are isoelectonic with primary silanes. We have failed to observe catalytic coupling with the test substrates L = diethylether, THF, thiophene, pyridine, or trimethylamine. The ether complexes give the relatively inert titanoceneborohydride while the amine complexes do not react at all. Titanoceneborohydride does not catalyze the coupling of silanes.

THE MECHANISMS OF SILANE AND GERMANE COUPLING.

By reaction s of dimethyltitanocene with phenylsilane in appropriate ratios we have been able to isolate and structurally characterize the two binuclear complexes 1 and 2 [10].

1   2

Although compound 2 spontaneously decomposes in solution to 1 and polysilane and 1 reacts with phenylsilane to give 2

110

and hydrogen, these reactions occur much too slowly to account for the catalytic reaction. We have studied the kinetics of the co-hydrogenation of olefins according to equation 2 [11] :

$$RSiH_3 + \overset{}{\underset{}{C}} = \overset{}{\underset{}{C} } \xrightarrow{Cp_2TiMe_2} H\overset{R}{\underset{H}{(Si)_n}}H + H-\overset{|}{\underset{|}{C}}-\overset{|}{\underset{|}{C}}-H \qquad (2$$

In this reaction the evolution of H_2 is completely suppressed, as are complications associated with the couping reaction occuring in a soution saturated with hydrogen. The co-hydrogenation system permits the study of kinetics and also extends reactivity to more sterically hindered silanes which do not react at a significant rate in the absence of olefin. These silanes (indicated by * below) give dimers rather than higher oligomers. The observed order of reactivity for several silanes is the following:

PhSiH_3 > p-tolylSiH_3 > PhMeSiH_2* > PhSiD_3 > n-HexylSiH_3 >
(13.2)   (9.8)   (4.6)   (3.6)   (1.0)
BenzylSiH_3 > CyclohexylSiH_3*
(1.0)   (0.5)

In the presence of cyclohexene,the polymerization of phenyl-silane follows the rate law:

$$Rate = k^{\prime} [catalyst]^{1/2}[silane]^{1/2}[olefin]^0 \qquad (3$$

We interpret this behaviour in terms of a pseudo-equilibrium in which 1 undergoes cleavage by reaction with $PhSiH_3$, thus:

$$Cp_2Ti(H)(HSiPhH)TiCp_2 + PhSiH_3 \longrightarrow 2 Cp_2Ti(H)SiH_2Ph \qquad (4$$
$$\underset{\text{1}}{} \qquad\qquad\qquad\qquad\qquad\qquad \underset{\text{3}}{}$$

The hydridosilyltitanium(IV) species is then expected to undergo α-hydride elimination leading to loss of H_2 and formation of a silylenetitanium intermediate. A number of routes for the homologation of the silylene can be imagined, one of which is shown in Scheme 1. This type of mechanism explains most of the observed features of the oligo-merization reaction. The necessity for an α-hydride elimination precludes reaction of secondary silanes beyond the dimer by this route. The strong steric effect on the rate for different silanes is explained by the necessity for the silyl ligand to assume an eclipsed configuration

Scheme 1. Proposed mechanism for $RSiH_3$ polymerization

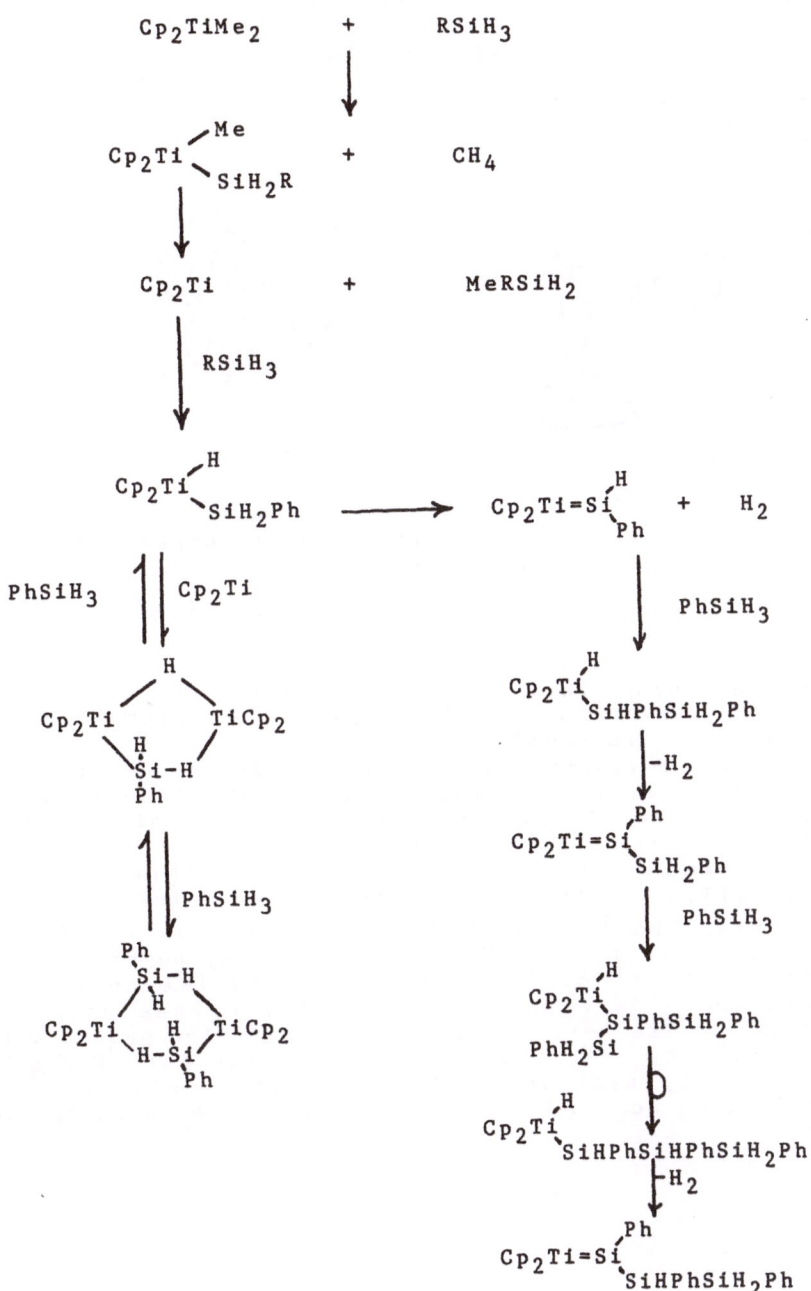

relative to the Cp$_2$TiH fragment in order for the Si-H to
transfer to the Ti-H as shown in fig.4. In the eclipsed
configuration, the Si-H overlaps optimally with the empty
titanocene LUMO. This overlap is viewed as critical and the
occupancy of this orbital by electrons, or ligands is viewed
as the most likely explanation for the unique activity of
the group 4 catalysts.

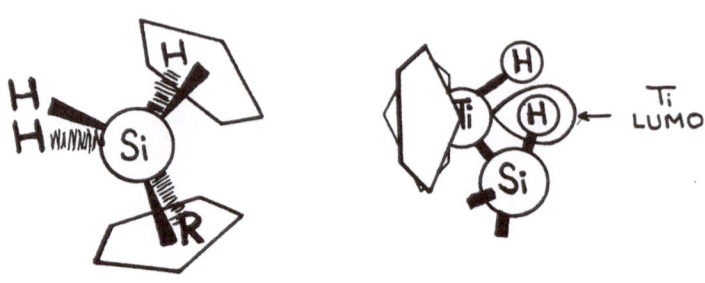

Figure 4. Role of conformation and the empty LUMO on Ti in
α-hydride transfer.

Although we have no direct evidence for the
participation of the species 3, the zirconium analogue
participates in the chemistry of the zirconocene catalyzed
silane polymerization. A stoichiometric reaction of dimethylzirconocene with phenylsilane gives the crystalline compound
4 [12]. This compound is a dimer of the zirconocene analogue
of 4 with a fragment which can be imagined to arise from the
insertion of a phenylsilylene moiety into a Zr-Me bond.
Phenylsilane also reacts with zirconocenedihydride to give
the symmetrical phenylsilylhydride 5 in solution, but we
have been unable to isolate this compound as a pure
crystalline solid. It is identified in solution from its NMR
spectra. The ^{1}H- and ^{29}Si-NMR parameters for 4 and 5 are
listed in table III. These compounds may be compared to the
known dihydrides [Cp$_2$ZrH$_2$]$_2$ [13] and [Me-Cp$_2$ZrH$_2$]$_2$ [14].
Compound 4 is a poor catalyst for silane coupling, due
partly to its very low solubility and also probably to a low
equilibrium constant for dissociation into the monomeric
phenylsilylhydride. We believe that rapid silane oligo-
merization occurs by the same route as shown in Scheme 1 for
Ti.

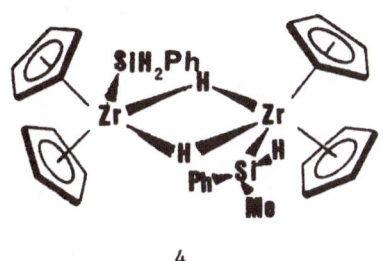

4

At higher temperatures (>60C) dimethylzirconocene also catalytically dimerizes methylphenylsilane and diphenylsilane. Again, we believe that this dimerization occurs via homolysis of Zr-Si bonds. Although zirconium(IV) is more difficult to reduce than titanium(IV), paramagnetic zirconium hydride species have been detected by ESR in silane/dimethylzirconocene solutions. The concentration of these species is greatly increased by heating or photolysis with UV light.

POLYORGANOSILYLENES AS NOVEL MATERIALS.

The polyorganosilylenes described in this work are unusual in that they cannot be prepared by classical Wurtz-type reactions. The presence of Si-H along the backbone and at the chain ends offers the opportunity for a wide variety of functionalizations. This dimension has thus far been very little explored. We have however demonstrated that all of the Si-H groups may be hydrosilated with olefins, or the end SiH$_2$ functions may be selectively hydrosilated under appropriate conditions. The latter reaction with dienes, or diynes, permits the concatenation of the oligosilanes to higher molecular weight materials.

Si-H funtions are known to react readily with a wide variety of transition metal complexes with formation of metal-Si bonds. These kinds of reactions allow the preparation of a large new range of polymeric compounds, either with heterometal atoms as part of the backbone of a copolymer, or with pendant heterometal substituents. Such materials would be expected to exhibit their own intersting photochemistry and electronic properties. They could also yieldinterestingpyrolysisproducts.

Table III ^{1}H-NMR data for some silylhydride and hydride complexes of zirconocene.

REACTION	NMR	ASSIGNMENT		
				SiMePhH
			H	
PhSiH$_3$ + Cp$_2$ZrMe$_2$	5.3-5.5	Cp$_2$Zr		ZrCp$_2$
	4.9(m)		H	
	4.8(s)	PhH$_2$Si		
	-5.0(ABq)			
				SiH$_2$Ph
			H	
PhSiH$_3$ + Cp$_2$ZrH$_2$	5.4(s)	Cp$_2$Zr		ZrCp$_2$
	4.8(S)		H	
	-4.9(s)	PhH$_2$Si		
				H
			H	
Ph$_2$SiH$_2$ + Cp$_2$ZrMe$_2$	3.8(dd)	Cp$_2$Zr		ZrCp$_2$
	-4.0(m)		H	
	-4.5(m)	X		
				H
			H	
Ref.13	3.85(t)	Cp$_2$Zr		ZrCp$_2$
	-3.45(t)		H	
			H	
				H
			H	
Ref.14	3.75(t)	Cp$_2'$Zr		ZrCp$_2'$
	-2.98(t)		H	
			H	

The polysilylenes with a single organic substituent are considerably more reactive than the peralkylated and perarylated analogues. For example the undergo autoxidation to poly(siloxanes) on standing in air, in solution ($t_{1/2}$ days) or the solid state ($t_{1/2}$ months). The oxidation occurs selectively at the Si-Si bond, suggesting that it is not a free radical process. Attempts to effect the analogous reaction with sulfur have been unsuccessful.

In conclusion we will mention the intriguing possiblity of catalytic coupling of SiH_4. Clearly, it would be of great technological interest to be able to catalytically decompose SiH_4, either to amorphous silicon or to (SiH) under mild conditions. We have successfully polymerized SiH_4 to $(SiH_2)_n$ using dimethyltitanocene, but for reasons that are not evident, dimethylzirconocene is not active. The stucture of the $(SiH_2)_n$ polymer is based on the value of   Si-H from the IR spectrum [15]. It is very insoluble in the organic solvent medium which probably accounts for the failure of the secondary SiH functions to react further to any significant degree. It is also pyrophoric and extremely difficult to handle.

ACKNOWLEDGEMENTS

The author thanks all those co-workers whose names are cited in the references. Financial support for the work from the Natural Sciences and Engineering Research Council of Canada, the Fonds FCAR du Québec, the Dow Corning Corporation and Esso Canada are gratefully acknowledged.

REFERENCES

1. Rösch,L; Altnau,G.;Erb,W.;Pickardt,J.; Bruncks,.N. J.- Organomet.Chem. (1980) 197, 51.
2. Hencken,G.; Weiss,E. Chem.Ber. (1973) 106, 1747.
3. Samuel,E; Harrod,J.F. J.Am.Chem.Soc. (1984) 106, 1859.
4. Aitken,C.; Harrod,J.F.; Samuel,E. J.Organomet.Chem. (1985) 279, C11.
5. Harrod,J.F.; Gill,U. unpublished results.
6. Hengge,E.; Lunzer,F. Monatshefte für Chemie, (1976) 107, 371.
7. Malek,A; Harrod,J.F., unpublished results.
8. Barry,J-P.; Harrod,J.F., unpublished results.
9. Aitken,C.; Harrod, J.F., unpublished results.
10. Aitken,C.; Harrod,J.F.; Samuel,E. J.Am.Chem.Soc. (1986) 108, 4059.
11. Yun,S.S.;Harrod,J.F. unpublished results.
12. Harrod,J.F.; Aitken,C; Samuel,E. Can.J.Chem. (1986) 64, 1677.
13. Bickley,D.G.; Nguyen,H.; Bougeard,P.; Burns,R.C.; McGlinchey, M.J. J.Organomet.Chem. (1983) 246, 257.
14. Jones,S.B.; Petersen,J.L. Inorg.Chem. (1985) 20, 2889.
15. Jeffrey,F.R.; Shanks,H.R. U.S.Patent, 4,353,788, Oct.12, 1982.

NEW POLYCARBOSILANE MODELS:

PREPARATION AND CHARACTERIZATION

OF A POLY(METHYLCHLORO)SILMETHYLENE

Eric BACQUÉ, Jean-Paul PILLOT,* Marc BIROT and Jacques DUNOGUÈS
Laboratoire de Chimie organique et organométallique (U.A.35 CNRS)
Université de Bordeaux I, 351 cours de la Libération,
33405 Talence cedex — France.

SYNOPSIS

In order to prepare new polycarbosilane models,
polydimethylsilmethylene $-(Me_2SiCH_2)_n$ chain was modified by a chemical
route. The involved aluminium chloride catalyzed interchange reaction
of methyl and chlorine on silicon, using trimethylchlorosilane, points
to a selective method of preparing a poly(methylchloro)-silmethylene
in high yield. This polymer constitutes a convenient intermediate for
the synthesis of others functional polycarbosilanes.

INTRODUCTION

Yajima's route for the obtention of SiC requires the pyrolysis of
polycarbosilanes (PCS) which are prepared by thermal rearrangement of
polydimethylsilane (PDMS), this last polymer resulting from the dechlorination
- polycondensation of dimethyldichlorosilane[1]. This strategy is represented
in Scheme 1.

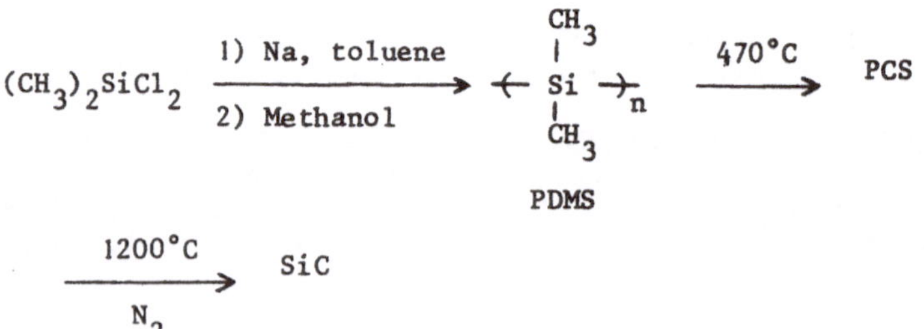

$$(CH_3)_2SiCl_2 \xrightarrow[\text{2) Methanol}]{\text{1) Na, toluene}} \left(\underset{\underset{CH_3}{|}}{\overset{\overset{CH_3}{|}}{Si}} \right)_n \xrightarrow{470°C} PCS$$

PDMS

$$\xrightarrow[N_2]{1200°C} SiC$$

Scheme 1

---

* Authors to whom correspondence should be adressed

116

R. M. Laine (ed.), Transformation of Organometallics into Common and Exotic Materials: Design and Activation, 116–132.
© 1988 by Martinus Nijhoff Publishers.

This approach suffers from the limitation that the structures of the involved polymers (PDMS and PCS) remain unclear.

The insolubility of PDMS hinders any detailed characterization of this material. In particular, the quantity and the structural influence of oxygen introduced by the methanolytic-hydrolytic workup is not determinable.

Furthermore the polycarbosilane structure is generally represented by the alternation of methylene and methylhydrogenosilylene units, on the asumption that the thermal transformation proceeds exclusively according to Kumada's rearrangement[2]. This reaction was first observed with disilanes and formally amounts to the insertion of a methylene group into an Si-Si bond according to :

$$
\begin{array}{ccc}
\underset{\displaystyle\overset{|}{\underset{|}{Me}}}{Me} \quad \underset{\displaystyle\overset{|}{\underset{|}{Me}}}{Me} & \overset{\Delta}{\longrightarrow} & \underset{\displaystyle\overset{|}{\underset{|}{Me}}}{H} \quad \underset{\displaystyle\overset{|}{\underset{|}{Me}}}{Me} \\
- Si - Si - & & - SiCH_2Si - \\
\end{array}
$$

Unfortunately the actual polycarbosilane cannot be entirely represented by the ideal structure $-(MeHSiCH_2)-_n$. Besides the presence of Si-O bonds due to the PDMS oxygenation, Yajima[3] and other authors[4] have reported that PCS possesses a certain degree of cross-linking which cannot be taken into account by Kumada's transposition.

Okamura has even suggested the existence of cyclic patterns resulting from unknown secondary reactions leading to a complex framework for Yajima's PCS.

Other than these results, only a paucity of information concerning the transformations polysilane-polycarbosilane and polycarbosilane-SiC can be found in the literature. In particular, for any given PCS precursor , the misappreciation of its backbone structure rules out establishing any correlation between its structure, its thermal behaviour and its ability to yield high SiC conversion rate .

In order to resolve these problems, the preparation of polycarbosilane models with clearly established structures, appeared necessary. So, we report herein the synthesis of an original linear polycarbosilane — the poly(methylchloro)silmethylene —with highly reactive Si-Cl bonds, thus allowing the preparation of further models for the organosilicon precursor-ceramic (SiC or SiC/Si_3N_4) transformation.

RESULTS AND DISCUSSION

Polydimethylsilomethylene (II) was known to possess an unambigous linear structure[5]. This soluble and unoxygenated polymer is readily obtained from 1,1,3,3-tetramethyl-1,3 disilacyclobutane (I), in high yields under mild conditions[6], according to eq. (1) :

$$\frac{n}{2} \quad \text{(I)} \quad \xrightarrow{\varepsilon H_2PtCl_6 \cdot 6H_2O} \quad \leftarrow Si - CH_2 \rightarrow_n \quad (1)$$

(I)   (II)

By bulk polymerization of (I) at 80°C, a polymer (II) was obtained as a brown, slightly tacky rubber, soluble in common organic solvents and flowing within a few days. GPC determination of its molecular weight distribution showed the presence of high molecular weight structures (up to values of several million) and gave $\overline{M}_n$ = 250,000.

(II) was reported to be stable up to 600°C in an inert atmosphere[7] but it left no residue after pyrolysis at 1000°C[8]. This property, attributed to its linearity, relegates this polymer to being a precursor of SiC.

On the other hand, substituted disilacyclobutanes are much less reactive than (I). In the same conditions $(ClMeSiCH_2)_2$ for instance, yielded only a viscous liquid with low molecular weight structures[5]. Moreover, because of the delicate synthesis involved for the preparation of this chlorinated compound, polymer (II) appeared to be a more convenient starting material for our purposes.

In order to take advantage of its interesting properties, it seemed logical to undertake its chemical modification, in the perspective of preparing suitable SiC precursor models. Our strategy involved the conversion of (II) into a chlorinated polymer (III). Then exploiting the reactivity of Si-Cl bonds, it would seem feasible to prepare new polycarbosilane models as shown in Scheme 2 :

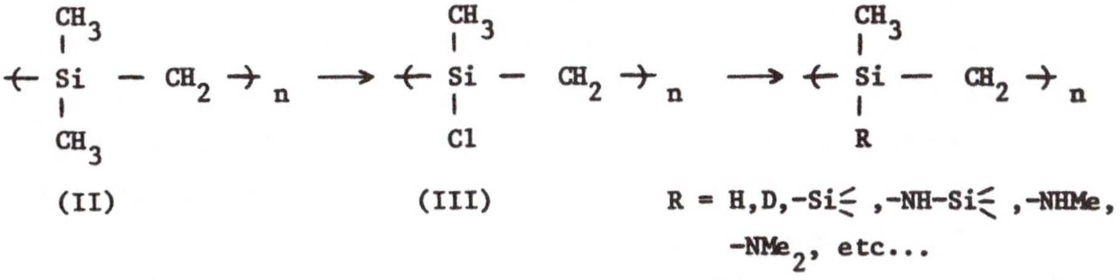

$$
\underset{(II)}{\left\langle\,\underset{\underset{CH_3}{|}}{\overset{\overset{CH_3}{|}}{Si}}-CH_2\,\right\rangle_n}\;\longrightarrow\;\underset{(III)}{\left\langle\,\underset{\underset{Cl}{|}}{\overset{\overset{CH_3}{|}}{Si}}-CH_2\,\right\rangle_n}\;\longrightarrow\;\left\langle\,\underset{\underset{R}{|}}{\overset{\overset{CH_3}{|}}{Si}}-CH_2\,\right\rangle_n
$$

R = H, D, -Si≤ , -NH-Si≤ , -NHMe,
-NMe$_2$, etc...

Scheme 2

1) CHLORINATION OF (II)

Many routes effecting the conversion of Si-CH$_3$ bonds into Si-Cl bonds have been reported for short chain molecules, but most of them are inapplicable in the case of polymers.

Furthermore, quantitative and selective reactions are needed to yield the model compounds required, in particular the substitution of two methyl groups on the same silicon must be avoided.

Two methods corresponding to our needs have been selected : the action of iodine chloride and the action of trimethylchlorosilane in the presence of AlCl$_3$.

a) *Action of ICl*

This reagent has already been used in our laboratories to carry out the selective substitution of alkyl groups in tetraalkylsilanes[9], according to eq. (2) :

$$
R_3SiCH_3 + 2\ ICl \xrightarrow{\quad CCl_4 \quad} R_3SiCl + I_2 + CH_3Cl \qquad (2)
$$

Those reactions are regarded to be selective so long as the associated R groups are bulky enough (for R = iPr, nBu, etc...). The progression of the reactions can be followed both by NMR spectroscopy and gas evolution. When reacted with a stoichiometric amount of ICl and after devolatilization under reduced pressure, polymer (II) led to a flowing and viscous purple oil (III$_a$).

The ^{1}H-NMR spectrum (60 MHz, δppm) showed deshielded resonance peaks centred at 0.5 ppm corresponding to the expected ($-CH_3SiCl-$) units, a large signal at 0.05 ppm revealing appreciable residual Me_2Si units, overlapping with the methylene protons and a weak peak at 2.8 ppm, assigned to $\geq SiCH_2Cl$ groups. Furthermore, the final weight of (III_a) and the volume of gas are less than expected (78% and 85% of their respective theoritical values). All these observations suggest not only that $SiCH_3$ bonds are cleaved by ICl, but that $SiCH_2$ bonds are also involved, leading to the formation of both Si-Cl and Si-CH_2Cl bonds. As ICl was stoichiometrically introduced, resulting side reactions decrease the amount of gas, and unreacted Me_2Si units remain. Weight deficiency of (III_a) may result from splitting of two neighbouring Si-CH_2 bonds, accompanied by formation of low b.p. by-products eliminated during the devolatilization.

Thus, polymer (III_a) cannot be considered as a suitable model because it possesses a complex structure. The poor selectivity of the reaction, coupled with the difficulties of handling large amounts of ICl, led us to investigate another route.

b) *Action of Me$_3$SiCl/AlCl$_3$*

M. KUMADA *et al.* found that the selective substitution of a methyl group per silicon atom was readily performed on molecules such as bis(trimethylsilyl)methane[10]. The principle of this reaction was applied to polymer (II) and the modified polymer (III_b) was obtained according to eq. (3) :

$$
\begin{array}{ccc}
\underset{|}{\overset{CH_3}{\underset{CH_3}{\overset{|}{Si}}}} - CH_2 \Big)_n
&
\xrightleftharpoons[AlCl_3]{(CH_3)_3SiCl, \text{ excess}}
&
\underset{|}{\overset{CH_3}{\underset{Cl}{\overset{|}{Si}}}} - CH_2 \Big)_n + n(CH_3)_4Si \quad (3)
\end{array}
$$

(II)   (III_b)'   TMS

In order to shift the equilibrium,   TMS must be removed by distillation.

A large excess of trimethylchlorosilane allowed the initial solubilization of (II), and a catalytic amount of $AlCl_3$ (5% by weight relative to (II)) was added to this solution, which was then warmed to reflux. Within a few hours, more than 95% of the expected yield of TMS was recovered upon distillation. After filtration and devolatilization under reduced pressure at 100°C, polymer (III_b) was obtained as a brown viscous oil in quasi-quantitative yield, with only a few drops of low-boiling point products (other than TMS) being detected.

Polymer (III_b) is in a fluid state at room temperature. It is soluble in polar solvents (ethers, aromatic and chlorinated solvents) but insoluble in non-polar solvents such as hexane. A comparison of the GPC curves of polymers (II) and (III_b) shows a marked decrease of the average molecular weights for the chlorinated polymer ($\overline{Mn}$ = 2330 instead of 250 000), revealing that a few $SiCH_2$ bonds have been broken (Fig.1). Thus it can be assumed that two reactions converge in the formation of (III_b), as shown in Scheme 3 :

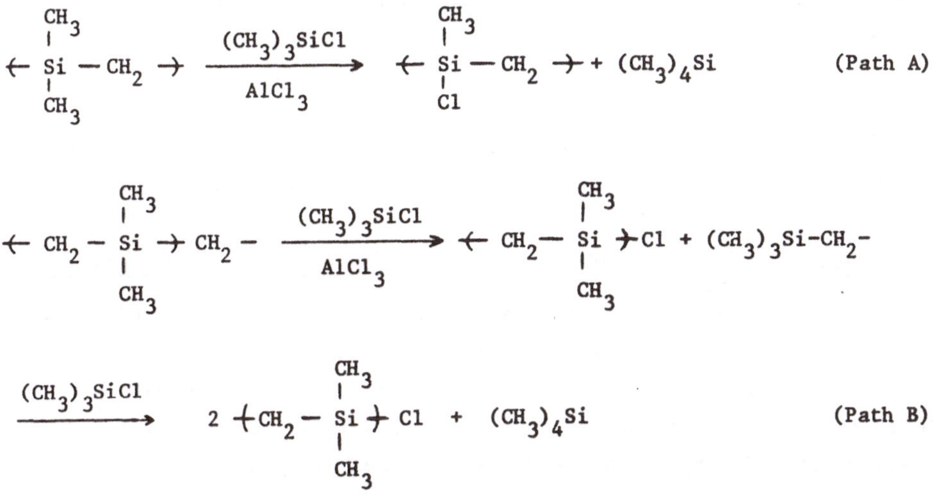

$$\begin{array}{ccc}
\overset{CH_3}{\underset{CH_3}{\overset{|}{\underset{|}{\leftarrow Si}}}} - CH_2 \rightarrow & \xrightarrow[\;AlCl_3\;]{(CH_3)_3SiCl} & \overset{CH_3}{\underset{Cl}{\overset{|}{\underset{|}{\leftarrow Si}}}} - CH_2 \rightarrow + (CH_3)_4Si \qquad \text{(Path A)}
\end{array}$$

$$\begin{array}{cc}
\overset{\qquad CH_3}{\leftarrow CH_2 - \underset{\underset{CH_3}{|}}{\overset{|}{Si}} \rightarrow CH_2 - } & \xrightarrow[\;AlCl_3\;]{(CH_3)_3SiCl} \qquad \overset{CH_3}{\leftarrow CH_2 - \underset{\underset{CH_3}{|}}{\overset{|}{Si}} \rightarrow Cl + (CH_3)_3Si-CH_2-}
\end{array}$$

$$\xrightarrow{(CH_3)_3SiCl} \qquad 2 \overset{\qquad CH_3}{\underset{\underset{CH_3}{|}}{\overset{|}{\leftarrow CH_2 - Si \rightarrow}}} Cl + (CH_3)_4Si \qquad \text{(Path B)}$$

Scheme 3

Looking at the $\overline{Mn}$ values, it can be inferred that path A is largely predominant, contributing by more than 90% to the formation of the polymer. The theoritical amount of TMS is not significantly modified by path B which, like path A, produces one TMS molecule per substituted silicon. The

first step of path B slightly increases the expected final polymer quantity.
Thus the action of $Me_3SiCl/AlCl_3$ led to a polycarbosilane with a linear
backbone, $-(MeClSiCH_2-)_n$ exhibiting $ClMe_2Si$ units at the chain terminals.
The structure of polymer (III_b) can therefore be written according to :

$$Cl - \underset{\underset{CH_3}{|}}{\overset{\overset{CH_3}{|}}{Si}} - CH_2 \overset{}{\overbrace{}} \underset{\underset{Cl}{|}}{\overset{\overset{CH_3}{|}}{Si}} - CH_2 \overset{}{\underbrace{}}_n \underset{\underset{CH_3}{|}}{\overset{\overset{CH_3}{|}}{Si}} - Cl \qquad \text{with } \bar{n} = 23 \qquad \text{(III b')}$$

The $^1H - {}^{29}Si$-and ^{13}C-NMR spectra of polymer (III_b) are depicted in
Figures 2-5, and IR data listed in Table 1. As it can be seen in Figure 2c,
its ^1H-NMR spectrum (60 MHz) shows the characteristic and reproducible
signals at 0.71, 0.63 and 0.53 ppm. No $SiCH_2Cl$ or $SiCH_3Cl_2$ groups were
observed. However, when the reaction did not reach completion, the spectrum
exhibited three other signals centered at 0.2 ppm (assigned to CH_2SiMe_2
remaining units), whose intensity decreases as the reaction proceeded
(Fig. 2b).
    ^{13}C-and ^{29}Si-NMR spectra of (III_b) both exhibit three separate signals
for each nucleus, which are believed to result from stereosensitivity effects
induced by asymetric silicon atoms (Fig. 4a and 5). Studies with regards to
short-chain models are in progress and will be soon reported.

c) *Synthesis of 2,4,6-trichloro, 2,4,6-trimethyl, 2,4,6-trisilaheptane (V)*

In order to explain the high-field ^1H-NMR spectra of (III_b) (Fig. 3a),
we have synthesized this novel short-chain model following a two-step
procedure. The first step involved the condensation of methyldichlorosilane
with the Grignard reagent prepared from dimethyl(chloromethyl)silane[11]
yielding 2,4,6-trimethyl, 2,4,6-trisilaheptane IV (eq. 4) :

$$CH_3SiHCl_2 + 2CH_3SiHCH_2MgCl \xrightarrow{THF} \underset{\underset{H}{|}}{\overset{\overset{CH_3}{|}}{CH_3-Si}}-CH_2-\underset{\underset{H}{|}}{\overset{\overset{CH_3}{|}}{Si}}-CH_2-\underset{\underset{H}{|}}{\overset{\overset{CH_3}{|}}{Si}}-CH_3 \qquad (4)$$

$$\text{(IV)} \qquad 66 \text{ \%}$$

Subsequent chlorination of (IV) by CCl_4/Pd-C according to L.H. SOMMER[12] led to compound (V) (eq. 5) :

$$IV \xrightarrow[\epsilon Pd/C]{CCl_4} CH_3-\underset{\underset{Cl}{|}}{\overset{\overset{CH_3}{|}}{Si}} - CH_2 - \underset{\underset{Cl}{|}}{\overset{\overset{CH_3}{|}}{Si}} - CH_2 - \underset{\underset{Cl}{|}}{\overset{\overset{CH_3}{|}}{Si}} - CH_3 \qquad (5)$$

$$(V) \qquad 83\ \%$$

200 MHz ^{1}H- and 39.8 MHz ^{29}Si-NMR spectra of compound V are given in Fig. 3b and 5. The comparison with the spectra of polycarbosilane (III_b) allowed an unambigous attribution of principal signals of this polymer. So the ^{1}H-NMR signals of the methylene protons and the internal methyl protons appeared at 0.76 and 0.62 ppm respectively, meanwhile the terminal methyl protons gave a signal at 0.52 ppm. Integration and subsequent calculation of the intensity ratios yielded an average value for n equal to 23, in good agreement with the GPC one.

Moreover, the ^{29}Si resonance peaks of (III_b) at 25.70 and 28.35 ppm were assigned respectively to the internal and chain-end silicon nuclei by comparison with the ^{29}Si chemical shifts of the carbosilane (V).

CONCLUSION

Reaction of trimethylchlorosilane with polydimethylsilmethylene $(-Me_2SiCH_2-)_n$, in the presence of a catalytic amount of $AlCl_3$ led readily and quantitatively to a novél chlorinated polycarbosilane, $-(MeSiCl-CH_2-)$. It has been demonstrated that this polymer possesses a linear skeleton with $SiMe_2Cl$ groups at the chain end terminals. It opens the way to new models of polycarbosilanes with a clearly defined structure,owing to the versatile reactivity of its SiCl bonds. Correlation between precursor structure and its capacity to give SiC ceramics will be pointed out for the first time.

EXPERIMENTAL

Chlorosilanes

Commercially available trimethylchlorosilane, dichloromethylsilane, and chloromethyldimethylchlorosilane, were purified before use by distillation over magnesium turnings, under nitrogen in a column filled with glass helices.

1,1,3,3-tetramethyldisilacyclobutane (I)

This compound, synthesized according to Kriner's method $(bp_{760}=119°C)$, was a generous gift from Rhône Poulenc Industries.

Chloromethyldimethylsilane[14]

Chloromethyldimethylsilane was prepared by reduction of $ClMe_2SiCH_2Cl$ with lithium aluminium hydride in ether (72% yield).

Solvents

The tetrahydrofuran (THF) used as the Grignard reaction solvent was dried before use by repeated refluxing over a sodium-benzophenone solution under a dry nitrogen atmosphere, followed by distillation.

Apparatus

All the reaction vessels were thoroughly dried and purged with dry nitrogen before use.

Characterization

Molecular weights of the copolymers were determined by gel permeation chromatography (GPC) with four Microstyragel columns calibrated by polystyrene standards (porosity ranges of 500, 10^3, 10^4 and 10^5 Å) and with THF as the eluent at a flow rate of 1mL/mn. The detection system used was a Waters Associates Differential Refractometer R 401.

60 MHz proton NMR spectra were checked on an HITACHI PERKIN-ELMER R 24B spectrometer, in CCl_4 or $DCCl_3$ solutions (δ ppm) and benzene as internal standard.

200 MHz proton and 39.76 MHz ^{29}Si-NMR spectra were checked with a
Bruker AC 200 model, on polymer solutions in deuterochloroform or benzene-d6.

Infrared (IR) spectra were examined in the region 4000-600 cm^{-1}
(product films between NaCl plates).

1 - Preparation of polydimethylsilmethylene (II)

Disilacyclobutane (I) (100g, 0.695 mol) were introduced in a 1-L flask
fitted with a reflux condenser connected to a vacuum-argon line.

After introducing H_2PtCl_6.6 H_2O (0.25g), the reaction mixture was twice
degassed by the freeze and thaw technique, evacuating to 0.1 mmHg. Finally
the flask was allowed to warm to room temperature, under argon atmosphere.

Then the reaction flask was lowered into an oil bath maintained at 80°C.
After 3 minutes, a violent polymerization accompanied by the evolution of
white smoke occured, yielding a brown gum. Addition of 1.5 L. of warm
cyclohexane under magnetic stirring resulted within a few hours in the
complete solubilization of the polymer.

Filtration eliminated colloïdal platinium and, after concentration on a
rotary evaporator and devolatilization (0.5 mmHg at 100°C), polymer (II) was
obtained (96.5g, 96.5 %) under the form of a brown rubber.

2- Preparation of (III$_a$)

A one-liter two-necked, round-bottomed flask equipped with a magnetic
bar was fitted with a pressure-equalizing addition funnel and a reflux
condenser connected to a $CaCl_2$ column, relayed to a safety flask and to a filled
graduated 6-L flask inverted into a saturated salt-water tank in order to
measure the evolution of methylchloride.

Polymer(II)(19 g), dissolved in 200 mL of CCl_4 was first introduced
into the flask. Then ICl (85.6 g, 0.527 mol) in CCl_4 (250 mL) was dropwise
added at 20°C over 60 mn with stirring to the reaction mixture. Gas evolved
immediatly, the volume totalling 1 liter after 20 mn; then the rate decreases
to zero after complete addition.

Then the temperature of the reaction mixture was slowly increased to
reflux, in order to complete the reaction. Gas was evolved once again and
then stopped when five liters were collected (5.9 liters expected).

After 6 hours, the reaction mixture was left without warming under argon; then the solvent and iodine were eliminated under reduced pressure (1 mmHg) at 50°C.

Work up yielded 19 g (expected 24.4g)of a fluid purple oil (77.8%).

3 - Preparation of (III$_b$)

Polymer (II) (22.6 g, 0.313 mol) and Me$_3$SiCl (170.9 g, 1.575 mol) were stirred under dry nitrogen in a 500 mL flask until complete dissolution.

After addition of anhydrous AlCl$_3$ (1.25 g), the flask was set under a spinning band column equipped with a condenser in which icewater was circulated.

Then the viscous brown solution was warmed under magnetic stirring. When reflux temperature was reached, the mixture began to froth. After a few minutes, a notable fluidization was observed accompanied by the appearance of an orange brown coloration, as well as the first drops of TMS condensing at the top of the column. The TMS was then immediatly distilled and two hours later, more than the half theoritical quantity had been collected. The rate of TMS formation then slowly decreased and its separation became more delicate. After 8 hours, the formation of TMS was complete (reflux temperature at the column top did not fall below 50°C) and the mixture was left under argon. Thus, TMS (44 mL, theoretical value: 42.6 mL) containing some Me$_3$SiCl fractions were collected by distillation : NMR titration allowed the TMS quantity collected to be estimated as 97 % of the theoritical value.

Filtration under nitrogen atmosphere followed by concentration with a rotary evaporator yielded a viscous liquid. Devolatilization under vacuum (1 mmHg) at 100°C, permitted the elimination of any remaining Me$_3$SiCl and most of the aluminium chloride. 28.7 g of a fluid brown oil was thus obtained after cooling under argon. Assuming that the formulation is (MeClSiCH$_2$)$_n$, the reaction yield was estimated to be quantitative.

Elemental analysis based on (III b')formula :

	C %	H %	Cl %	Si %
calculated	26.28	5.58	38.03	30.13
found	25.78	5.60	37.20	29.60

4 – <u>Synthesis of</u> $(HMe_2SiCH_2)_2SiHMe$ (IV)

In a 500 ml three-necked flask equipped with a magnetic bar, fitted with a dropping funnel, a condenser, a thermometric well and protected from atmospheric moisture by a $CaCl_2$ column, magnesium powder (8.7 g, 0.35 mol) and dried THF (15 mL) were successivly introduced. The mixture was warmed to 60°C and a few mL of a solution consisting of THF (85 mL) and HMe_2SiCH_2Cl (27.2 g, 0.25 mol) was added.

Addition of a few drops of 1,2-dichloroethane initiated the reaction. The remaining solution was added dropwise in 90 mn to the dark mixture, with stirring.

The reaction mixture was refluxed for three days. After cooling $MeSiHCl_2$ (14.5 g, 0.126 mol) in THF (100 mL) was added in 30 minutes. A grey color soon appeared as the internal temperature increased to 60°C.

Then the mixture was refluxed for 24 hours. THF was eliminated under reduced pressure and the remaining white solid was washed with pentane (80 mL) before filtration. Addition of pentane and subsequent filtration were twice more and the resulting pale yellow solution was distilled, to afford (IV) (15.8 g, 66 %), b.p. 82°C/30 mmHg.

5 – <u>Synthesis of (V)</u> : $(ClMe_2SiCH_2)_2SiMeCl$

In a 250 ml three-necked flash equipped with a magnetic bar, a dropping funnel, a thermometric well and a condenser protected from atmospheric moisture by a $CaCl_2$ column, Pd/C (0.3 g) in CCl_4 (30 mL), was introduced. Then IV (6 g) (0.0315 mol) in CCl_4 (30 mL) was added dropwise over 25 mn with stirring. The temperature of the mixture rose to 50°C and the flask wwas cooled to 20°C.

After the addition was complete, the reaction mixture was left at room temperature for 20 h. At that time no Si-H bonds were detected by IR spectroscopy and the mixture was filtered off under argon atmosphere; the resulting colourless liquid was concentrated and then distilled under reduced pressure to give 7.8 g of compound (V) (7.8g, 83%, b.p. 80°C/1mmHg).

REFERENCES

(1) S. Yajima, K. Ohamura, J. Hayashi and M. Omori
    J. Am. Cer. Soc., $\underline{59}$ (7-8), 324(1976).

(2) K. Shiina and M. Kumada
    J. Org. Chem. $\underline{23}$, 139(1958).

(3) S. Yajima, Y. Hasegawa, J. Havashi and M. Iimura
    J. Mater. Sci., $\underline{13}$, 2569(1978).

(4) Y. Hasegawa and K. Okamura
    J. Mater. Sci., $\underline{21}$, 321(1986).

(5) W.A. Kriner
    J. Polymer Sci. A, $\underline{4}$, 444(1966).

(6) D.R. Weyenberg and L.E. Nelson
    J. Org. Chem., $\underline{30}$, 2618(1965).

(7) V.M. Vdovin, K.S. Puschchevaya and A.D. Petrov
    Dokl. Akad. Nauk SSSR, $\underline{141}$, 843(1961).

(8) C.L. Schilling, J.-P. Wesson and T.C. Williams
    Am. Ceram. Soc. Bull., $\underline{62}$(8), 912(1983).

(9) M. Bordeau, M. Djamei, J. Dunoguès and R. Calas
    Bull. Chem. Soc. Fr., $\underline{2}$ (5-6), 159(1982).

(10) M. Ishikawa, M. Kumada and H. Sakurai
    J. Organometal. Chem., $\underline{23}$, 63(1970).

(11)   G. Greber and G. Degler
    Macromol. Chem., $\underline{52}$, 199(1962).

(12) J.D. Citron, J.E. Lyons and L.H. Sommer
    J. Org. Chem. $\underline{34}$, 638(1969).

(13) W.A. Kriner
    J. Org. Chem., $\underline{29}$, 1601(1964).

(14) O.W. Steward and O.R. Pierce
    J. Am. Chem. Soc., $\underline{80}$, 4932(1961).

Acknowledgements.

    We are indebted to Rhône Poulenc Industries for their generous gift
of trimethylchlorosilane, chromethyldimethylchlorosilane and
1,1,3,3-tetramethyldisilacyclobutane. We are also grateful to Société
Européenne de Propulsion (SEP) and Centre National de la Recherche
Scientifique for their grant (E.B.).

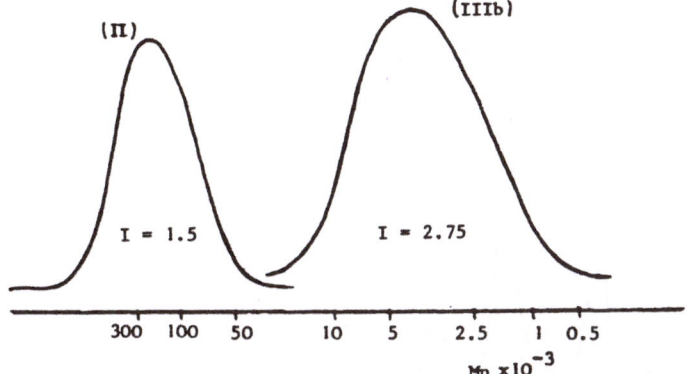

I = Polydispersity index

Fig. 1 - GPC elution curves of polycarbosilanes(II)and(IIIb)
Molecular weights are relative to polystyrene standards.

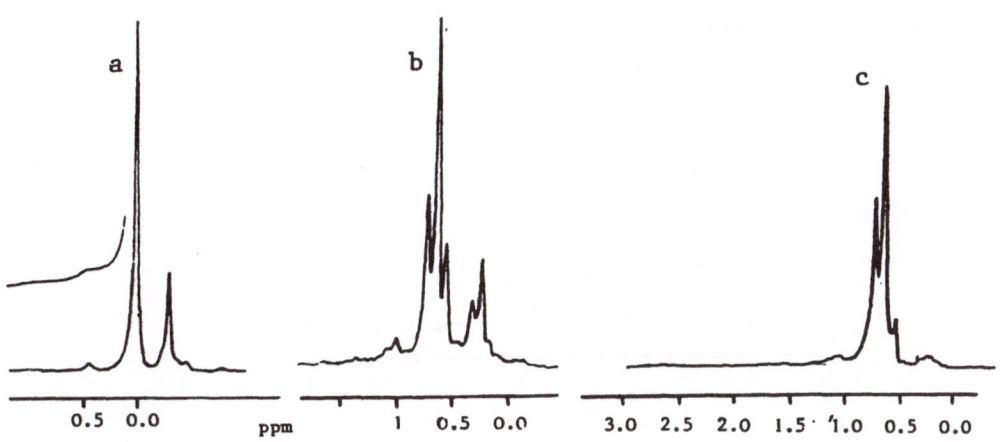

Fig. 2. 60 MHz ^{1}H NMR spectra. a) Starting $\text{-(Me}_2\text{SiCH}_2\text{)}_{\overline{n}}$(II); b) After
uncomplete reaction time ; c) Pure $\text{-(MeSiClCH}_2\text{)}_{\overline{n}}$(IIIb)

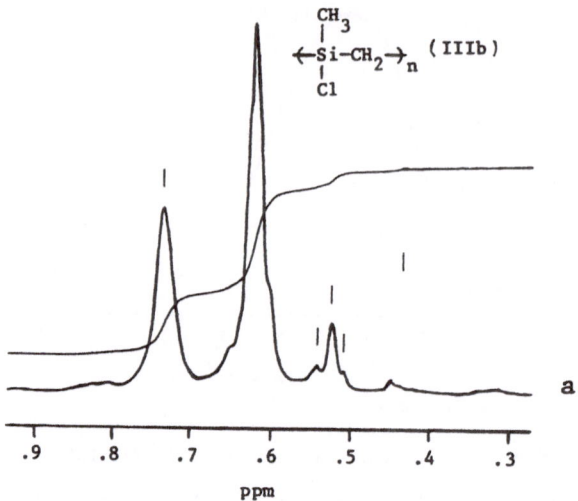

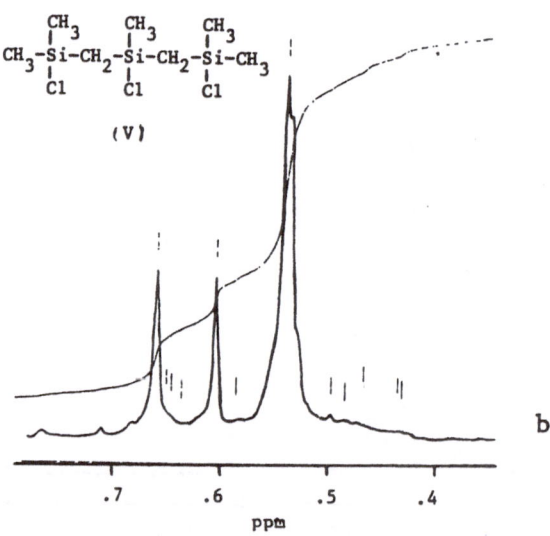

Fig.3. ^{1}H-200 MHz NMR spectra of polycarbosilanes(IIIb) and (V)

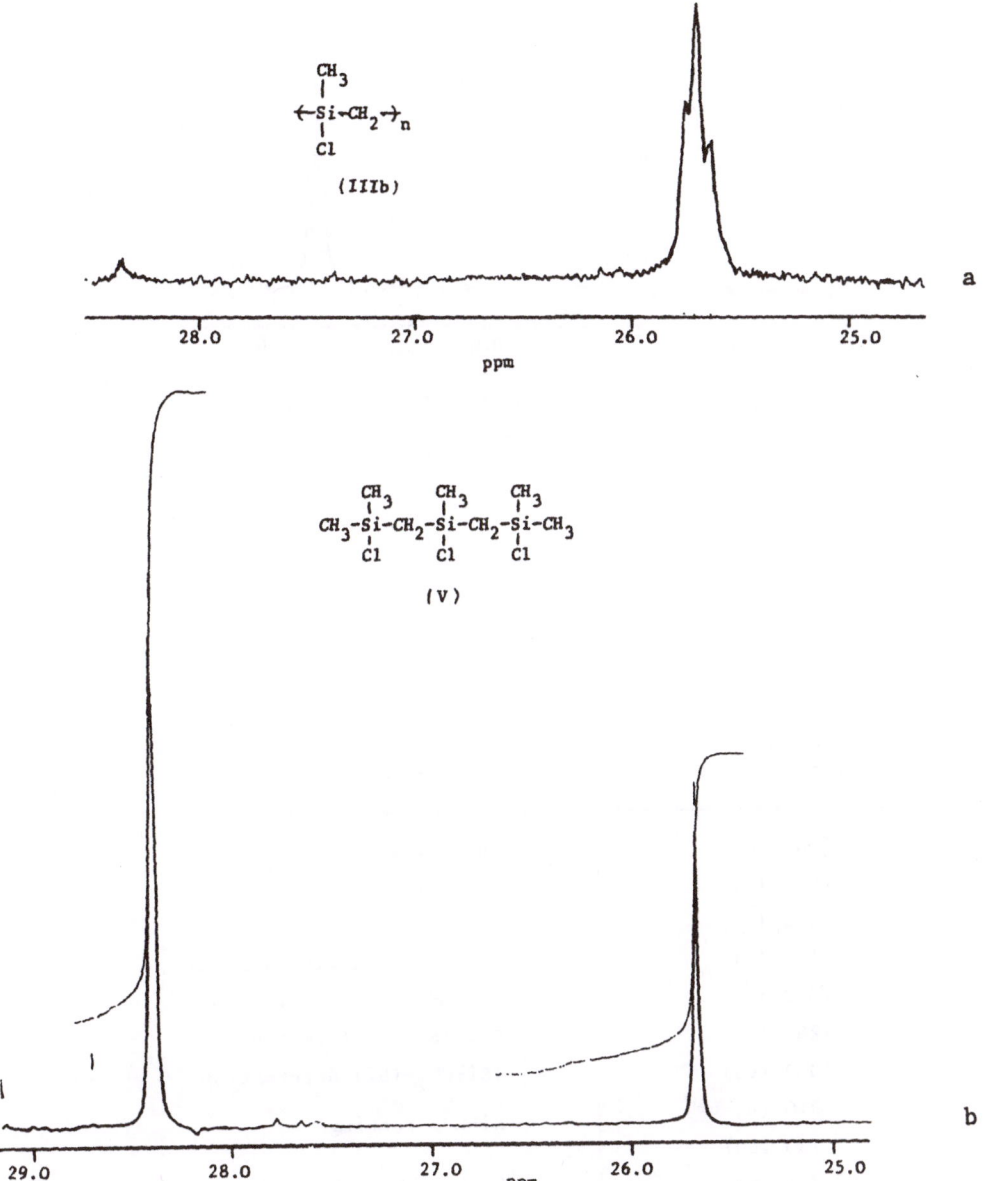

Fig.4. ^{29}Si NMR spectra of polycarbosilanes(IIIb)and(V)

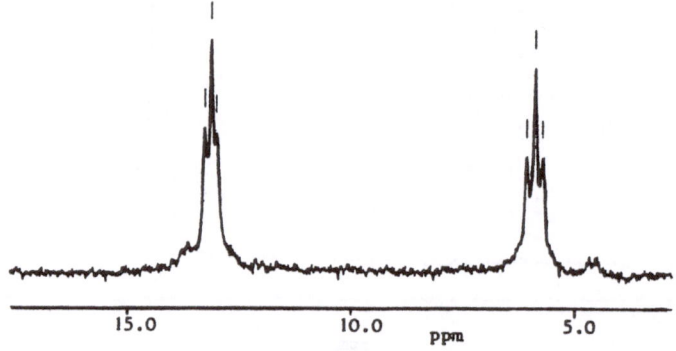

Fig. 5. Polycarbosilane(IIIb): ^{13}C NMR spectrum
broad band decoupling.

TABLE I

Infrared Absorption : Polycarbosilane(III b)

Wave number (cm^{-1})	Assignments
2965 (m)	CH_3 stretch (a)
2900 (w)	CH_2 stretch (a)
2880 (w)	CH_3 stretch (s)
1410 (m)	CH_3 (-Si) deformation (a)
1350 (m)	(Si)CH_2-(Si) deformation
1260 (s)	CH_3-(Si) deformation
1060 (vs)	(Si)CH_2-(Si) deformation
810 (b, vs)	Si-CH_3 rocking
750 (sh)	
720 (m)	Si-C stretch (a)
660 (m)	Si-C stretch (s)

ORGANOSILICON POLYMERS AS PRECURSORS FOR SILICON-CONTAINING CERAMICS

DIETMAR SEYFERTH
Massachusetts Institute of Technology, Dept. of Chemistry,
Cambridge, MA 02139

ABSTRACT

Following general comments about the preceramic polymer approach to the preparation of ceramic materials, we describe the preparation of a novel polysilazane by dehydrocyclodimerization of the ammonolysis product of CH_3SiHCl_2, the conversion of such polymers to ceramic products and their use in upgrading polysilanes of type $[(CH_3SiH)_x(CH_3Si)_y]_n$ and $[CH_3Si(H)CH_2]_n$ to more useful ceramic precursors.

INTRODUCTION AND GENERAL COMMENTS

Silicon-containing ceramics include the oxide materials, silica and the silicates; the binary compounds of silicon with non-metals, principally silicon carbide and silicon nitride; silicon oxynitride and the sialons; main group and transition metal silicides, and, finally, elemental silicon itself. There is a vigorous research activity throughout the world on the preparation of all of these classes of solid silicon compounds by the newer preparative techniques. In this report we will focus on silicon carbide and silicon nitride.

Silicon carbide, SiC [1] and silicon nitride, Si_3N_4 [2], have been known for some time. Their properties, especially their high thermal and chemical stability, their hardness, their high strength, as well as other properties have led to useful applications for both of these materials. Silicon carbide has been an article of commerce since the development of the Acheson process for its manufacture just

R. M. Laine (ed.), Transformation of Organometallics into Common and Exotic Materials: Design and Activation, 133–154.
© *1988 by Martinus Nijhoff Publishers.*

before the turn of the century, but silicon nitride is a
relative newcomer as far as commercial utilization goes [3].

The "conventional" methods for the preparation of SiC
and Si_3N_4, the high temperature reaction of fine grade sand
and coke (with additions of sawdust and NaCl) in an electric
furnace (the Acheson process) for the former and usually the
direct nitridation of elemental silicon or the reaction of
silicon tetrachloride with ammonia (in the gas phase or in
solution) for the latter, do not involve soluble or fusible
intermediates. For many applications of these materials
this is not necessarily a disadvantage (e.g., for the
application of SiC as an abrasive), but for some of the more
recent desired applications soluble or fusible (i.e.,
processable) intermediates are required.

The need for soluble or fusible precursors whose
pyrolysis will give the desired ceramic material has led to
a new area of macromolecular science, that of preceramic
polymers [4]. Such polymers are needed for a number of
different applications. Ceramic powders by themselves are
difficult to form into bulk bodies of complex shape.
Although ceramists have addressed this problem using the
more conventional ceramics techniques with some success,
preceramic polymers could, in principle, serve in such
applications, either as the sole material from which the
shaped body is made or as a binder for the ceramic powder
from which the shaped body is to be made. In either case,
pyrolysis of the green body would then convert the polymer
to a ceramic material, hopefully of the desired composition.
In the latter alternative, shrinkage during pyrolysis should
not be great, but when the green body is made entirely of
preceramic polymer, shrinkage on pyrolysis could be
considerable.

Ceramic fibers of diverse chemical compositions are
sought for application in the production of metal-, ceramic-
glass- and polymer-matrix composites [5]. The presence of
such ceramic fibers in a matrix, provided they have the
right length-to-diameter ratio and are distributed uniformly
throughout the matrix, can result in very considerable
increases in the strength (i.e., fracture toughness) of the
resulting material. To prepare such ceramic fibers, a

suitable polymeric precursor is needed (which can be spun by melt-spinning, dry-spinning, or wet-spinning techniques) to fibers which then can be pyrolyzed (with or without a prior cure step).

Some materials with otherwise very useful properties such as high thermal stability and great strength and toughness are unstable with respect to oxidation at high temperatures. A notable example of such a class of materials is that of the carbon-carbon composites. If these materials could be protected against oxidation by infiltration of their pores and the effective coating of their surface by a polymer whose pyrolysis gives an oxidation-resistant ceramic material, then one would have available new dimensions of applicability of such carbon-carbon composite materials.

In order to have a _useful_ preceramic polymer, considerations of structure and reactivity are of paramount importance. Not every inorganic or organometallic polymer will be a useful preceramic polymer. Some more general considerations merit discussion. Although preceramic polymers are potentially "high value" products _if_ the desired properties result from their use, the more generally useful and practical systems will be those based on commercially available, relatively cheap starting monomers. Preferably, the polymer synthesis should involve simple, easily effected chemistry which proceeds in high yield. The preceramic polymer itself should be liquid or, if a solid, it should be fusible and/or soluble in at least some organic solvents, i.e., it should be processable. It would simplify matters if the polymer were stable on storage at room temperature and stable toward atmospheric oxygen and moisture. Its pyrolysis should provide a high yield of ceramic residue and the pyrolysis volatiles preferably should be non-hazardous and non-toxic. In the requirement of high ceramic yield, economic considerations are only secondary. If the weight loss on pyrolysis is low, shrinkage will be minimized as will be the destructive effects of the gases evolved during the pyrolysis.

There are important considerations as far as the chemistry is concerned. First, the design of the preceramic

polymer is of crucial importance. Many linear
organometallic and inorganic polymers, even if they are of
high molecular weight, decompose thermally by formation and
evolution of small cyclic molecules, and thus the ceramic
yield is low. In such thermolyses, chain scission is
followed by "back-biting" of the reactive terminus thus
generated at a bond further along the chain. Thus high
molecular weight, linear poly(dimethylsiloxanes) decompose
thermally principally by extruding small cyclic oligomers,
$(Me_2SiO)_n$, n = 3,4,5 . . . When a polymer is characterized
by this type of thermal decomposition, the ceramic yield
will be low and it will be necessary to convert the linear
polymer structure to a cross-linked one by suitable chemical
reactions prior to its pyrolysis. In terms of the high
ceramic yield requirement, the ideal preceramic polymer is
one which has functional substituent groups which will give
an efficient thermal cross-linking process so that on
pyrolysis non-volatile, three-dimensional networks-- which
lead to maximum weight retention--are formed. Thus,
preceramic polymer design requires the introduction of
reactive or potentially reactive functionality.

In the design of preceramic polymers, achievement of the
desired elemental composition in the ceramic obtained from
them (SiC and Si_3N_4 in the present cases) is a major
problem. For instance, in the case of polymers aimed at the
production of SiC on pyrolysis, it is more usual than not to
obtain solid residues after pyrolysis which, in addition to
SiC, contain an excess either of free carbon or free
silicon. In order to get close to the desired elemental
composition, two approaches have been found useful in our
research: (1) The use of two comonomers in the appropriate
ratio in preparation of the polymer, and (2) the use of
chemical or physical combinations of two different polymers
in the appropriate ratio.

Preceramic polymers intended for melt-spinning require a
compromise. If the thermal cross-linking process is too
effective at relatively low temperatures (100-200°C), then
melt-spinning will not be possible since heating will induce
cross-linking and will produce an infusible material prior
to the spinning. A less effective cross-linking process is

required so that the polymer forms a stable melt which can be extruded through the holes of the spinneret. The resulting polymer fiber, however, must then be "cured", i.e., cross-linked, chemically or by irradiation, to render it infusible so that the fiber form is retained on pyrolysis. Finally, there still are chemical options in the pyrolysis step. Certainly, the rate of pyrolysis, i.e., the time/temperature profile of the pyrolysis, is extremely important. However, the gas stream used in the pyrolysis also is of great importance. One may carry out "inert" or "reactive" gas pyrolyses. An example of how one may in this way change the nature of the ceramic product is provided by one of our preceramic polymers which will be discussed in more detail later in this paper. This polymer, of composition $[(CH_3SiHNH)_a(CH_3SiN)_b]_m$, gives a <u>black</u> solid, a mixture of SiC, Si_3N_4, and some free carbon, on pyrolysis to 1000°C in an inert gas stream (nitrogen or argon). However, when the pyrolysis is carried out in a stream of ammonia, a <u>white</u> solid remains which usually contains less than 0.5% total carbon and is essentially pure silicon nitride. At higher temperatures (>400°C), the NH_3 molecules effect nucleophilic cleavage of the Si-C bonds present in the polymer and the methyl groups are lost as CH_4. Such chemistry at higher temperatures can be an important and sometimes useful part of the pyrolysis process.

The first useful organosilicon preceramic polymer, a silicon carbide precursor, was developed by S. Yajima and his coworkers at Tohoku University in Japan [6]. The chemistry leading to the Yajima polycarbosilane has been described in the paper of Bacqué, Pillot, Birot and Dunoguès. As might be expected on the basis of the 2 C/1 Si ratio of the $(CH_3)_2SiCl_2$ starting material, the ceramic fibers contain free carbon as well as silicon carbide. A typical analysis [6] showed a composition 1 SiC/0.78 C/0.22 SiO_2. (The latter is introduced in the oxidative cure step of the polycarbosilane fiber).

The Yajima polycarbosilane, while it was one of the first, is not the only polymeric precursor to silicon carbide which has been developed. Another useful system which merits mention is the polycarbosilane which resulted

from research carried out by C.L. Schilling and his
coworkers in the Union Carbide Laboratories in Tarrytown,
New York [7].

NEW PRECERAMIC POLYMER SYSTEMS: RESEARCH AT M.I.T.

Our own research has been aimed at polymer precursors
for silicon nitride, silicon carbide as well as SiC/Si_3N_4
combinations.

Our initial work involved the preparation of
polysilazanes by the ammonolysis of dichlorosilane [8], but
a useful product did not result. Our further studies of the
polysilazanes obtained using methyldichlorosilane as
starting material were more successful and useful preceramic
polymers were obtained [9].

Methyldichlorosilane is commerically available; it is a
by-product of the "Direct Process", the high temperature,
copper-catalyzed reaction of methyl chloride with elemental
silicon and it is potentially inexpensive. The ammonolysis
of CH_3SiHCl_2 has been reported to give a mixture of cyclic
and (possibly) linear oligomers, $[CH_3SiHNH]_x$ [10]. The
ammonolysis product, after removal of the precipitated NH_4Cl
which also is produced, can be isolated as a clear, mobile
liquid in high yield. Its C, H and N analysis and its
spectroscopic (^1H NMR, IR) data are in agreement with the
$[CH_3SiHNH]_x$ formulation. Molecular weight determinations
(cryoscopy in benzene) of several preparations ranged from
280-320 g/mol (x = 4.7-5.4). The product is quite stable at
room temperature, but it is sensitive to moisture and must
be protected from the atmosphere. This mixture of
$[CH_3SiHNH]_x$ oligomers is not suitable for ceramics
preparation without further processing. On pyrolysis to
1000°C in a stream of nitrogen the ceramic yield is only
20%. Clearly, it is necessary to convert these cyclic
$[CH_3SiHNH]_x$ oligomers to material of higher molecular
weight.

The conversion of the cyclopolysilazanes obtained by
ammonolysis of diorganodichlorosilanes was investigated by
Rochow and his coworkers some years ago when there was

interest in polysilazanes as polymers in their own right
[11]. Their procedure, the ammonium salt-induced
polymerization, which in the case of hexamethyl-
cyclotrisilazane appears to give polymers containing both
linear and cyclic components, was applied to the CH_3SiHCl_2
ammonolysis product. It produced a very viscous oil of
higher molecular weight, but the ceramic yield obtained in
pyrolysis was a disappointing 36%. The $Ru_3(CO)_{12}$-catalyzed
ring-opening polymerization of cyclo-$[(CH_3)_2SiNH]_4$, reported
recently by Zoeckler and Laine [12], could not be adapted to
the conversion of the $[CH_3SiHNH]_x$ cyclics to a soluble
polymer. An insoluble, rubbery solid was formed, which
suggests that Si-H bonds as well as the Si-N bonds were
activated by the transition metal catalyst.

The solution to our problem of converting the
$[CH_3SiHNH]_x$ cyclics to a useful preceramic polymer was
provided by earlier workers [13] who had described the
conversion of silylamines of type 1 to cyclodisilazanes, 2,
in high yield by the action of potassium in di-n-butyl ether
(eq. 1). In this dehydrocyclodimerization reaction, the

$$2 \; \underset{\substack{| \; | \\ H \; H}}{R_2SiNR'} \xrightarrow[\text{Bu}_2\text{O}]{K} 2 \; H_2 \; + \; R_2Si\diamond SiR_2$$

1   2   (1)

potassium serves to metallate the NH functions to give 3.
This then either eliminates H^- from silicon to give an

R₂Si——NR' structure (3): R_2Si—NR' with H below Si, minus sign, K^+

Structure (4): R'—N at top, bonded to R_2Si and SiR_2 (with H below SiR₂), and N^- with R' below.

$$\underline{3} \qquad\qquad \underline{4}$$

intermediate with a silicon-nitrogen double bond, $R_2Si=NR'$, which then undergoes head-to-tail dimerization to form $\underline{2}$, or alternatively, reacts with a molecule of $R_2Si(H)NHR'$ to give intermediate $\underline{4}$ which undergoes cyclization to $\underline{2}$ with displacement of H^-. This interesting mechanistic question still needs to be resolved.

The repeating unit in the $[CH_3SiHNH]_x$ cyclics is $\underline{5}$. For

Structure 5: $-Si-N-$ with CH_3 above Si, and H H below.

Structure 6: 8-membered ring with alternating Si and N, methyl and H substituents.

$$\underline{5} \qquad\qquad \underline{6}$$

example, the cyclic tetramer is the 8-membered ring compound $\underline{6}$. On the basis of equation 1, the adjacent NH and SiH groups provide the functionality which permits the molecular weight of the $[CH_3SiHNH]_x$ cyclics to be increased. It may be expected that a sheet polymer will be formed.

Treatment of the CH_3SiHCl_2 ammonolysis product, cyclo-$[CH_3SiHNH]_x$, with a catalytic amount of a base (generally an alkali metal base) strong enough to deprotonate the N-H function in a suitable solvent results in evolution of hydrogen. The resulting solution contains polymeric basic species and these are quenched by addition of methyl iodide or a chlorosilane. After filtration to remove alkali metal halide, evaporation of the filtrate gives the product in essentially quantitative yield. In our experiments we generally have used catalytic amounts of potassium hydride as the base. When THF was used as reaction solvent, the product was isolated in the form of a white powder (average molecular weight, 1180) which was found to be soluble in hexane, benzene, diethyl ether, THF, and other common organic solvents. It is imperative to exclude atmospheric moisture since the $[CH_3SiHNH]_x$ cyclics are readily hydrolyzed. The product polymer on the other hand, is of greatly diminished sensitivity to hydrolysis.

The composition of the polysilazane product of the experiment detailed above, ascertained by proton NMR spectroscopy, was $(CH_3SiHNH)_{0.39}(CH_3SiHNCH_3)_{0.04}(CH_3SiN)_{0.57}$ and its combustion analysis (C,H,N,Si) agreed with this formulation. These results are compatible with a process in which $(CH_3SiHNH)_n$ rings are linked together via Si_2N_2 bridges. Thus, if for example, the silazane 6 were to be polymerized by such a dehydrocyclodimerization process, the eight-membered rings could be linked as shown in 7. The ex-

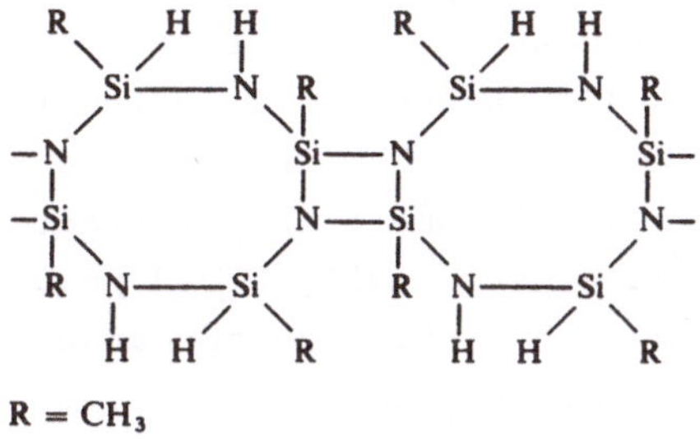

R = CH$_3$

7

perimental $(CH_3SiN/(CH_3SiHNH)$ ratio of ~1.3 indicates that further linking via Si_2N_2 rings must have taken place, probably to give a sheet-like structure. High polymers obviously are not formed, but the molecular weight is increased sufficiently so that pyrolysis proceeds satisfactorily.

A better understanding of the chemistry leading to these polysilazanes and of their structure is needed. For instance, in our various preparations the average molecular weights of the products obtained after the methyl iodide quench varied between 800 and 2000. Before the methyl iodide quench, silylamide functions (i.e., catalytically active functions) still were present, yet growth to higher molecular weights did not occur. The reason for this must be connected with the solution structure and conformation of the polymers.

Whatever the structure of the silazane polymers obtained by KH treatment of the $[CH_3SiHNH]_x$ cyclics, these polymers are excellent ceramic precursors. Examination of the polymers from various preparations by TGA showed the weight loss on pyrolysis to be only between 15 and 20%. The pyrolysis appears to take place in three steps: a 5% weight loss (involving evolution of H_2) from 100°C to 350°C; a 2% weight loss from 350-550°C and a 9% weight loss from 550°C to 900°C. During the 550-900°C stage a mixture of H_2 and methane was evolved. A trace of ammonia, in addition to H_2, was lost between 350 and 550°C.

In a typical bulk pyrolysis experiment, pyrolysis was conducted under a slow stream of nitrogen. The sample was heated quickly to 500°C and then slowly (over 8 h) to 1420°C and was held at 1420°C for 2 h. The ceramic powder was a single body and black. Powder X-ray diffraction (CuK_α with Ni filter) showed only very small, broad peaks for α-Si_3N_4. (At 1490°C, lines due to β-SiC also appeared). Scanning electron microscopy analysis showed little discernible microstructure with only a few very fine grains appearing at high magnification. The bulk appearance of the ceramic suggested that pyrolysis took place after the polymer had melted. There were many large holes and craters where the liquid bubbles apparently had burst. In such experiments ceramic yields usually were between 80 and 85%. The polymer used in the experiment above had been prepared in diethyl

ether. It had a molecular weight of around 900 and went
through a melt phase when it was heated. This could be
shown when it was heated in a sealed capillary: it began to
soften around 65°C, becoming more fluid with increasing
temperature. The polymer prepared in THF is of higher
molecular weight (MW = 1700-2000) and does not soften when
heated to 350°C. It gave an 83% yield of a black ceramic
material on pyrolysis under nitrogen to 1000°C.

The pyrolysis of the silazane polymer may be represented
by eq. 2. Here the ceramic yield (Si_3N_4 + SiC + C)

$$2 \ (CH_3SiHNH)(CH_3SiN) \longrightarrow Si_3N_4 + SiC + C + 2 \ CH_4 + 4 \ H_2 \quad (2)$$

would be 83 wt%. An analysis of such a ceramic product gave
12.87% C, 26.07% N, and 59.52% Si. This analysis is
compatible with eq. 2 and leads to a ceramic constitution,
based on the 4 Si of eq. 2, of 0.88 Si_3N_4, 1.27 SiC, 0.75 C.

Thus the chemistry leading to the desired ceramic
product is quite satisfactory: most of the requirements
mentioned earlier are met.

Initial ceramic studies have been promising.
Isostatically pressed (40,000 psi) bars of the polymer on
pyrolysis to 1100°C gave a coherent, rectangular ceramic bar
which had not cracked or bloated and could not be broken by
hand.

In collaboration with ceramists at the Celanese Research
Company it was found that our meltable polysilazane cannot
be melt-spun. Apparently, the thermal cross-linking process
which is so effective in giving a high ceramic yield on
pyrolysis takes place in the heated chamber of the spinning
machine and quickly gives infusible polymer. However, the
infusible polysilazane which is obtained when the
preparation is carried out in THF solution can be dry-spun.
In this process the solid polysilazane is dissolved in an
appropriate solvent and then is extruded through a spinneret
into a heated drying chamber in which the solvent is
volatilized, leaving the solid polymer. These polymer
fibers could be pyrolyzed to give ceramic fibers. In
simpler experiments, it was shown that fibers one to two

feet in length could be drawn from the sticky, waxy solid which remained when a toluene solution of the polysilazane was evaporated. Pyrolysis of these fibers under nitrogen produced long, flexible black fibers. This aspect of our work is in only at a very early stage, but these initial observations are encouraging.

Our polysilazane also serves well as a binder for ceramic powders. The preparation of such composites (using commercial samples of fine α-SiC, β-SiC and α-Si$_3$N$_4$ of 0.36-0.4 μm mean particle size) required appropriate dispersion studies. The ceramic powder was dispersed in a solution of toluene containing the appropriate weight of polysilazane and then the toluene was evaporated using a rotary evaporator, leaving a waxy residue. Vacuum distillation removed the remaining solvent and left chunks of solid material. These were finely ground and pressed into a bar at 5000 psi. Isostatic pressing to 40,000 psi followed and then the bars were pyrolyzed in a tube furnace under nitrogen (10°C a minute, to 1100°C). The maximum density (~2.4 g/cc) was achieved in these experiments with a polymer loading of 30%.

Our research also has been directed at SiC precursors. It began with an examination of a potential starting material in which the C:Si ratio was 1, the ratio desired in the derived ceramic product.

Available methylsilicon compounds with a 1 C/1 Si stoichiometry are CH_3SiCl_3 and CH_3SiHCl_2. The former gives highly cross-linked, insoluble products on treatment with an alkali metal in a suitable diluent. The latter, in principle, could give $[CH_3SiH]_n$ cyclic oligomers and linear polymers on reaction with an alkali metal. In practice, the Si-H linkages also are reactive toward alkali metals. Thus, mixed organochlorosilane systems containing some CH_3SiHCl_2 have been treated with metallic potassium by Schilling and Williams [14]. It was reported that the CH_3SiHCl_2-based contribution to the final product was $(CH_3SiH)_{0.2}(CH_3Si)_{0.8}$, i.e., about 80% of the available Si-H bonds had reacted. Such Si-H reactions lead to cross-linking in the product, or to formation of polycyclic species if cyclic products are preferred. Nevertheless, we have used this known reaction of CH_3SiHCl_2 with an alkali metal as an entry to new preceramic polymers.

When the reaction of CH_3SiHCl_2 with sodium pieces was carried out in tetrahydrofuran medium, a white solid was isolated in 48% yield. This solid was poorly soluble in hexane, somewhat soluble in benzene, and quite soluble in THF. Its molecular weight could not be determined by cryoscopy in benzene because of its limited solubility in that solvent. Its ^{1}H NMR spectrum (in $CDCl_3$) indicated that extensive reaction of Si-H bonds had occurred. The $\delta(SiH)/\delta SiCH_3$ integration led to a constitution $[(CH_3SiH)_{0.4}(CH_3Si)_{0.6}]_n$. Here the CH_3SiH units are ring and chain members which are not branching sites; the CH_3Si units are ring and chain members which are branching sites. In our reactions it is expected that mixtures of polycyclic and linear (possibly cross-linked) polysilanes will be formed. (Attempts to distill out pure compounds from our preparations were not successful. Less than 10% of the product was volatile at higher temperatures at 10^{-4} torr.)

The ceramic yield obtained when the $[(CH_3SiH)_{0.4}(CH_3Si)_{0.6}]_n$ polymer was pyrolyzed (TGA to 1000°C) was 60%; a gray-black solid was obtained whose analysis indicated a composition 1.0 SiC + 0.49 Si.

The reaction of methyldichlorosilane with sodium in a solvent system composed of six parts of hexane and one of THF gave a higher yield of product which was soluble in organic solvents. Such reactions give a colorless, cloudy oil in 75 to over 80% yield which is soluble in many organic solvents. In various experiments the molecular weight (cryoscopic in benzene) averaged 520-740 and the constitution (by ^{1}H NMR) $[(CH_3SiH)_{0.76}(CH_3Si)_{0.24}]_n$ to $[(CH_3Si)_{0.9}(CH_3Si)_{0.1}]_n$. This less cross-linked material (compared to the product obtained in THF alone) gave much lower yields of ceramic product on pyrolysis to 1000°C (TGA yields ranging from 12-27% in various runs). Again, the product was (by analysis) a mixture of SiC and elemental silicon, 1.0 SiC + 0.42 Si being a typical composition.

The results described above are not especially promising. The CH_3SiHCl_2/Na product which on pyrolysis gives a reasonable ceramic yield is of limited solubility in organic solvents and its conversion to ceramic fibers requires a photolysis/oxidation cure step. The CH_3SiHCl_2/Na product in which crosslinking is not as extensive and which is very soluble in organic solvents gives unacceptably low ceramic yields on pyrolysis. Furthermore, only in the case of the preparations carried out in 6-7/1 hexane/THF solvent medium were the yields of soluble product satisfactory. Finally, in all cases the ceramic product contained a considerable excess of "free" silicon over the ideal SiC composition. It was obvious that further chemical modification of the $[(CH_3SiH)_x(CH_3Si)_y]_n$ products obtained in the CH_3SiHCl_2/Na reactions was required.

A number of approaches which we tried did not lead to success, but during the course of our studies we found that treatment of the $[(CH_3SiH)_x(CH_3Si)_y]_n$ products with alkali metal amides (catalytic quantities) serves to convert them to materials of higher molecular weight whose pyrolysis gives significantly higher ceramic yields. Thus, in one example, to 0.05 mol of liquid $[(CH_3SiH)_{0.85}(CH_3Si)_{0.15}]$ in THF was added, under nitrogen, a solution of about 1.25 mmol (2.5 mol%) of $[(CH_3)_3Si]_2NK$ in THF. The resulting red solution was treated with methyl iodide. Subsequent nonhydrolytic workup gave a soluble white powder in 68% yield, molecular weight 1000, whose pyrolysis to 1000°C gave a ceramic yield of 63%.

The proton NMR spectra of these products showed only broad resonances in the Si-H and Si-CH_3 regions. In the starting $[(CH_3SiH)_x(CH_3Si)_y]_n$ materials observed proton NMR integration ratios, $SiCH_3$/SiH, ranged from 3.27-3.74. This ratio was quite different in the case of the products of the silylamide-catalyzed processes, ranging from 8.8 to 14. Both Si-H and Si-Si bonds are reactive toward nucleophilic reagents. In the case of the alkali metal silylamides, the following processes can be envisioned:

$$(R_3Si)_2NK + -\overset{|}{\underset{|}{Si}}H \longrightarrow -\overset{|}{\underset{|}{Si}}N(SiR_3)_2 + KH \qquad (3)$$

$$(R_3Si)_2NK + -\overset{|}{\underset{|}{Si}}H \longrightarrow (R_3Si)_2NH + -\overset{|}{\underset{|}{Si}}K \qquad (4)$$

$$(R_3Si)_2NK + -\overset{|}{\underset{|}{Si}}-\overset{|}{\underset{|}{Si}}- \longrightarrow (R_3Si)_2N-\overset{|}{\underset{|}{Si}}- + -\overset{|}{\underset{|}{Si}}K \qquad (5)$$

In each process, a new reactive nucleophile is generated: KH in equation 2, a silyl alkali metal function in reactions 3 and 4. These also could undergo nucleophilic attack on the $[(CH_3SiH)_x(CH_3Si)_y]_n$ system and during these reactions some of the oligomeric species which comprise the starting material would be linked together, giving products of higher molecular weight. Other processes are possible as well, e.g., a silylene process as shown in equation 5. Thus not only anionic species but also neutral silylenes could be involved as intermediates. In any case, extensive loss of Si-H takes place during this catalyzed process: it is more

$$(R_3Si)_2NK \ + \ \ -\overset{|}{\underset{H}{Si}}-\overset{|}{\underset{|}{Si}}- \ \longrightarrow \ (R_3Si)_2N\overset{|}{\underset{|}{Si}}- \ + \ -\overset{H}{\underset{|}{Si}}-K$$

$$\searrow Si: \ + \ KH \quad (6)$$

than a simple redistribution reaction. Further studies relating to the mechanism of this process must be carried out.

While these silylamide-catalyzed reactions have provided a good way to solve the problem of the low ceramic yield in the pyrolysis of $[(CH_3SiH)_x(CH_3Si)_y]_n$, the problem of the elemental composition of the ceramic product remained (i.e., the problem of Si/C ratios greater than one) since only catalytic quantities of the silylamide were used.

As noted above, KH-catalyzed polymerization of the CH_3SiHCl_2 ammonolysis product gives a polymeric silylamide of type $[(CH_3SiHNH)_a(CH_3SiN)_b(CH_3SiHNK)_c]$. In a typical example, a ~0.9, b~1.3, c~0.04, so there is only a low concentration of silylamide functions in the polymer. This polymeric silylamide reacts with electrophiles other than methyl iodide, e.g., with diverse chlorosilanes, and it has been isolated and analyzed. Since it is a silylamide, we expected that it also would react with $[(CH_3SiH)_x(CH_3Si)_y]_n$ polysilane-type materials. Not only would it be expected to convert the latter into material of higher molecular weight, but it also would be expected to improve the Si/C ratio (i.e., bring it closer to 1). As noted above, pyrolysis of $[(CH_3SiHNH)_a(CH_3SiN)_b(CH_3SiHNCH_3)_c]_n$ gives a ceramic product in 80–85% yield containing Si_3N_4, SiC and excess carbon. Thus, combination of the two species in the appropriate stoichiometry, i.e., of $[(CH_3SiH)_x(CH_3Si)_y]_n$ and $[(CH_3SiHNH)_a(CH_3SiN)_b(CH_3SiHNK)_c]_n$, and pyrolysis of the product (which we will call a "graft" polymer) after CH_3I quench could, in principle, lead to a ceramic product in which the excess Si obtained in pyrolysis of the former and the excess C obtained in the pyrolysis of the latter combine to give SiC. A further benefit might be expected from such a combination in the formation of ceramic fibers: The $[(CH_3SiHNH)_a(CH_3SiN)_b(CH_3SiHNCH_3)_c]_2$ system is self-curing as the temperature is raised on the way to the production of a ceramic material; the $[(CH_3SiH)_x(CH_3Si)_y]_n$ system, as described above, is not. It might be expected that a combination of the two would give a self-curing system.

With these ideas in mind, experiments were carried out in which the two polymer systems, $[(CH_3SiH)_x(CH_3Si)_y]_n$ and the "living" polymer-silyl amide, $[(CH_3SiHNH)_a(CH_3SiN)_b(CH_3SiHNK)_c]_n$, were mixed in THF solution in varying proportions (2.4:1 to 1:2 mole ratio) and allowed to react at room temperature for 1 h and at reflux for 1 h. (Such experiments were carried out with the $[(CH_3SiH)_x(CH_3Si)_y]_n$ materials prepared in hexane/THF as well as with those prepared in THF alone.) After quenching with methyl iodide, nonhydrolytic workup gave a new polymer in nearly quantitative yield (based on weight of material charged). The molecular weight of these products was in the 1800–2500 range. Their pyrolysis under nitrogen gave ceramic products in 74–83% yield. Thus, the reaction of the two polymer systems gives a new polymer in close to quantitative yield which seems to be an excellent new preceramic polymer in terms of ceramic yield.

In an alternative method of synthesis of $[(CH_3SiH)_x(CH_3Si)_y][(CH_3SiHNH)_a(CH_3SiN)_b]$ "combined" polymers, the polysilyl amide was generated <u>in situ</u> in the presence of $[(CH_3SiH)_x(CH_3Si)_y]_n$. This, however, gave materials that were somewhat different. In one such experiment, a mixture of $(CH_3SiHNH)_n$ cyclics (as obtained in the ammonolysis of CH_3SiHCl_2 in THF) and the $[(CH_3SiH)_x(CH_3Si)_y]_n$ material (x = 0.76; y = 0.26) in THF was treated with a catalytic amount of KH. After the reaction mixture had been treated with methyl iodide, the usual workup gave an 89% yield of hexane-soluble white powder, molecular weight ~2750. On pyrolysis, this material gave a 73% yield of a black ceramic.

The "combined" polymer prepared in this way ("in situ polymer") was in some ways different from the "combined" polymer prepared by the first method ("graft" polymer). Principal differences were observed in their proton NMR spectra and in the form of their TGA curves. This suggests that the two differently prepared polymers have different structures. It is likely that in the "in situ" preparation intermediates formed by the action of KH on the $(CH_3SiHNH)_n$ cyclics are intercepted by reaction with the $[(CH_3SiH)_x(CH_3Si)_y]_n$ also present before the $[(CH_3SiHNH)_a(CH_3SiH)_b(CH_3SiHNK)_c]_n$ polymer (which is the starting reactant used in the "graft" procedure) has a chance to be formed to the extent of its usual molecular weight. Thus, less of the original CH₃SiHNH protons are lost and/or more of those of the $[(CH_3SiH)_x(CH_3Si)_y]_n$ system are reacted.

The TGA curves of the "graft" polymer and the "in situ" polymer are different as well. Noteworthy in the former is a small weight loss between 100°C and 200°C, which begins at around 100°C. This initial small weight loss occurs only at higher temperature (beginning at ~175°C) in the case of the "in situ" polymer. This difference in initial thermal stability could well have chemical consequences of importance with respect to ceramics and both kinds of polymers may be useful as preceramic materials.

Physical blends of $[(CH_3SiH)_x(CH_3Si)_y]_n$ (solid polymer, THF preparation) and $[(CH_3SiHNH)_a(CH_3SiN)_b(CH_3SiHNCH_3)_c]_n$ were also examined. When about equimolar quantities of each were mixed and finely ground together, pyrolysis to 1000°C gave a 70% ceramic yield (TGA). It appears that a reaction between the two polymers already occurs at lower temperatures. When such mixtures were heated, either in the absence of a solvent at 100°C under nitrogen or in toluene solution at reflux, white powders were obtained which were insoluble in hexane, benzene, and THF. The ceramic yields (by TGA) were 67& and 75%, respectively.

Further experiments showed that the "combined" polymers may be converted to black ceramic fibers. Pyrolysis of pressed bars of the "combined polymer to 1000°C gave a black, foam product of irregular shape (74-76% ceramic yield). In other experiments, SiC powder was dispersed in toluene containing 20% by weight of the "combined" polymer. The solution was evaporated and the residue, a fine powder of SiC with the "combined" polymer binder, was pressed into bars and pyrolyzed at 1000°C. A ceramic bar (6% weight loss, slightly shrunk in size) was obtained.

The ceramic products obtained in the pyrolysis of the "combined" polymers have not been studied in detail, but some of them have been analyzed for C, N, and Si. The compositions of the ceramic materials obtained cover the range 1 Si_3N_4 + 3.3 to 6.6 SiC + 0.74 to 0.85 C. Thus, as expected, they are rich in silicon carbide and the excess Si which is obtained in the pyrolysis of the $[(CH_3SiH)_x(CH_3Si)_y)_n$ materials alone is not present, so that objective has been achieved. By proper adjustment of starting material ratios, we find that the excess carbon content can be minimized.

The Yajima polycarbosilane discussed earlier, as obtained by thermal rearrangement of the poly(dimethylsilylene) is a polymeric silicon hydride, with $[(CH_3)(H)SiCH_2]$ as the main repeating unit. As such, it also might be expected to react with our $[(CH_3SiHNH)_a(CH_3SiN)_b(CH_3SiHNK)_c]$ poly(silylamide). This was found to be the case. The commercially available Yajima polycarbosilane (sold in the U.S. by Dow Corning Corporation; our sample, a white solid, had a molecular weight of 1210 and a ceramic yield of 58% was obtained on pyrolysis) was found to react with our poly(silylamide). A reaction carried out in THF solution, initially at room temperature, then at reflux, followed by treatment of the reaction mixture with methyl iodide, gave after appropriate workup a nearly quantitative yield of a white solid which was very soluble in common organic solvents including hexane, benzene, and THF. When the polycarbosilane-to-polysilylamide ratio was approximately one, pyrolysis of the product polymer gave a black ceramic solid in 84% yield

which analysis showed to have a composition (1 SiC + 0.22
Si_3N_4 + 0.7 C). When the polycarbosilane/polysilylamide
ratio was ~5, the ceramic yield was lower (67%). In these
experiments the cyclo-$(CH_3SiHNH)_n$ starting materials used to
synthesize the polysilylamide had been prepared by CH_3SiHCl_2
ammonolysis in diethyl ether. When this preparation was
carried out in THF, the final ceramic yields obtained by
pyrolysis of the polycarbosilane/polysilylamide hybrid
polymer were 88% (1:1 reactant ratio) and 64% (5:1 reactant
ratio).

The "in situ" procedure in which the CH_3SiHCl_2
ammonolysis product, cyclo-$(CH_3SiHNH)_m$, was treated with a
catalytic quantity of KH in the presence of the
polycarbosilane, followed by a CH_3I quench and the usual
work-up, gave equally good results in terms of high final
ceramic yields, whether the starting cyclo-$(CH_3SiHNH)_m$ was
prepared in Et_2O or THF.

It is clear from these results that a new polymer is
formed when the polycarbosilane and the polymeric silylamide
are heated together in solution and then quenched with
methyl iodide. Proton NMR spectroscopy brought further
evidence of such a reaction.

A physical mixture of the polycarbosilane and the
polysilazane $[(CH_3SiHNH)_a(CH_3SiN)_b(CH_3SiHNCH_3)_c]_m$, also was
found to react when heated to 1000°C, giving good yields of
ceramic product. That appreciable reaction had occurred by
200°C was shown in an experiment in which a 1:1 by weight
mixture of the initially soluble polymers was converted to a
white, foamy solid which no longer was soluble in organic
solvents by such thermal treatment.

The combined polymers obtained by the "graft", "in
situ", and physical blend methods can be converted to black
ceramic fibers. Pyrolysis of pressed bars of the combined
polymers to 1000°C provides a black solid product. In other
experiments, SiC powder was dispersed in a toluene solution
containing 25% by weight of the combined powders. The
solvent was evaporated and the residue, a fine powder of
silicon carbide with combined polymer binder, was pressed
into bars and pyrolyzed at 1000°C. A ceramic bar was
obtained showing a low weight loss and slightly shrunken

size. Thus, the usual ceramic applications seem indicated. While this approach in which chemical combination of the Yajima polycarbosilane and our poly(silylamide) leads to new hybrid polymers successfully addresses the problem of ceramic yield, it does not deal with the problem of chemical composition. Pyrolysis of the polycarbosilane and of the polysilazane separately gives ceramic materials which in each case contain an excess of free carbon. As expected, the hybrid polymers produced by reaction of the polycarbosilane with the poly(silylamide) by either the "graft" or "in situ" procedures gives ceramic products which contain an excess of free carbon.

CONCLUSIONS

We have shown in this paper that useful preceramic organosilicon polymers can be prepared and that their design is an exercise in functional group chemistry. Furthermore, we have shown that an organosilicon polymer which seemed quite unpromising as far as application as a preceramic polymer is concerned could, through further chemistry, be incorporated into new polymers whose properties in terms of ceramic yield and elemental composition were quite acceptable for use as precursors for ceramic materials. It is obvious that the chemist can make a significant impact on this area of ceramics. However, it should be stressed that the useful applications of this chemistry can only be developed by close collaboration between the chemist and the ceramist.

ACKNOWLEDGMENTS

The work reported in this paper was carried out with generous support of the Office of Naval Research and the Air Force Office of Scientific Research. The results presented here derive from the Ph.D. dissertation of Gary H. Wiseman and the postdoctoral research of Dr. Yuan-Fu Yu. I thank these coworkers for their skillful and dedicated efforts.

REFERENCES

1. Gmelin Handbook of Inorganic Chemistry, 8th Edition,
    Springer-Verlag: Berlin, Silicon, Supplement Volumes B2,
    1984, and B3, 1986.

2. Messier, D.R.; Croft, W.J. in "Preparation and
    Properties of Solid-State Materials", Vol. 7, Wilcox,
    W.R., ed.; Dekker: New York, 1982, Chapter 2.

3. As recently as 1975, the statement "Silicon nitride has
    no established use." was applicable: Rochow, E.G., "The
    Chemistry of Silicon", Pergamon Texts in Inorganic
    Chemistry, Volume 9, Pergamon: Oxford, 1975, p. 1417.

4. (a) Wynne, K.J.; Rice, R.W. Ann. Rev. Mater. Sci.
         (1984) 14, 297.
    (b) Rice, R.W. Am. Ceram. Soc. Bull. (1983) 62, 889.

5. Rice, R.W. Chem. Tech. (1983) 230.

6. Yajima, S. Am. Ceram. Soc. Bull. (1983) 62, 893.

7. Schilling, C.L., Jr.; Wesson, J.P.; Williams, T.C. Am.
    Ceram. Soc. Bull. (1983) 62, 912.

8. (a) Seyferth, D.; Wiseman, G.H. and Prud'homme, C.C.
         J. Am Ceram. Soc. (1983) 66, C-13.
    (b) U.S. Patent 4,397,828 (August 9, 1983).

9. (a) Seyferth, D.; Wiseman, G.H. J. Am. Ceram. Soc.
         (1984) 67, C-132.
    (b) U.S. Patent 4,482,669 (13 Nov. 1984).
    (c) Seyferth, D.; Wiseman, G.H. in "Ultrastructure
         Processing of Ceramics, Glasses and Composites", 2,
         edited by L.L. Hench and D.R. Ulrich, Wiley: New
         York, 1986, Chapter 38.

10. Brewer, S.D.; Haber, C.P. J. Am. Chem. Soc. (1948)
    70, 3888.

11. (a) Krüger, C.; Rochow, E.G.   J. Polymer Sci., Part A
        (1964) 2, 3179.
    (b) Rochow, E.G.   Monatsh. Chem. (1964) 95, 750.

12. Zoeckler, M.T.; Laine, R.M.   J. Org. Chem. (1983) 48,
    2539.

13. Monsanto Company, Neth. Appl. 6,507,996 (December 23,
    1965); Chem. Abstr. (1966) 64, 19677d.

14. (a) Schilling, C.L., Jr.; Williams, T.C., Report 1983,
    TR-83-1, Order No. AD-A141546; Chem. Abstr. 101. 196820p
    (b) U.S. Patent 4,472,591 (18 Sept. 1984).

SECTION C

CHEMICAL VAPOR DEPOSITION

CHEMICAL VAPOR DEPOSITION OF Fe-Co THIN FILMS FROM Fe-Co ORGANOMETALLIC CLUSTERS

CORINNA L. CZEKAJ-KORN AND GREGORY L. GEOFFROY
Department of Chemistry, The Pennsylvania State University, University
Park, PA 16801

ABSTRACT:

HFeCo₃(CO)₁₂ and CpFeCo(CO)₆ have been employed as vapor phase
reagents in the chemical vapor deposition of amorphous, homogeneous,
coherent mixed-metal and mixed-metal oxide thin films of Fe and Co. Film
thickness can be controlled as a function of time with the stoichiometry of
the films reflecting the metal ratio of the starting organo-transition metal
complexes.

INTRODUCTION

Chemical Vapor Deposition (CVD) is a chemical process which typically
involves the reaction of vapor phase constituents very near or on the
surface of a heated substrate to produce a solid film, formed atom by
atom, and gaseous product species. The utility of CVD stems from the
ability to vary the morphology of a film of given composition by changing
the nature and surface structure of the substrate, the deposition
conditions, and the treatment of the film after deposition. This potential
to vary the deposit morphology provides a means of modifying the physical,
mechanical, electrical, magnetic and optical properties of a film and
ideally designing a material "structure for function".[1]

There are two major disadvantages in the presently practiced art of
CVD which limit the usefulness of the technique. The first involves the
high temperatures, generally greater than 750°C, that are necessary to
achieve useful reaction rates for most vapor combinations. This limits the
support materials on which the films can be deposited to only those
supports stable at these temperatures. For example, it is difficult to
deposit films by CVD on organic polymeric media, such as polyurethanes and
polyesters.[2] High temperatures also promote the diffusion of elements
within the film and between layers.[3] This has posed a serious problem in
the deposition of homogeneous metal alloy films and in the fabrication of
electronic microcircuits where high-purity individual layers are
essential. Also, study of the kinetics and mechanisms of the chemical
reactions occuring in the gas phase and/or on the substrate surface is
difficult in these temperature ranges because the reactions are limited by
thermodynamics rather than kinetics.[4]

The second disadvantage arises from the lack of suitable vapor phase
reactions for the formation of many desired phases. This is particularly
true in the formation of films of more than two components, such as
mixed-metal alloy and oxide films.

An alternative approach to the preparation of thin films which
circumvents some of the above difficulties involves the use of
organotransition metal complexes as the vapor phase reagents. Many
organotransition metal compounds have sufficient vapor pressures at
temperatures well below their decomposition points to allow them to be
vaporized into a reaction chamber. Normally, decomposition of the

157

R. M. Laine (ed.), Transformation of Organometallics into Common and Exotic Materials: Design and Activation, 157–164.
© *1988 by Martinus Nijhoff Publishers.*

organometallic would occur at temperatures below 500°C. Thus, these compounds should be excellent reagents for the preparation of thin films at temperatures significantly lower than those needed in conventional methods of preparation. Furthermore, at these lower temperatures, the decomposition of discreet organo-transition metal cluster complexes containing a variety of metal and ligand combinations should provide a method for depositing pure, homogeneous mixed-metal alloy and refractory films. The low temperatures would also facilitate kinetic and mechanistic studies of the deposition process.

To explore the utility of organo-transition metal cluster compounds as vapor phase reagents in CVD, we embarked on a study of mixed-metal cluster compounds which should yield homogeneous mixed-metal alloy and oxide films containing Fe, Co, Mn, and Ni. Polycrystalline and single crystal mixed-metal alloy and oxide films of these metals exhibit unique electronic and magnetic properties, which have been exploited in the production and magnetic switching and recording media, as well as micro electron circuitry.[5] Chemical vapor deposition has been employed in the production of magnetic films, but difficulties in controlling process variables and compositions to produce a consistent product has limited the successful application of this technique.[6-10] In these studies, determination of the exact chemical composition and structure of the magnetic films and the relationships between process variables, composition, structure, and magnetic properties was not typically addressed.

Herein we demonstrate that mixed-metal alloy films of Fe and Co can be prepared by the controlled thermal decomposition of suitable organo-transition metal complexes. An in-depth investigation shows that homogeneous, pure, metallic mixed-metal films of controlled composition and morphology can be prepared by varying process variables and the metal ratio of the starting compounds. In addition, the presence of oxygen during deposition or post-treatment with oxygen yields Co-Fe oxide materials.

EXPERIMENTAL

Synthesis and Characterization of Organo-transition Metal Compounds. The metal complexes $HFeCo_3(CO)_{12}$[10], mp = 85°C, and $CpFe(Co(CO)_6$[11], mp = 52-53° employed in this study were prepared by published procedures. Carbon monoxide (CP Grade) and hydrogen (Research Grade) were purchased from Linde Specialty Gases and used without further purification.

CVD Apparatus and Procedure. The CVD reactor and accompanying flow system employed in the initial studies are similar to those designed by other workers and is illustrated in Figure 1.[13] For each experimental run, the glass substrate was positioned in the deposition zone so as to be in line with the inlet tube. The system was evacuated to 10^{-3} mm Hg while heating the deposition chamber and inlet line to their respective temperatures. At this time the carrier gas was introduced into the system at the chosen flowrate. When the system was filled with carrier gas and the temperature had stabilized, the organo-transition metal compound was introduced into the flask. After a brief evacuation, the carrier gas was metered into the system, and the reaction flask was heated to the vaporization temperature. The vaporization and deposition temperatures and the flowrate were adjusted as necessary. Each run lasted between 6 and 12 hours. Sample weights of 50-100 mg were generally employed. Additional reactant or diluting gases were introduced into the deposition chamber only after consistent flowrate and temperatures were maintained. Post-treatment of the films was carried out in the same apparatus.

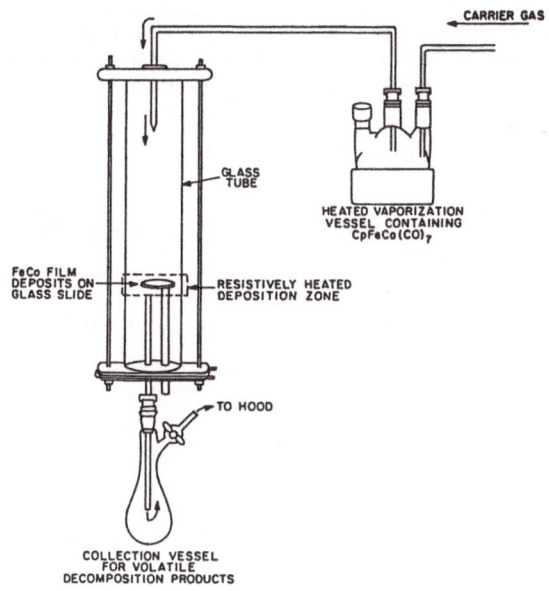

Figure 1. CVD apparatus

Determination of the Chemical Composition and Physical Properties of the Films. Film morphology, coherence, and thickness were analyzed using an ISI-DS13 Dual Stage Scanning Electron Microscope in both the secondary and back-scattered electron modes. The EDAX (Energy Dispersive Analysis of X-rays) attachment on this instrument was employed to determine the elements present, as well as their relative abundance and distribution within the sample. A Talysurf 10 Profilometer was used in film thickness determination. Scratch testing provided information on the adherence and coherence of the films. Metal analyses were conducted using a Perkin Elmer 703 Atomic Absorption Spectrometer. Standard solutions were prepared by dissolving $Co_2(CO)_8$, $[CpFe(CO)_2]_2$ (Pressure Chemical), and Fe and Co powders (Aldrich) in a mixture of 5 mL HCl and 10 mL HNO_3, their diluting to 1000 mL. The CVD prepared Fe/Co films were dissolved from the glass slide by soaking in the same HNO_3/HCl mixture. Carbon analyses were conducted by Schwartzkopf Analytical Laboratories. The presence of oxide phases was studied by electron microprobe analysis.

RESULTS AND DISCUSSION

Preparation of Fe/Co Films

Our initial work in exploring the utility of organo-transition metal cluster complexes to form thin films has focused on studying the influence of process variables on the chemical vapor deposition of $HFeCo_3(CO)_{12}$ and $CpFeCo(CO)_6$. These air stable complexes, with melting points of 80°C and 53°C respectively, were chosen for this study because they were expected to exhibit volatilities representative of polynuclear transition metal compounds, to have the potential to undergo a variety of decomposition modes, and to demonstrate that film stoichiometry could be tuned by changing the metal ratio in the organometallic precursor compounds. The influence of vaporization temperature, deposition temperature, carrier gas, flow rate, concentration, and deposition time on the formation, composition, morphology, and coherence of the cold deposits has been studied. The results are summarized in Table I.

Table I.   Study of Deposition Parameters Using $HFeCo_3(CO)_{12}$ and $CpFeCo(CO)_6$

Parameter	Optimum Value	Comments
Vaporization Temperature	5–10°C below the melting pt.	If significantly < melting pt., vapor pressure is insufficient. If ≥ melting pt, compounds decompose.
Deposition	300–350°C	If < 300°C, deposition rate is too slow. If > 350°C, gas phase nucleation occurs.
Flowrate	30–50 cc/min	If > 50 cc/min, deposition occurs only on the edges of the substrate or in the collection flask.
Time	6–10 hours	6 hours is optimum for $HFeCo_3(Co)_{12}$, 10 hours is optimum for $CpFeCo(CO)_6$, compound decomposition occurs after the optimum time.
Concentration	Saturated	Addition of N_2 to dilute concentration decreases the deposition rate.
Deposition Rate	5–10 mg/hour or 10–15 μm/ hour	Rates apply for optimum conditions outlined above and are variable due to absence of mass flow control.

The data in Table I indicate that decomposition of the organometallic precursors in the <u>vaporization flask</u> can be a problem if the vaporization temperature, carrier gas, and deposition time are not chosen carefully. Since such decomposition can significantly alter the gas phase concentration of Fe and Co and therefore the metal ratio and morphology of the film, the thermal decomposition of $HFeCo_3(CO)_{12}$ and $CpFeCo(CO)_6$ was studied. Both compounds were heated at their respective melting points in an atmosphere of CO for 8 hours. Infrared data during and after the procedure suggests the decomposition pathways summarized in equations 1 and 2. Chini and co-workers [11] described the

$$4\ HFeCo_3(CO)_{12}(s)\ +\ 8CO\ \xrightarrow{80°C}\ 4\ Fe(CO)_5(g)\ +\ 3\ Co_4(CO)_{12}(s)\ +\ 2H_2(g)\quad (1)$$

$$2\ CpFeCo(CO)_6(s)\ +\ xCO\ \xrightarrow{52°C}\ [Cp_2Fe(CO)_2]_2(s)\ +\ Co_y(CO)_x(s)\qquad (2)$$

thermal decomposition of solutions of $HFeCo_3(CO)_{12}$ to yield $Co_4(CO)_{12}$ but did not detect an Fe species. Because $CpFeCo(CO)_6$ is in equilibrium with $[CpFe(CO)_2]$, and $Co_2(CO)_8$ in solution [12], the decomposition pathway in equation 2 is also likely. The cobalt containing product is most a decomposition product of $Co_2(CO)_8$, which is thermal and light sensitive.

In addition, it was observed that the presence of air in the system during deposition resulted in the formation of highly-colored orange films. Also, the introduction of oxygen while heating the metallic film at the deposition temperature caused conversion of the silvery gray metallic film to an orange, presumably metal oxide, film.

Characterization of Fe/Co Films

Representative SEM micrographs of films prepared by the chemical vapor deposition of $HFeCo_3(CO)_{12}$ (Figure 2) and $CpFeCo(Co)_6$ (Figure 3) indicate

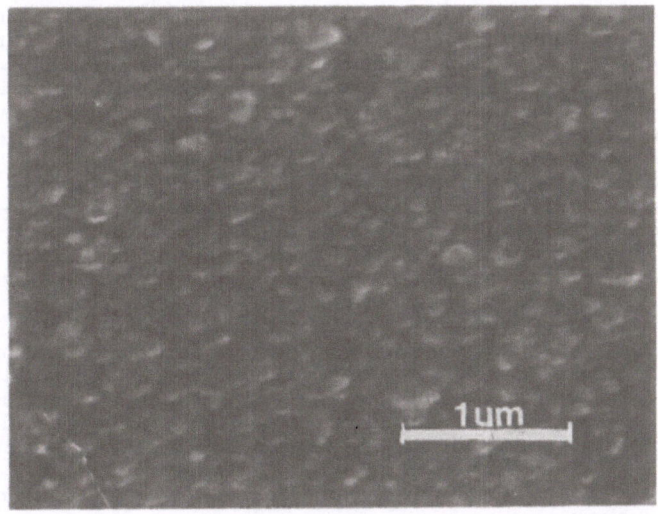

Figure 2. SEM of $HFeCo_3(CO)_{12}$ derived FeCo thin film at 57.8 Kx.

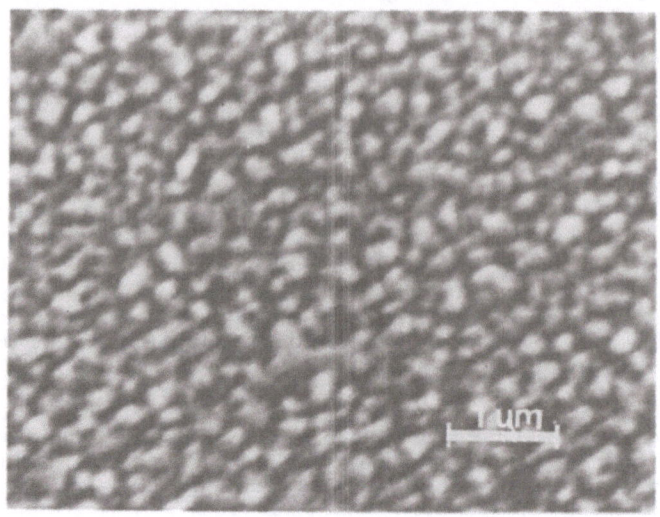

Figure 3.   SEM of CpFeCo(CO)$_6$ derived FeCo thin film at 19.7 Kx.

the presence of a coherent film formed by decomposition of the gas phase organometallic compound on the substrate surface. EDAX analysis showed that both Fe and Co were present in the films, and metal mapping suggested an even dispersion of these elements throughout the films. Film thickness was determined from electron micrographs taken in the back-scattered electron mode and ranged from 0.8-1.0μm for HFeCo$_3$(CO)$_{12}$ derived films and 1-1.5μm for CpFeCo(CO)$_6$ derived films. The films are amorphous with a granular surface composed of particles 700-1000Å in size in the HFeCo$_3$(CO)$_{12}$ derived films and 3000-5000Å in size in the CpFeCo(CO)$_6$ derived films. The amorphous nature of the films was verified by attempts to study the crystallinity and composition of these films by x-ray powder diffraction which gave no observable lines in the x-ray spectra. Although metallic in appearance, the films were only slightly conducting. Scratch and Scotch tape testing of the surface of the films showed them to be fairly coherent and adherent.

Attempts to investigate the morphology of the <u>oxidized</u> films by SEM and X-ray powder diffraction were unsuccessful because of their amorphous nature. EDAX analysis of these oxidized films suggested that both Fe and Co were present. Although oxidation of pre-deposited films did not affect film thickness, the introduction of oxygen or air <u>during</u> deposition appeared to decrease the thickness of the resulting films by ~25.0%. These amorphous films were metallic in appearance, coherent, adherent, and non-conducting.

The metal ratios of two sets of the films, as determined by atomic absorption spectroscopy, are summarized in Table II. Note that both the

Table II. Ratio of Fe:Co as Determined by Atomic Absorption
           Spectroscopy

Complex	Film	Weight Fe	Weight Co	Fe/Co Ratio
$CpFeCo(Co)_6$	1	0.0785mg	0.0693mg	1.20
	2	0.0778mg	0.0702mg	1.17
$HFeCo_3(CO)_{12}$	1	0.0382mg	0.0700mg	0.50
	2	0.0362mg	0.0832mg	0.45

$CpFeCo(CO)_6$ and $HFeCo_3(CO)_{12}$ derived films are iron rich by approximately
20% and 50% respectively. These percentages reflect the tendency of the
starting compounds to thermally decompose to yield volatile organo-iron
complexes, eqs 1 and 2. However, the observed ratios qualitatively
reflect the stoichiometry in the starting compounds. Carbon analyses
showed that carbon was not a significant impurity in these films, with no
carbon detected in a $HFeCo_3(CO)_{12}$ derived film and only 0.05%
detected in a $CpCoFe(CO)_6$ derived film. Attempts to investigate the oxygen
content of the metallic films by electron microprobe analysis showed that
no detectable oxide phases were present. This is in contrast to the films
exposed to oxygen, which showed the presence of both Fe^{3+} and Co^{2+} oxide
phases. Further investigation into the stoichiometry of and phases present
in these oxide films is necessary.

Once the characterization of the mixed metal oxide films is complete,
an investigation into the magnetic properties of both Fe_xCo_y and $Fe_xCo_yO_z$
films will commence. It is anticipated that these amorphous films, with
relatively small grain sizes, should exhibit magnetic behavior unlike
analogous polycrystalline and single crystal films prepared previously.

Preliminary results suggest that this approach can be extended to
other mixed-metal systems. To date, MnCo and HnFe films have been
deposited by the chemical vapor deposition of $MnCo(CO)_9$ and $CpMnFe(CO)_7$
respectively under similar conditions. SEM analyses of these films suggest
that grain size, morphology and thickness of these films suggest that
grain size, morphology and thickness of these films are in the range of
the Fe_xCo_y films. These systems, as well as the potential fabrication of
Fe_xNi_y, Mn_xNi_y, and Co_xNi_y films from heteronuclear organo-transition
metal complexes are currently under investigation.

CONCLUSION

Chemical vapor deposition has been employed in the production of
magnetic Fe_xCo_y films; however, variability in film thickness, morphology,
and composition has limited the successful application of this approach
[6-10]. By starting with an organo-transition metal complex which already
contains both metals and optimizing the deposition conditions, we have
succeeded in reproducibly forming amorphous, homogeneous, pure, coherent
mixed-metal and oxide films of Fe and Co. Film thickness can be controlled
as a function of time. The stoichiometry of the films reflects, in part,
the metal ratio of the starting organo-transition metal complex. This
suggests that the complex arrives at the substrate essentially intact
where at the higher temperatures it decomposes to yield Fe_xCo_y films.
Here, the use of Co as a carrier gas is crucial in stabilizing the cluster
in the gas phase prior to its arrival in the deposition zone. In addition,
the controlled oxidation of these films to yield $Fe_xCo_yO_x$ phases has been
demonstrated.

164

REFERENCES

[1] Luberhoff, B. J. in opening remarks at meeting of Industrial
     Engineering Section A during Fall 1985 National Meeting of the
     American Chemical Society.

[2] Bryant, W. A.; J. J. Mater. Sci. 12, 1285 (1977).

[3] Gupta, D. C.; Wang, P. Solid State Technol. 11, 48 (1977).

[4] Spear, K. private communication.

[5] Thompson, J. E. The Magnetic Properties of Materials, Great Britain:
     CRC Press (1968).

[6] Edelman, F. H. "The Preparation and Characterization of Thin
     Ferromagnetic Films, U. S. Departments of Commerce, Report No.
     PB151525, 1958.

[7] Halaby, S. A.; Kenny, N. S.; Murphy, J. A.   U.S. Patent 3 892 888.

[8] Chen, S.-L.; Murphy, J. A. U.S. Patent 3 859 129.

[9] For example, Linare, R. C.; McGraw, R. B.; Schroeder, J. B.
     J. Appl. Phys. 36, 2884 (1965).

[10] For example, Nagasawa, Ki, Banda, Y.; Takada, T. Japn. J. Appl.
     Phys. 7, 174 (1978).

[11] Chini, P.; Colli, L.; Deraldo, M.   Gazz. Chim. Ital. 90, 1005 (1960).

[12] Madach, T.; Vahrenkamp, H.   Chem. Ber. 113, 2675 (1980).

[13] Kaplin, Y. A.; Belysheva, G. V.; Zhipsov, S. F.; Domrachev, G. A.;
     Chernyshava, L. S. Zhurnal. Obschei Khimi 50, 118 (1980).

HOW TO MAKE METAL SILICIDE THIN FILMS FROM MOLECULAR SILICON-METAL COMPOUNDS —
AND HOW NOT TO.

B.J. AYLETT
Chemistry Department, Queen Mary College, London   E1 4NS, U.K.

ABSTRACT

The role of transition metal silicides in microelectronics devices is
explained, and current routes to their formation as thin films are
critically examined. It is shown how volatile molecular precursors with
prevenient Si-metal bonds can be used to prepare silicide films of a variety
of metals by chemical vapour deposition (CVD) at low pressure with an inert
carrier gas, under mild conditions. The scope of the method is discussed,
some mechanistic features of the CVD process are considered, and possible
future developments are examined.

THE IMPORTANCE OF METAL SILICIDE THIN FILMS

Silicides form an essential part of modern silicon-based semiconductor
fabrication technology [1,2]. Their chief uses in this area are as follows,
given in roughly ascending order of sophistication.

(a) To provide a barrier between the silicon substrate and an attached
metal conductor, often Au or Al (fig.1). This barrier inhibits, to a
great extent, the interdiffusion of silicon and metal atoms across the
interface, which would otherwise occur rapidly at quite modest temperatures (600-700 K) and slowly even at room temperature. The effect of
such mutual diffusion is to degrade the device and cause eventual
failure. These barrier layers must obviously be fairly good electrical
conductors, as well as being chemically inert; PtSi and WSi_2 are
examples of effective materials.

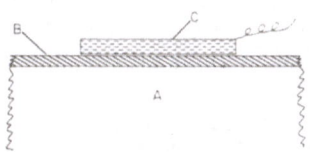

FIG. 1 Barrier layer in silicon device manufacture.

A, silicon substrate; B, silicide diffusion barrier; C, metal

contact and external connection.

R. M. Laine (ed.), Transformation of Organometallics into Common and Exotic Materials: Design and Activation, 165–177.
© 1988 by Martinus Nijhoff Publishers.

(b) To form Schottky barriers when deposited on suitably doped silicon substrates [3]. These are used as rectifying contacts and switching devices.

(c) To make interconnects and gates in very large-scale integration (VLSI) devices [1,4]. Here the requirement is for a material of low resistivity which can be used to provide circuit features of 1 μm (or less) in size, and yet retain fast operation. Fig.2 shows a typical metal oxide-silicon field effect transistor (MOSFET): the essential thin insulating layer of SiO_2 is easily formed on the silicide surface [5] by reaction with oxygen at about 1000 K. Complex devices with 10^5 - 10^6 circuit elements per chip are currently being produced, and still higher densities are being sought.

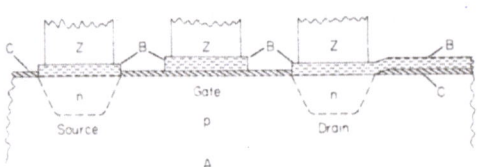

FIG. 2   Schematic view of a MOSFET device.

A, silicon substrate; B, silicide layer; C, thin SiO_2 layer; n, p, n-type and p-type regions; Z, metal conductors.

(d) To produce specialised devices employing silicides as the active elements. Some silicides (e.g. β-$FeSi_2$ and WSi_2) are semiconductors, and have been n- and p- doped; they can be used in place of (or together with) silicon, to produce different band gaps and electron mobilities [6].

(e) To develop possible new multiple quantum well (MQW) devices [7]. These consist of a succession of accurately-defined thin layers, each being of different yet accurately-controlled composition, deposited on a semiconducting substrate. Typical examples might have 10-20 layers, each of 20-50 Å thickness. Within such layers, electron propagation is essentially 2-dimensional in nature; the device as a whole shows radically new physical effects which promise to be of wide application.

Other non-electronic applications [8] of metal silicide films include protective coatings on metals for avionics and space applications, on turbine blades, and on fuel cell electrodes. Also electrodes for photo-electrochemical cells, when coated with metal silicides, show catalytic activity and could find use in solar energy conversion [9].

TYPICAL CHARACTERISTICS OF SILICIDES.

Only a brief summary is given here [10].

Stoicheiometry

There is a wide range of possibilities, e.g. Cr_3Si, Co_2Si, Mn_5Si_3, NiSi, $Mn_{27}Si_{47}$, $MoSi_2$, $IrSi_3$. A number of non-stoicheiometric phases are also known e.g. MnSi, α-$FeSi_2$, Ni_3Si (approximately). Many ternary silicides, silicide-nitrides, silicide-carbides etc. are also known.

Structures

These are frequently complex, with large unit cells of an unusual or unique kind. Metal-rich silicides are akin to alloys, with close M-M interactions. On the other hand, silicon-rich silicides often have chains, layers, or networks of linked silicon atoms, cemented together by metal atoms/ions. Because there is undoubtedly electron transfer from metal to silicon, the Si_n arrangements in such structures have been described [2] as charge-stabilised homopolymers.

In all cases, the M-Si interactions are strong. Some structural patterns are exemplified in Table I.

TABLE I
Structural Features of Some Silicides

	Feature	Example
(i)	Si_2	U_3Si_2
(ii)	Si_n chains	USi
(iii)	Si_n 6-ring layers	β-USi_2
(iv)	Si_n 3-d network	$ThSi_2$
(v)	Close-packed layers, M and Si in each	$CrSi_2$
(vi)	3-d lattice, like CaF_2	$NiSi_2$
(vii)	3-d lattice, like $PbCl_2$	Co_2Si
(viii)	3-d lattice, like β-W, Si[12] by V	V_3Si

Colour

Commercial bulk silicides are typically dark brown or black, but thin films of pure materials are commonly grey, silvery, or iridescent, depending on thickness.

Melting Point

Typically melting points are high (e.g. CoSi, 1690 K; WSi_2 2440 K).

Electrical properties

Usually they are rather poor metallic-type conductors (e.g. Ti_5Si_3, 3×10^3 ohm^{-1} cm^{-1}/298 K). A small number of silicon-rich silicides are semiconductors (e.g. β-$FeSi_2$; intrinsic band gap, 0.8 eV). Some metal rich silicides are superconducting at relatively high temperatures (e.g. V_3Si; T (transition), 17.1 K).

Chemical properties

Generally very inert: they are attached by fused alkali (especially with oxidising agents present), by fluorine, and also chlorine when heated. Superficial oxidation occurs on heating in air, but complete reaction requires very high temperatures (e.g. $MoSi_2$: > 1700 K).

Silicides are typically hard (e.g. Cr_3Si, 1000 kg mm^{-2}/298 K), of moderate densities (e.g. $TiSi_2$, 4.4 g cm^{-3}), and can show a variety of magnetic behaviour (e.g. $ZrSi_2$ diamagnetic; CoSi, paramagnetic; Fe_5Si_2, ferromagnetic.

PRESENT PREPARATIVE METHODS AND THEIR LIMITATIONS

Thin films are usually made by the following methods [1,2].

(a) Sputtering. A thin metal layer is deposited on the silicon substrate, then heated. Interdiffusion occurs, giving a metal silicide. The composition of the product often varies, depending on the temperature and duration of heating. For example, Ni on Si yielded Ni_2Si at 570 K, NiSi at 670 K, and $NiSi_2$ above 950 K. Refractory silicides such as $MoSi_2$ may need annealing at up to 1150 K.

(b) Molecular beam epitaxy (MBE). This method can produce films of high purity and accurate composition in the laboratory, but is less well suited to production processes.

(c) Plasma transport. Rapid deposition is possible but imperfections may be produced in the substrate and/or in the thin film, due to the energetic nature of the discharge.

(d) Chemical vapour deposition (CVD). This has the advantage of producing uniform deposits, even on irregular surfaces: it is also attractive for large-scale production, and can easily produce a succession of different deposits. However, the processes so far described, such as that in eqn. (1) [11], lead to halide contamination of the film.

$$WF_6 + SiH_4 \quad \xrightarrow[\text{plasma}]{570 \text{ K}} \quad WSi_2, \text{ etc.} \qquad (1)$$

Thus only methods (a) and (b) above are really suitable for microelectronics device fabrication. They also have the great advantage of being able to produce epitaxial deposits under favourable circumstances [12].

It might be supposed that pyrolysis of a SiH_4/metal carbonyl mixture under CVD conditions would lead to good metal silicide films. In the first place, however, the number of volatile metal carbonyls is quite limited (those of V, Fe, Ni are very volatile, those of Cr, Mo, W, Mn, Re, Ru, Os, Co, Rh, Ir, less so), and related metal precursors are essentially limited to the carbonyl hydrides of Mn, Re, Co and Fe. Moreover, the maintenance of a correct Si:metal ratio between two precursors of very different volatilities in the gas stream is not a trivial matter; even if this is done, there remains the problem of how to maintain even deposition with a constant Si:metal ratio at each point of the surface. For these reasons, we turned our attention to precursors with PREVENIENT Si-METAL BONDS, namely volatile molecular compounds that already contained the required silicon and metal linked together in the correct ratio to give the required metal silicide. Provided that no homogeneous decomposition of a harmful kind supervenes (i.e. one resulting in the selective loss of either volatile silicon or volatile metal compounds), there should be a fixed Si:metal ratio at the moment of surface deposition.

MOLECULAR SILICON-METAL COMPOUNDS AS POTENTIAL PRECURSORS

A wide range of molecular compounds is known in which one or more silicon atoms are linked to one or more transition metal atoms [3]. Table II shows the metals concerned.

TABLE II
Transition metals forming molecular compounds with silicon

Ti	V	Cr	Mn	Fe	Co	Ni	Cu
Zr	Nb?	Mo		Ru	Rh	Pd	
Hf	Ta?	W	Re	Os	Ir	Pt	Au

Some of these derivatives are simple volatile carbonyl derivatives such as $H_3SiCo(CO)_4$; many alkyl- and aryl-silyl examples are also known (e.g. $Me_3SiMn(CO)_5$. Less volatile derivatives include metal cyclopentadienyl derivatives such as $Cl_3SiMo(CO)_3(C_5H_5)$ and clusters such as $(OC)_3Ru(\mu-SiMe_2)_3Ru(CO)_3(Ru-Ru)$. As examples of simple compounds with differing Si:metal ratios, we have $(H_3Si)_2Fe(CO)_4$, $H_2Si[Co(CO)_4]_2$, $ISi[Mn(CO)_5]_3$ and $Si[Re(CO)_5]_4$.

Out of all these compounds, the hydridosilicon derivatives appear to be the most attractive, since it is well-known that the Si-H bond is usually cleaved thermally at temperatures below 750 K; although the Si-I bond is also thermally rather labile, there is the distinct possibility of silicon-to-metal transfer of iodine. Alkyl- and aryl-silicon derivatives are likely to yield films containing carbon, since the Si-C bond normally persists to high temperatures. Considering the other end of the molecule, CO is more likely to be lost cleanly from the metal than a cyclopentadienyl ligand. Finally, the relatively high volatility of the simple $H_3SiM(CO)_n$ derivatives makes them the best choice for CVD purposes.

THERMOLYSIS OF $H_3SiM(CO)_n$ DERIVATIVES

Initial experiments showed that approximate orders of thermal stability were:

$$H_3SiV(CO)_6 \; < \; H_3SiFeH(CO)_4 \; < \; H_3SiCo(CO)_4 \; < \; H_3SiMn(CO)_5$$
$$(300 \text{ K}) \qquad\qquad (340 \text{ K}) \qquad\qquad (375 \text{ K}) \qquad\qquad (400 \text{ K})$$

$$< \; (H_2Si)_2Fe(CO)_4 \; < \; H_3SiRe(CO)_5 \; < \; H_2Si[Re(CO)_5]_2$$
$$(420 \text{ K}) \qquad\qquad (430 \text{ K}) \qquad\qquad (450 \text{ K})$$

$$H_3SiCo(CO)_4 \; < \; H_3SiIr(CO)_4$$

$$H_2Si[Mn(CO)_5]_2 \; < \; H_2Si[Re(CO)_5]_2$$

Where known, temperatures at which decomposition begins are indicated.

Possible decomposition routes

Reported modes of decomposition of molecular silicon-metal compounds include the following [13].

1. Disproportionation about silicon (eqns. 2 and 3) and about metal (eqn. 4).

$$X_3SiCo(CO)_4 \longrightarrow SiX_4 + X_2Si[Co(CO)_4]_2 \qquad (2)$$
$$(X = H, F)$$

$$H_3SiMn(CO)_5 \longrightarrow SiH_4 + H_2Si[Mn(CO)_5]_2 \qquad (3)$$

$$\underline{cis}\text{-}Cl_3SiFeH(CO)_4 \xrightarrow{\Delta} (Cl_3Si)_2Fe(CO)_4 + \ldots\ldots \qquad (4)$$

Disproportionation about silicon is very common for hydrido- or fluoro-derivatives of the type X_nSiY_{4-n} (X = H, F: n = 1-3), and is associated particularly with liquid phase systems under mild conditions; it is believed to occur via a bimolecular intermediate with Y-Si interactions.

2. On heating, many alkyl- or aryl-silicon derivatives yield disiloxanes (e.g. eqns. 5 and 6): the hydrido-derivative $H_3SiV(CO)_6$ decomposes almost quantitatively as shown in eqn. 7.

$$R_3SiCo(CO)_4 \xrightarrow{\Delta} (R_3Si)_2O + \ldots\ldots \qquad (5)$$
$$(R = Et, Ph)$$

$$Me_3SiMo(CO)_3(C_5H_5) \xrightarrow{\Delta} (Me_3Si)_2O + \ldots\ldots \qquad (6)$$

$$H_3SiV(CO)_6 \xrightarrow{300 \text{ K}} (H_3Si)_2O + V(CO)_6 + [V(CO)_5C]_n \qquad (7)$$

No doubt these reactions proceed via electron donation from oxygen (of a CO group) to silicon in another molecule, followed by Si-metal bond cleavage and Si-O bond formation. This conclusion is strengthened by the observation of "half-way" products in the controlled pyrolysis of $Me_3SiCo(CO)_4$ (eqn. 8).

$$Me_3SiCo(CO)_4 \xrightarrow[50 \text{ hr.}]{380 \text{ K}} Me_3SiOCCo_3(CO)_9 \tag{8}$$
$$+ (Me_3SiOC)_4Co_2(CO)_4 + \cdots\cdots$$

3. Finally, removal of hydrogen as H_2 and of CO as such may occur in a "stripping" reaction, analogous to commonly observed mass spectrometric fragmentation patterns (e.g. eqn. 9).

$$H_3SiCo(CO)_4 \xrightarrow{\hspace{2cm}} H_mSiCo(CO)_n \tag{9}$$
$$(m = 3-0; \; n = 4-0)$$

Note, however, that other Si-metal compounds do not always fragment in this simple way; for example, metal cyclopentadienyl derivatives often give rise to abundant $SiC_5H_5^+$ ions (believed to be $_n{}^5$-species)[14] and consequent $SiCH^+$ fragments.

Initial experiments in which $H_3SiCo(CO)_4$ was heated in a sealed tube so that some liquid phase was present revealed a wide range of volatile products, from which it was inferred that disproportionation, silicon transfer from metal to oxygen, "stripping", and possibly also silylene extrusion were taking place (eqn. 10).

$$H_3SiCo(CO)_4 \xrightarrow{\Delta} H_2 + CO + SiH_4 + (H_3Si)_2O \tag{10}$$
$$+ HCo(CO)_4 + H_2Si[Co(CO)_4]_2 + \cdots\cdots$$

Later, more precise analysis of the volatile products from pyrolysis in sealed tubes of gaseous SiH_3-metal compounds, notably $H_3SiMn(CO)_5$, showed that H_2, CO, SiH_4 and CH_4 were chiefly formed: the last arose from subsequent reaction between H_2 and CO[13, 15, 16]. In all cases, a dark-coloured involatile solid was also formed, partly as an electrically conducting thin film on the tube walls. Analysis showed that the films contained C, H and O in addition to Si and metal: a typical composition, resulting from the pyrolysis of $H_3SiMn(CO)_5$ at 720 K for 1 hr., was $C_{1.36}H_{0.97}Mn_{1.00}O_{1.36}Si_{1.00}$, approximating to "$[HSiMn(CO)_{1.4}]$". There is conflicting evidence as to the nature of this solid: it is tempting to suppose that the hydrogen remains attached to silicon and that the carbon and oxygen are present as CO attached to metal. In support of this, alkaline hydrolysis of the solids yields H_2, and in some cases treatment with bromine has displaced up to 90% of the putative "combined CO" as a $CO/COBr_2$ mixture. In other cases, however, powder X-ray evidence has been found for SiO_2 and manganese carbide phases in the solid. The most likely explanation is that much of the solid is indeed an amorphous network polymer of linked (SiH) and $[Mn(CO)_n]$ units, but that a small proportion is the product of O-to-Si migration and consequent metal carbide formation, as discussed earlier (cf. eqn. 7).

Clearly an effective route to pure metal silicides must suppress disproportionation and migration of silicon to oxygen (both associated with close intermolecular contacts), while encouraging Si-H and M-CO cleavage (which can occur in isolated molecules). This suggests the use either of gas flow methods with low partial pressures of the Si-metal precursors or of dilute solutions of the precursor in inert solvents; in both cases, bimolecular processes are disfavoured. In practice, it is the first of these possibilities which has received most study, although, as indicated later, it may be necessary to pay more attention to solution techniques in the future.

CVD processes [13, 15, 16]

The apparatus is shown diagrammatically in Fig. 3. The carrier gas (He or Ar), with a flow rate of 50–150 cm^3s^{-1} at a pressure of 50–200 Pa, incorporates a small amount of the volatile precursor from a sample reservoir; the temperature of the the reservoir is such as to give a precursor partial pressure of 5–10 Pa. The resulting mixture passes through a cooled inlet tube and then emerges from a nozzle into the furnace zone. It impinges on the substrate plate, typically some 0.5 cm from the nozzle and independently heated. With furnace and/or substrate temperatures in the range 670–770 K, the precursor completely decomposes, and a thin, often shiny, layer of metal silicide is formed on the substrate and surrounding surfaces. The deposits are then characterised by X-ray analysis, electron microprobe analysis, and atomic absorption spectroscopy.

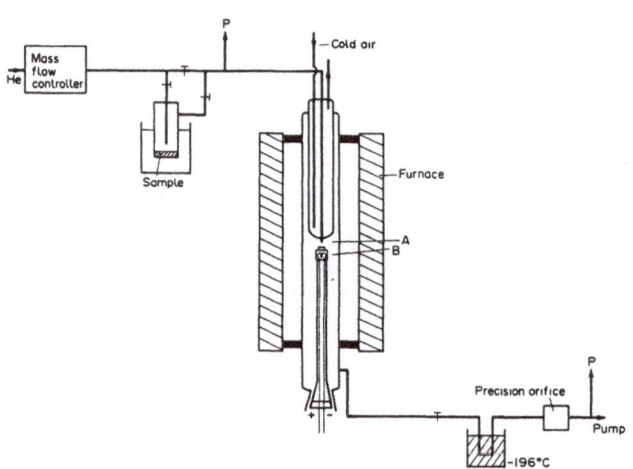

FIG. 3 The CVD apparatus.

A, vapour exit nozzle from cooled inlet tube; B, substrate plate on

holder with heater; P, pressure-monitoring device.

Volatile precursors so far employed include those shown in Table III and also $H_2Si[Re(CO)_5]_2$.

TABLE III

Products of CVD processes

Precursor	Product(s)
$H_3SiCo(CO)_4$	CoSi
$H_3SiMn(CO)_5$	$Mn_5Si_3 + MnSi_x (x \simeq 1.25)$
$H_3SiRe(CO)_5$	$Re_5Si_3 + \ldots\ldots$
$(H_3Si)_2Fe(CO)_4$	$\beta\text{-}FeSi_2$

In all cases, the overall Si:metal ratio in the thin film is the same as that in the precursor. However, it can be seen that in some cases phase disproportionation occurs, yielding a metal-rich and a silicon-rich phase; this is no doubt attributable to the favourable stability of M_5Si_3 phases [10]. Satisfactory films with thicknesses ranging from about 10 nm to 1 μm can be deposited on a variety of substrates, including Si, SiO_2, GaAs, InP, and metals.

Mechanistic considerations

Various stages in the process can be considered [16].

1) Activation of the precursor. This must occur as a result of (i) direct thermal excitation, and (ii) kinetic energy transfer from colliding noble gas atoms. The role of (ii) is clearly vital, since it is a matter of experience that good deposits are formed only when a large excess of noble gas atoms is present.

The effect of the colliding atoms is two-fold. In the first place, they may break bonds directly by simply knocking off atoms or groups present in the precursor. Space-filling models show that H atoms and CO goups are exposed on the periphery of the molecule, and are thus more likely to be removed in this way: the Si-metal unit is relatively hidden. Secondly, the colliding atoms will cause excitation of the precursor molecule, the latter's additional energy largely appearing as excess vibrational energy. Rapid partitioning of this energy around the molecule can than lead to bond rupture. Estimates [13, 17] of bond dissociation energies [D, kJ mol^{-1}: Si-H, ~250; M-CO, ~150; Si-M, 250-300] suggest that the Si-metal bond will largely remain intact.

It is not possible at present to estimate the relative importance of these various effects, although analysis of the changes produced by altering the mass of the carrier gas atoms (and hence their momentum and kinetic energy) may well prove helpful.

2) Later stages. The resulting $H_xSiM(CO)_y$ species may, if their partial pressure is sufficiently high, coalesce to form small clusters $H_bSi_aM_a(CO)_c$: there are precedents for this behaviour [16]. These clusters (or the monomer) continue to lose CO and hydrogen by the processes outlined in 1) above until simple $(SiM)_p$ molecules are produced (where p = 1 or a small integer). It is important that all these species remain in the gas phase: too great a degree of aggregation into clusters will result in the formation of small solid particles, which then fall to the substrate as a "snowstorm", giving rise to a powdery deposit.

The final stage begins with the deposition of $(SiM)_p$ molecules on the surface, yielding an amorphous phase. Rearrangement then occurs to give a micro-crystalline silicide phase (or phases): recent work [18] has shown that similar rearrangements of metal silicide phases on surfaces can occur readily at temperatures as low as 415 K.

This sequence of events then corresponds to a series of essentially homogenous gas-phase reactions as the main decomposition route, and will be favoured when the ambient temperature T_1 is similar to the substrate temperature T_2 (see Fig. 4). It has the great advantage that volatiles (H_2, CO) will not be evolved from the thin film material itself, as they will have been removed in earlier stages: such evolution tends to disrupt the thin film structure. Consistent with this, it is known to be harder to produce good films under "heterogeneous" conditions, i.e. when heating is provided solely at the substrate $(T_2 \gg T_1$, Fig.4); even under these conditions, it is possible that some pre-pyrolysis occurs in the gas-phase, just above the surface of the substrate.

It may be noted that this sequence, involving the retention of the strong prevenient Si-metal bond throughout the process, is fundamentally different from the CVD processes noted earlier (eqn. 1, for example), which use separate silicon and metal precursors. It also differs from the more familiar MOCVD processes currently used for production of III/V semiconductors such as GaAs; here, the III-V bond must, during the deposition process, first be formed and then strengthened.

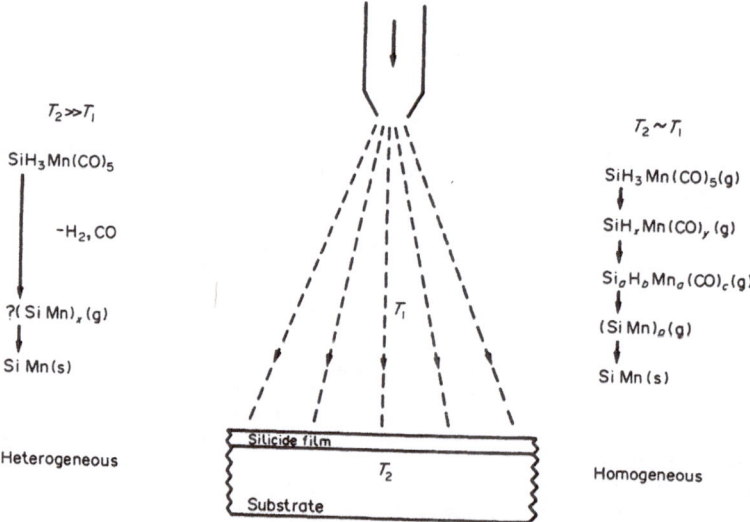

FIG. 4   The decomposition/deposition zone in the CVD apparatus (region A-B in Fig. 3) and possible reaction paths (see text).

FUTURE OUTLOOK

Precursors .

Suitable volatile $H_3Si-M(CO)_n$ precursors are not too plentiful. Apart from those mentioned in Table III, $H_3SiV(CO)_6$, $H_3SiFeH(CO)_4$ and $H_3SiIr(CO)_4$ [13] are known, and no doubt Nb and Ta analogues of the first, Ru and Os analogues of the second, and a Rh analogue of the third could be synthesised. Not unexpectedly, no compounds of this kind are known for Ti, Zr and Hf. More surprisingly however, it has not so far proved possible to make compounds such as $(H_3Si)_2M(CO)_n$ (M = Cr, Mo, W; n = ?4, ?5) or $(H_3Si)_2$-$Ni(CO)_n$ (n = ?2). When metal ligands other than Co are considered, PF_3 is promising, and $H_3SiM(PF_3)_4$ (M = Co, Rh, Ir) is an obvious possibility (R_3Si analogues are already known) Other Group V donors (e.g. PPh_3) will reduce the volatility markedly. Nitrosyl ligands on the metal are probably best avoided, since although they might confer volatility, the risk of NO$\rightarrow$Si attack is too great. On the other hand, π-coordinated alkenes and alkynes are likely to be removed cleanly.

Another possible class is represented by $H_3SiM(CO)_n(cp^*)$, where cp^* represents $\eta^5-C_5H_5$ or a substituted variant. A few examples are known, such as $H_3SiFe(CO)_2(C_5H_5)$ and $H_3SiMo(CO)_3(C_5H_4SiMe_3)$ [19]; However, they are of rather low volatility and pose the threat of incomplete removal of the cyclopentadienyl group under CVD conditions. Resulting incorporation of carbon into the thin film could be harmful for device applications.

The general class $H_2Si[M(CO)_n]_2$ is known for M = Co, n = 4 or M = Mn, Re, n = 5, and all of these are potential precursors for "M_2Si" deposition. Variants with other metal ligands, such as $H_2Si[Fe(CO)_2cp]_2$ [14], are of marginal volatility, while it has not so far proved possible to synthesise $H_2Si[M(CO)_3cp^*]_2$ (M = Cr, Mo, W) [19].

Many other types of Si-metal molecular compounds are known [13], such as the tris and tetrakis derivatives $HSi[M(CO)_5]_3$ and $Si[M(CO)_5]_4$ (M = Mn, Re) [20], clusters of various types e.g. $[(OC)_4Co]SiCo_3(CO)_9$, and cyclic derivatives such as $L_2PtSiR_2CH:CHSiR_2$. Besides the undesirable presence of Si-C bonds in the last example, these classes are all too involatile to be used as precursors under normal CVD conditions, and other deposition techniques would need to be developed.

Deposition techniques

It is clear from the preceding section that the number of Si-metal precursors that are suitable for use with existing techniques is rather limited. This is because for many Si-metal compounds their decomposition temperatures lie only just above or are lower than the temperatures at which they are appreciably volatile.

Future possibilites include the use of spray deposition, in which a dilute solution of the precursor in an inert solvent is introduced to the reaction vessel and rapidly heated; the solvent is quickly removed, leaving particles of the solute, with virtually molecular dimensions, in the gas phase. These particles then travel to the substrate, undergoing decomposition en route. Photo-induced decomposition, either in the gas phase or on the substrate surface, will also repay investigation; it is known [13,16] that simple silyl metal carbonyls undergo UV photolysis, yielding products generally similar to those form thermally-induced reactions. Also plasma-enhanced deposition has yet to be evaluated in this area.

In summary, it appears that a great deal of difficult synthetic work, together with careful mechanistic studies and improved deposition technology are needed in order to realise the potential of metal silicides. These must be linked, of course, with the most sophisticated application of physical characterization techniques such as LEED, Auger spectroscopy, SIMS, and Rutherford back-scattering. But the rewards are great, not least because similar ideas can be applied to many other transition metal/main group systems, giving rise to a very wide range of new materials.

Acknowledgement. I thank SERC and British Telecom for their support of some of this work.

References

1. J.M. Poate, K.N. Tu and J.W. Mayer, "Thin Films," Wiley, New York, 1978; S.P. Murarka, "Silicides for VLSI applications," Academic Press, New York, 1983; M. Wittmer, Thin Solid Films, 107, 99, (1983); K.Y. Ahn and S. Basavaiah, ibid., 118, 163 (1984).

2. B.J. Aylett, British Polymer J., 1986, in press.

3. R.T. Tung, J. Vacuum Sci. Technology, B2, 465 (1984).

4. S.P. Murarka, J. Vacuum Sci. Technology, B2, 693 (1984); S. Furukawa (ed.) "Layered Structures & Interface Kinetics," Reidel, Amsterdam, 1985.

5. See, e.g. L.N. Lie, W.A. Tiller and K.C. Sarawat, J. Appl. Phys., 56, 2127 (1984); A. Cros, R.A. Pollak and K.N. Tu, ibid., 57, 2253 (1985); W.J. Strydom, J.C. Lombard, and R. Pretorius, Thin Solid Films, 131, 215, (1985).

6. B.-Y. Tsaur and C.H. Anderson, Appl. Physics letters, 47, 527 (1985).

7. G. Bauer (ed.), "Two-dimensional Systems, Hetero-structures, and Superlattices," Springer, Berlin, 1984.

8. B. Aronsson, T.Lundstrom and S. Rundqvist, "Borides, Silicides and Phosphides," Methuen, London, 1965; Kirk-Othmer Encyclopaedia of Chemical Technology, 3rd. edition, Interscience, New York, 1982/83.

9. A.J. Bard, F.-R.F. Fan, G.A. Hope and R.G. Keil, in "Inorganic Chemistry: Towards the 21st Century," ed. M.H. Chisholm, A.C.S. Symposium Series, 211, 93, (1983); M.S. Wrighton, ibid., p.59.

10. For further details see refs. cited in refs. 2 and 8 above.

11. M.J. Cooke, Vacuum, 35, 67 (1985)., and refs cited there. See also: P.K. Tedrow, V.Ilderem and R.Reif, Appl. Physics letters, 46, 189 (1985).

12. A. Zur, T.C. McGill and M.-A. Nicolet, J. Appl. Phys., 57, 600 (1985); C.J. Chien, H.G. Cheng, C.W. Nieh and L.J. Chen, ibid., 57, 1887 (1985).

13. B.J. Aylett, Adv. Inorg. Chem. Radiochem., 25, 1 (1982.

14. B.J. Aylett and H.M. Colquhoun, J. Chem. Research, (S) P.148; (M) P.1677 (1977).

15. B.J. Aylett and H.M. Colquhoun, J. Chem. Soc. Dalton Trans., 2058 (1977).

16. B.J. Aylett and A.A. Tannahill, Vacuum, 35, 435, (1985).

17. B.J. Aylett, "Organometallic Compounds" 4th. edn., Vol.1, Part 2, Chapman & Hall, London, 1979.

18. C.A. Hewitt, S.S. Lau, I. Suni and L.S. Hung, J. Appl. Physics, 57, 1089 (1985).

19. B.J. Aylett and M. Hampden-Smith, unpublished work.

20. B.J. Aylett and M.T. Taghipour, J. Organometal. Chem., 249, 55, (1983).

CHEMICAL VAPOR DEPOSITION OF BORON NITRIDE USING ORGANO-METALLIC ADDUCTS

D. M. SCHLEICH[*], W. Y. F. LAI AND A. LAM
Polytechnic University, 333 Jay Street, Brooklyn, N.Y. 11201

ABSTRACT

We have studied the preparation of boron nitride thin films by the chemical vapor deposition of adducts. We have observed that the adduct decomposition does not yield stoichiometric films. The addition of ammonia to the reaction mixture lowers the B/N ratio, however the films are still not stoichiometric. We have also looked at the effect of ammonia addition on hydrogen incorporation in our boron nitride films.

INTRODUCTION

Boron nitride is the first member in the III-V semiconductor family. As a bulk material it is usually of interest for its ceramic properties; however, it has other applications in the semiconductor field. Because of its transparency to X-rays, BN thin films are being exploited in X-ray lithographic treatment of wafers,[1] in addition, because of the III-V nature of the material it can be used for growth crucibles for semiconducting III-V compounds - specifically GaAs and InP.

Our initial interest in BN was to develop an inexpensive and relatively non-toxic technique to coat silica boats for the Bridgman growth of GaAs single crystals. Since gallium metal may react chemically at high temperatures with silica to introduce silicon impurities within a crystal, we wished to prevent contact between the precrystalline melt and the silica boat. Although pyrolytic BN crucibles are routinely used in Czochrolski growth of GaAs, the necessity to freeze the entire melt within a Bridgman boat would greatly increase the risk of breaking the extremely expensive BN boat. Our approach was to use standard silica Bridgman boats and place a thick protective layer on the surface. The added advantage of this approach was to limit the high thermal conductivity associated with BN which creates problems in directional solidification.

The literature indicates a wide variety of reagents which have been used for the chemical vapor deposition of BN. The first reports of BN thin film deposition were in 1963 when Haberect et al. performed static low pressure deposition of trichloroborazine [2]. The materials prepared by this approach were reported to be polycrystalline. This work was followed by several researchers in the field trying to prepare boron nitride films for various tasks. The general approach has been to introduce a boron containing compound, usually B_2H_6,[3,4] or BCl_3,[5] into a reactor where it would combine with a nitrogen containing compound, usually NH_3,[3,4,5] or methyl amine [6]. The exact product of this reaction would depend on the temperature of the substrate and the relative concentration of nitrogen to boron compounds. In all of these reactions one definite problem exists. When any of the above mentioned boron compounds come into contact with the above mentioned nitrogen compounds solid adducts are formed. These solid adducts will cause homogeneous gas phase precipitations which can create particulate contamination of an optical quality film. In addition, these reactions can result in depletion of the gas phase species and subsequent inhomogeneities in deposited film thickness.

Although the field of MOCVD has been widely studied in the semiconducting members of the III-V group, little work has been directed to the MOCVD growth of BN. A distinct advantage of the MOCVD approach to BN growth is

178

R. M. Laine (ed.), Transformation of Organometallics into Common and Exotic Materials: Design and Activation, 178–184.

the existence of relatively volatile, liquid adducts which contain both nitrogen and boron. An additional advantage of the use of organo-metallic adducts is to stabilize otherwise highly reactive boron complexes. It has previously been shown that this approach could be used to prepare thin films of the semiconducting members of the III-V family.[7]

EXPERIMENTAL RESULTS

We chose to study specifically low pressure chemical vapor deposition of the triethylamine:borane complex. We have since expanded this work to the diethylamine:borane complex with very similar results. The main attractions of these compounds are relatively high volatility (~4-10 torr at 65°C), substantial stability and only moderate flammability, low-cost and easy accessibility.

Two types of reactors have been used to prepare thin BN films. A simple hot walled reactor (Figure 1) specifically prepared to coat silica boats, and a cold wall reactor equipped with more sophisticated controls (see Figure 2). Our initial results using the simple hot walled reactor indicated that independent of the temperature, pressure or carrier gas within the reactor, using only the adduct we were not able to prepare stoichiometric BN. This was evident because the films that we prepared were always colored, ranging from yellow to brownish black. Since stoichiometric BN has a very large optical gap (~5.8 eV) [8], and the boron rich compounds have progressively lower optical gaps eventually falling to about 1 eV; we felt that our compounds were boron rich. XPS measurements indicated that these films had B/N ratios as high as 10/1 with very low carbon and oxygen impurity concentration. In order to understand better the processes occurring we performed a series of tests using our cold wall reactor. In these tests we varied the substrate growth temperature from 450°C - 800°C. Within this range all films prepared were amorphous to X-rays. In fact, we had previously observed in our hot walled reactor that even at temperatures as high as 1100°C, all our films were amorphous to X-rays. The in situ quadrupole mass spectrometer immediately indicated that even at very low temperatures (400° C) the borane:triethylamine complex was breaking apart. In addition we observed a large number of amine fragments but very few if any boron containing fragments. It is not at all surprising that the adduct, with a bond energy of only about 30 Kcal/mole [9], cleaves when exposed to a high temperature. It further appears that the cleaved amine can then escape from the surface whereas the highly reactive borane polymerizes.

Therefore, although we initially start with a gas phase adduct containing exactly the correct B/N stoichiometry this ratio is destroyed during the decomposition process. In order to compensate for this regular depletion in nitrogen we decided to introduce an ammonia stream into the system. A simple nitrogen carrier gas is not adequate to introduce active nitrogen atoms into the reaction zone. In order to control the relative ratio of borane:triethylamine complex to ammonia we decided to maintain the flow of ammonia constant at 50 SCCM, (standard cubic centimeters per minute, Brooks mass flow control meter) and vary the temperature of the borane:triethylamine complex in order to increase its vapor pressure. We prepared films on two substrates during each run, a silica substrate and a silicon substrate. The silica substrate allowed us to observe immediately color changes of the film and the silicon film allowed us to perform FT-IR analysis (Digilab-FT 60). The mass spectrometer immediately indicated the existence of the ammonia, and we could estimate the ratio of complex to ammonia by comparing the intensities of ammonia to amine (see Figures 3a & b).

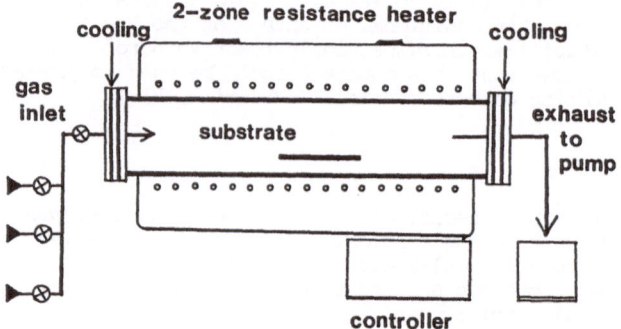

Figure1 Hot-Wall LP-CVD Reactor

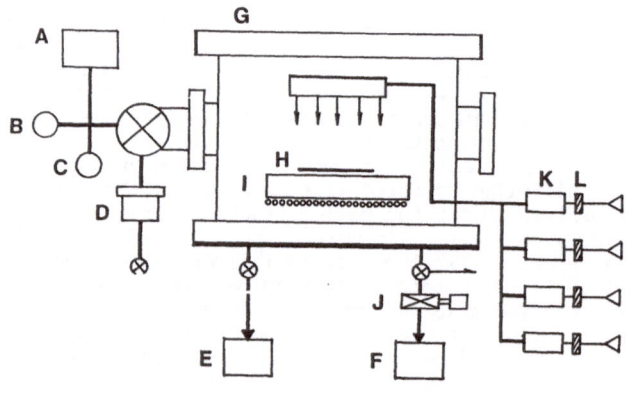

A. mass spectrometer
B. ionization gauge
C. capacitance monometer
D. turbo pump
E. rough pump
F. processing pump

G. reactor
H. substrate
I. resistance heater
J. throttle valve
K. flow controller
L. filter

Figure 2 Schematic Representation of the CVD Reactor

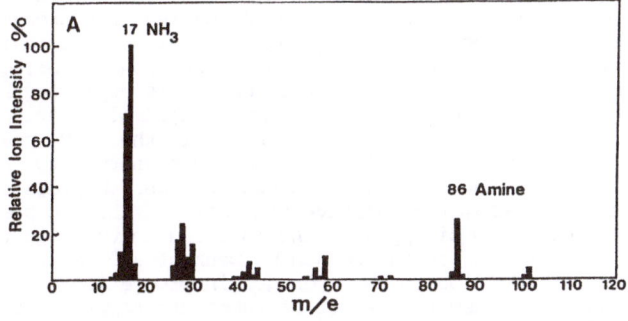

Figure 3a Ammonia Conc. in Reactor

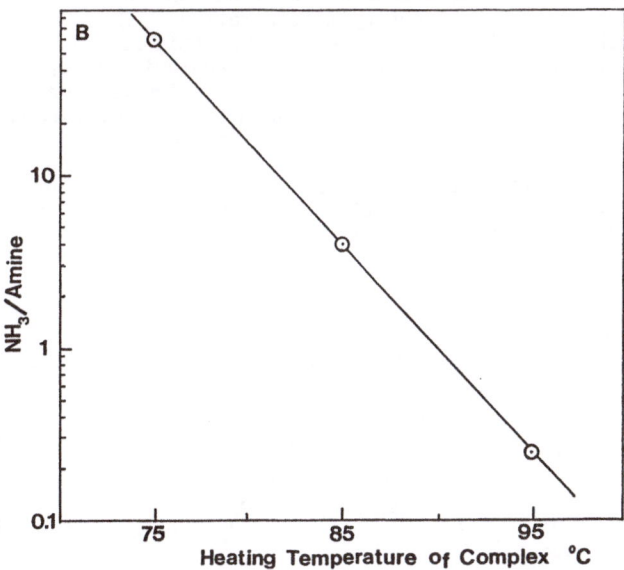

Figure 3b Ammonia/Amine Ratio as a Function
Adduct Temperature

RESULTS AND CONCLUSIONS

Several films were prepared at complex temperatures ranging from 60-95 C, while the substrate temperature was maintained at temperatures from 450 - 800 C. As the complex temperature was lowered, i.e. as the active N to B ratio increased the films became more transparent. At 65 C the films deposited on the quartz show only a very slight yellow hue. At temperatures below 60 C the deposition rate becomes too slow for reasonable film preparation.

A second property that we observed in the films was hydrogen incorporation. The FT-IR of the silicon substrate films allowed us to see probable N-H (~ 3400 cm^{-1}), B-H (~ 2500 cm^{-1}) and B-N (~ 1370 cm^{-1}) vibrations (see Figure 4). The assignment of the B-H and B-N is unambiguous, however it is possible that the 3400 cm^{-1} peak could be hydroxyl. We have interpreted the peak as N-H since there are no other peaks indicative of hydroxyl, the peak does not broaden on exposure to air, and the peak does not exist without ammonia being added to the system. In Figure 5 we show the effect of liquid adduct temperature at a fixed ammonia flow on the incorporated hydrogen at a substrate temperature of 600 C. When the ammonia/complex ratio is high (at low complex temperatures) we observe a minimum in the B-H peak, whereas simultaneously the N-H peak is almost equal in magnitude to the B-H peak. As the temperature of the complex is increased, and the relative ammonia concentration is decreased, the relative intensity of the B-H peak increases and the N-H peak begins to disappear. As indicted in the figure, when no ammonia is added to the reactor we see no peak at 3400 and a relatively large B-H peak. In addition to the hydrogen incorporation as a function of mixture, we also examined the effect of substrate temperature at a fixed complex to ammonia ratio. In Figure 6 we see that as the substrate tempera-ture is increased the B-H peak decreases in intensity more rapidly than the N-H peak decreases.

X-Ray photo electron spectroscopy (XPS) was performed on the films to determine stoichiometry and impurity incorporation. Prior to argon sput-tering some carbon was observed in the film but this rapidly disappeared to trace amounts after sputtering. The chemical analysis for B/N ratio indi-cated that the major factor determining stoichiometry was the active nitrogen concentration in the gas phase. When the ammonia concentration was very low (ammonia/amine ratio from mass spectral data <1) the B/N ratio for films deposited at 600 C was ~9/1. When the ammonia concentration was increased (ammonia/amine ratio from mass spectrum ~5/1) the B/N ratio falls to 3/1. The value of the B/N ratio continues to drop but even at 10/1 ammonia/amine ratios the B/N ratio has still not reached 1.

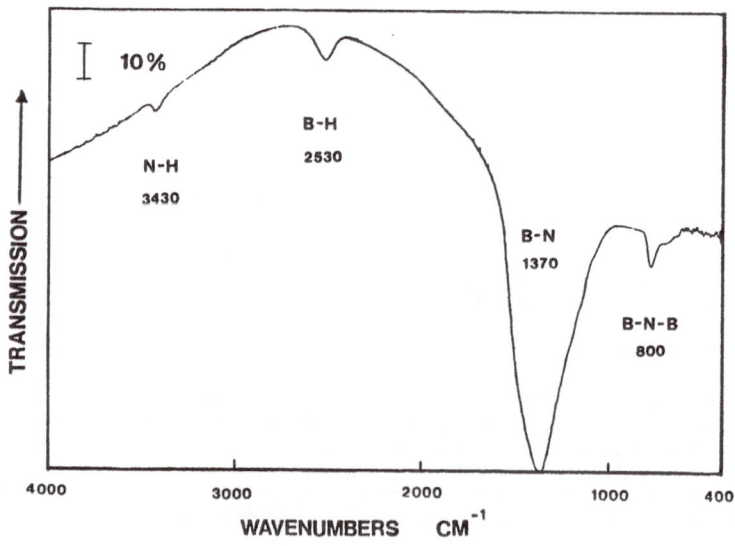

Figure 4   FTIR Spectrum of BN Film

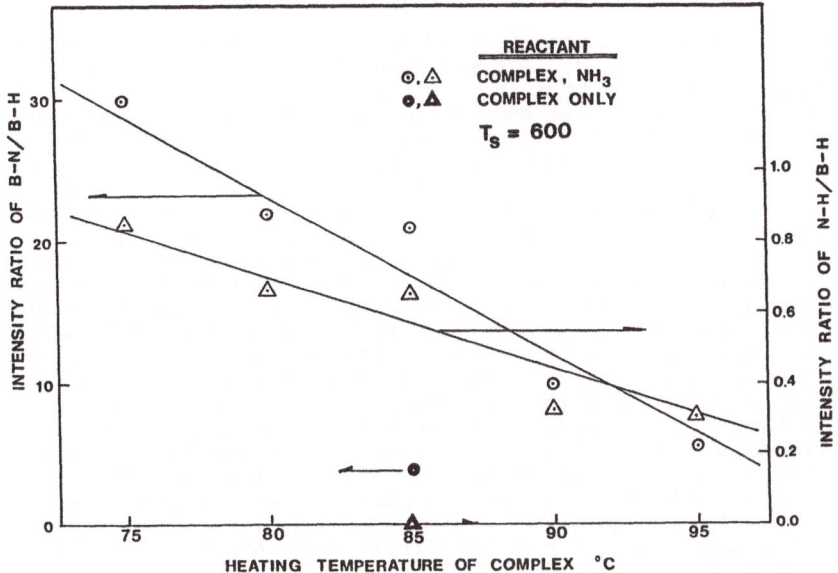

Figure 5 The Residual Hydrogen Conc. as a Function of Temperature

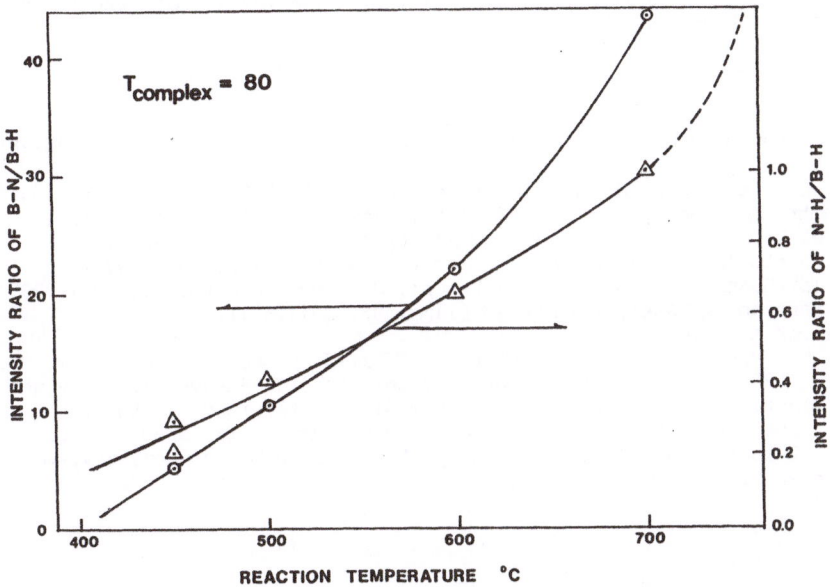

Figure 6 The Residual Hydrogen Conc. as a Function of Complex / NH_3

CONCLUSION

We have observed that it is possible to prepare B N films by MOCVD using an adduct. However, without adding additional active nitrogen to the triethylamine borane complex it is not possible to obtain stoichiometric BN. If we raise the temperature of deposition (>700 C) we can eliminate residual hydrogen as observed by FT-IR spectroscopy. The films obtained either in a hot walled or cold walled reactor are amorphous to X-rays even when prepared at substrate temperatures as high as 1100 C. The films have good adhesion to both silica and silicon. Subsequent heating in a vacuum of the films prepared at 700 C or greater shows no degradation or flaking even at 1300 C. The films show no particulate contamination, since the adduct is a liquid. In order to transform the reaction towards stoichiometric boron nitride, ammonia can be added to the reactor. When this is done at low temperatures (<800 C) we always observe some N-H vibration but a lower intensity of B-H vibration. The films formed are now highly transparent, indicative of a lower B/N ratio in the film. As the active N/B ratio in the gas decreases so does the transparency of the film formed, and XPS data indicates that the N/B ratio decreases. It appears that at low temperatures, although we can substantially decrease the B-H concentration we compensate by increasing the N-H concentration. Therefore the strength of the residual N-H requires high temperature to obtain hydrogen free samples. We will be continuing this work by observing the effects of adding either triethylamine or trimethyl-amine as the additional excess active nitrogen compound. In addition we will be studying the mechanism of decomposition. However, it does appear that one may use the borane adducts to facilitate the addition of boron into the reactor zones without the difficulties associated with diborane.

ACKNOWLEDGEMENT

This work was sponsored by the Joint Services Electronics Program under Contract No. F49620-85-C-0078.

REFERENCES

1. H. J. Levinstein, S. P. Murarka and D. S. Williams, U.S. Patent 4522842, June 1985.
2. R. J. Patterson, R. D. Humphries and R. R. Haberecht, Proceedings of Electrochem. Soc., April 15-18, 1963, Pittsburgh, Abstract No. 103.
3. M. J. Rand and J. F. Roberts, J. Electrochem. Soc., 115, 423(1968).
4. H. Makoto and S. Katsufuso, J. Electrochem. Soc. 122, 1671(1975).
5. T. Takahashi, H. Itoh and A. Takeuchi, J. Cryst. Growth, 47, 245(1979).
6. M. Sano and M. Aoki, Thin Solid Films, 83, 247(1981).
7. See for example
   H. M. Manasevit, F. M. Erdman and W. I. Simpson, J. Electrochem. Soc., 118, 1864(1971), or A. Zaouk, E. Salvetat, J. Sakaya, F. Maury and G. Constant, J. Cryst. Growth, 55, 135(1981) and references within.
8. A. Zunger, A. Katzir and A. Halperin, Phys. Rev. B 13, 5560(1976).
9. H. Steinberg and A. L. McCloskey, Progress in Boron Chemistry, Vol. 1. (The MacMillan Company, N.Y., 1964).

ORGANOMETALLIC COMPOUNDS FOR III-V DEVICES

J. HAIGH
British Telecom Research Laboratories, Martlesham Heath, Suffolk, UK.

ABSTRACT

The chemical characteristics of the Organo-Metallic Vapour Phase Epitaxy (OMVPE) of III-V compounds are reviewed briefly, as are the mechanistics of the pyrolytic component processes using simple precursors. Significant trends in the development of new precursors are sketched out. Chemical trends emerging from current work on photo-initiated deposition from organometallics are then isolated and discussed.

INTRODUCTION

Some ten to fifteen years ago, the epitaxial chemical vapour deposition of semiconductors was purely the province of inorganic chemists. Silicon could be deposited as single crystal epitaxial layers by the reduction of $SiCl_4$ with H_2, and gallium arsenide similarly by the reduction of gaseous mixtures of GaCl and $AsCl_3$ or GaCl and AsH_3. These techniques were later successfully developed to produce InP (from InCl +PCl_3 or PH_3) and the solid solutions (Ga,In)As and (Ga,In)(As,P). Because the reactions take place in a kinetically facile manner it was necessary to arrange the temperatures and partial pressures in the reactor in such a way that the reactants and products were close to equilibrium. This constraint applied at both ends of the process, namely the production of the vapour feeds from their solid or liquid sources, and the generation of the layer from the vapour. Thus even for a system as complex as the growth of (Ga,In)As, which is described by the following equations;

$$Ga + HCl + AsCl_3 + 3/2 \ H_2 \quad \rightarrow \quad GaAs + 4 \ HCl$$
$$In + HCl + AsCl_3 + 3/2 \ H_2 \quad \rightarrow \quad InAs + 4 \ HCl$$
$$GaAs + InAs \quad \rightarrow \quad (Ga,In)As$$

- "pseudo-equilibrium" constants could be calculated which agreed not only with those calculated from another, totally different, growth technique [1] but also with the true equilibrium constants, so closely did the system lie to equilibrium. In systems such as (Ga,In)As, very high purity material could be produced, but in others, such as (Ga,Al)As, the equilibria are highly unfavourable and attempts to force the reaction tend to mean the adoption of extreme conditions and lead to contamination problems.

The advent, at about this time, of organometallic (or metallo-organic) vapour-phase epitaxy (OMVPE or MOVPE; both names are used) for III-V compounds provided the practitioners with an alternative technique that was based on a totally different chemistry. Gallium arsenide, for example, could be produced by a reaction involving trimethylgallium and arsine;

$$Ga(CH_3)_3 + AsH_3 \quad \rightarrow \quad GaAs + 3 \ CH_4$$

185

R. M. Laine (ed.), Transformation of Organometallics into Common and Exotic Materials: Design and Activation, 185–194.
© *1988 by Martinus Nijhoff Publishers.*

With this reaction, calculations indicate that the experimental conditions are far from equilibrium, with negative free energy changes of many 10's of kilojoules per mole. The great attraction of the process, indeed, is its freedom from the constraints of operation near equilibrium. Thus the temperatures and vapour pressures do not require great accuracy of control, and since kinetics inhibit the occurrence of the reactions at low temperature, cold-wall reactors can be used, making equipment design easier and solving some of the contamination problems attendant upon the older technique (although, as I will show, introducing others).

As the process was applied to increasing numbers of semiconductor materials, and as more groups of workers independently developed it to their own requirements, so the range of experimental conditions under which it could be carried out expanded. To date, gallium arsenide has been reported as grown from $Ga(CH_3)_3$ (or $Ga(C_2H_5)_3$) and AsH_3 under the following sets of conditions.

a) 1 atmosphere total pressure, with partial pressures of the active species around 0.001 - 0.1 torr and the balance consisting of H_2, He or a mixture of the two [2],

b) 10 - 100 torr total pressure, with approximately the same partial pressures of the active species as above [3],

c) 10^{-3} torr total pressure, with a wide (often unstated) range of partial pressures of the active species [4].

Similarly the temperature of growth has varied from 450 to 700 degC [5].

At present, several groups are engaged in optimising the chemistry of the process, in order to solve problems that have arisen under one or other of these variants. In the first part of the paper, some of this work will be surveyed.

The great versatility of organometallic pyrolysis reactions, and the success of OMVPE and OMCVD, have tempted workers to investigate them for the deposition of other components of the semiconductor device structure. Deposition of SiO_2 from SiH_4 and oxidising agents, of Si_3N_4 from SiH_4 and NH_3 have been extensively used of course, the former as a thermal process and both as plasma-assisted processes. Thermal (pyrolytic) deposition of W from $W(CO)_6$ is also well-known [6]. However there exists a range of elements which have potential for, or are presently used in, device fabrication, and for which some form of organometallic deposition would find ready applications.

The second part of this paper will be concerned with aspects not of the pyrolytic but of the photolytic deposition of some of these elements. Photolysis as a deposition tool has developed rapidly since the advent of reliable lasers. (Over 400 papers have appeared on the subject since 1982.) On the assumption that some order will eventually appear in this field and that a predictive capacity would be of value, an attempt will be made to draw out some of the common features of the photochemistry and to suggest guidelines for the development of new organometallic precursors.

MECHANISTICS OF THE OMVPE REACTIONS FORMING III-V COMPOUNDS.

The body of work on the mechanisms of these reactions is fragmentary and scattered, and no systematic investigation of, for example , the surface component, has been reported. The following, however, is an attempt to draw out some key points.

1. The reaction is nominally a co-pyrolysis, since it is generally possible to perform the two component parts - Group III and Group V pyrolyses - separately, and under the same temperature conditions as the combined reaction;

$$MR_3 \quad \rightarrow \quad M + \text{hydrocarbons}$$
$$PnH_3 \quad \rightarrow \quad 1/2\ Pn_2 + 3/2\ H_2$$

(M = Gp III metal; Pn = Gp V [pnictide])

However this classification is misleading. Only with In alkyls is the metal obtained. Thus the solo pyrolysis of $Ga(CH_3)_3$ - in the absence of H_2 - produces a compound $(GaCH_3)_n$ and not metallic Ga (7,8). (This difference may reflect the fact that the Ga-C bond is stronger than, for example In-C. Reference 8 gives values of 245 and 162 kJ/mol).

2. The two component reactions differ in their susceptibility to surface catalysis. Thus it has been shown that the Group V pyrolyses are markedly catalysed by III-V crystal surfaces [9] , but that the Group III pyrolyses appear not to be, although sensitive to glass surfaces [10,11].

3. In the usual cold-wall deposition system, the heated susceptor which holds the substrate crystal is the only route for heat into the chamber. This heat is transmitted through it to the incoming reactant gas flow, and a thermal gas-phase boundary zone is established. One might expect some degree of pyrolysis to take place as the reactant species diffuse through this zone, and this appears to be borne out by changes that occur in the gas-phase infrared spectra, both of MR_3 [12] and of PnH_3 [13]. However there appears to be no firm evidence that their decomposition routes differ when mixed from the routes followed when separate.

4. The extent to which the two species bond together before surface attachment, and whether such bonded species are intermediates, are obviously questions of great importance, but unanswered. There is evidence, with some reactants, for the transient formation of donor-acceptor complexes, although the free energies of formation of these complexes appear to be too low for them to exist in appreciable concentrations at growth temperatures, and when deliberately introduced [15] they seem to decompose to the simple reactants [10] before surface attachment. Various relatively strongly-bonded Ga-As species have been used a precursors in their own right. However further investigation is needed here and it is possible that these may also break up to simpler species before surface attachment.

5. During OMVPE of solid solutions of the type (Ga,In)(As,P), where the species $Ga(CH_3)_3$, $In(CH_3)_3$, AsH_3 and PH_3 are all present, the incorporation of As is markedly favoured over that of P. This is in line with the greater ease of pyrolysis of AsH_3 over PH_3 [14], and again tends to indicate that the Group V hydride pyrolysis preserves its individual characteristics.

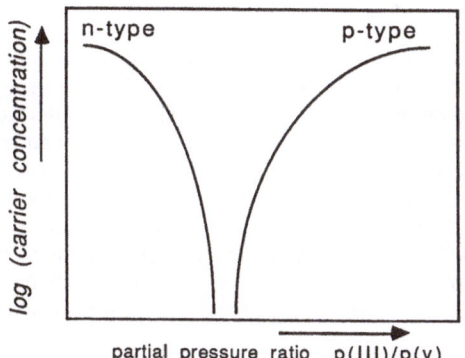

Figure 1. The carrier concentration of nominally undoped GaAs as a function of the reactant partial pressure ratio.

The general drift of these mechanistic features indicates that for a considerable portion of their routes the two pyrolyses have their independent characteristics and proceed separately. There clearly must be an interaction however, and direct evidence that this occurs at the attachment stage is provided by the following observation. Unintentional electrical doping of the semiconductor occurs in OMVPE as in all growth methods; it may be due to contaminant incorporation or to native defects, depending on the details of the growth procedure. The concentration and type of doping tends to show a strong dependence on the ratio of Group III to Group V reactant species, as shown schematically in Figure 1. Low temperature luminescence studies indicate that the excess p-doping can be associated with carbon incorporation on the sub-ppm scale (see ref. 4, for example). An excess of Group V hydride is found to be able to suppress the carbon incorporation. The only source of carbon is the Group III reactant, so that this suppression seems to indicate that the incorporation of Group III, Group V and carbon in the lattice are concerted processes. It might be objected that the carbon could arise from hydrocarbon species liberated before surface attachment but finding their way to, and pyrolysing upon, the substrate as shown below; this is possible, but there is spectroscopic evidence that species with Ga-C and In-C bonds are genuine intermediates at the surface attachment stage [16], and hence that there is Ga-C or In-C bond-breaking at the surface. Thus there is a strong inference that the concerted stage corresponds to the surface reaction.

NEW PRECURSORS IN OMVPE OF III-V's

In view of the somewhat fragmentary nature of the evidence on mechanisms using the present precursors, it might seem premature and even rash to try to suggest ways of developing new ones along systematic lines, or to pontificate upon the work of others who adopt a semi-empirical approach to the problem. This is especially so bearing in mind the huge number of organometallic species that might conceivably be synthesised and tried out. To take another point, much work has been done on for example donor-acceptor complexes [15] of the type $(CH_3)_3In-P(C_2H_5)_3$, which offer considerable advantages technically but which seem to pyrolyse via the parent species such as $(CH_3)_3In$ and so do not introduce new chemistry. Some of these pitfalls may be avoided if attention is confined to two somewhat related technological imperatives which have provided urgent motivation for the investigations.

The first of these is safety. Arsine and phosphine are highly toxic gases. Alkylarsines and phosphines, although probably just as toxic on a weight-for-weight basis as the parent compounds, are less volatile and therefore safer to handle. Growth of (Ga,In)As using $As(CH_3)_3$ was first reported some time ago [17]. Because arsine is less stable to pyrolysis than phosphine, as already mentioned, and alkylarsines in general are more stable than their parent compound, combinations of phosphine and alkylarsines are potentially especially attractive for arsenide-phosphide alloys.

Alkylphosphines have an analogous stability relative to phosphine, but in this case the stability presents rather than solves problems, especially if it is desired to use them to produce alloys. Recently however iso- and tertiary- butylphosphines have been used [18] with some success in InP growth. In these compounds steric crowding appears to weaken the P-C bond.

The second motivation is to provide MR_3 species which lead to less of the undesirable carbon incorporation discussed above. Here some success has been obtained using the triethyls (see below). With these species a beta-elimination is possible, giving Et_2MH as intermediate, and later presumably $EtMH_2$ and MH_3. At the same time the carbon species eliminated is of a non-radical nature and therefore less inclined to polymerise to carbon. This development appears to be of special significance in the ultra-low pressure growth regime around 10^{-3} torr , where it has been postulated [19] that the absence of hydrogen in the gas flow is deleterious, since it would normally serve to getter CH_3 radicals. Carbon incorporation is invariably higher in this regime than with the higher-pressure growth, however, and it remains to be shown whether the level can be reduced sufficiently to meet the most stringent of the device requirements (probably below 10^{15} cm^{-3} or 10 - 100 ppb).

$$CH_2\text{---}CH_2\text{---}In \longrightarrow H\text{---}In + C_2H_4$$
$$|$$
$$H$$

DEPOSITION OF ELEMENTAL MATERIALS BY UV PHOTOLYSIS.

Although some distinguished earlier work is of significance, the efforts of Ehrlich, Osgood, Deutsch, Boyer and Collins [20] and a few others have primarily been responsible for bringing the technique almost to a practical technology with several elements. Figure 2 shows the elements whose photolytic generation has been exploited for semiconductor use - including the group Cd,Hg,Ga,In,Si,Ge,As,P,Te which form the semiconductors, and the dopants B and Zn. The advantage of the photolytic method over other deposition techniques such as evaporation, CVD, plasma deposition are firstly the relative gentleness and secondly the possibilities - elegantly explored in the work of Ehrlich with Tsao [21] and other colleagues - of deposits spatially delineated with a definition down to 1 micron or better.

For the purposes of this discussion uv is defined, somewhat unusually, as any radiation capable of bringing about electronic transitions in organometallic molecules. In practice this differs only slightly from the conventional definition. Apart from a little work in the vacuum uv - mostly with the Hg lamp line at 185 nm - the bulk of the reported work uses radiation in the band 193 - 500 nm (Fig.3). This armoury encompasses the ArF, KrF, XeF and XeCl excimer laser lines, the various lines of the Ar$^+$ laser around 400 - 500 nm , and the 257 nm line obtained by frequency doubling the 514 nm Ar$^+$ laser line. In addition there is the 254 nm Hg lamp line, and various broad band uv lamp sources. The significance of the cut-off at 193 nm is that this is the shortest uv laser line transmitted by silica optics, which become opaque below 180-190nm.

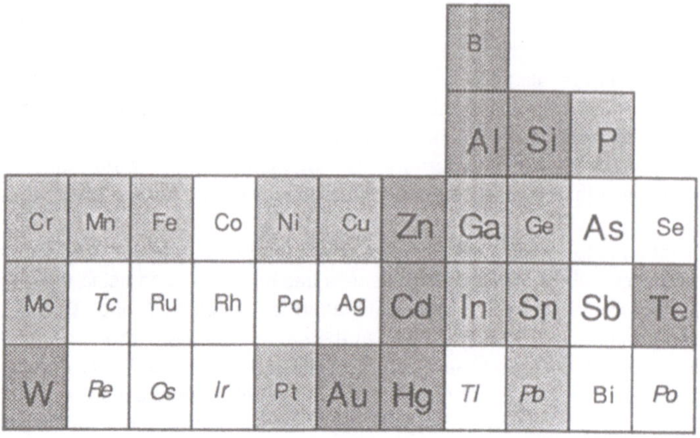

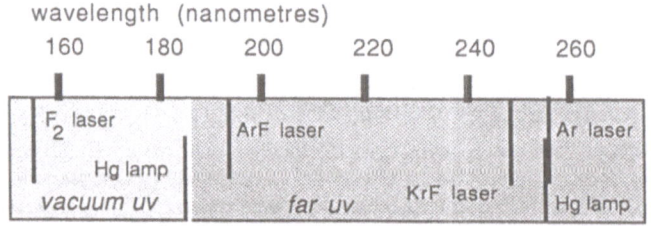

Figure 2. Elements photodeposited to semiconductor standard (heavy shading) and to preliminary standard (lighter shading). Heavy type - elements of importance in present semiconductor technology. (The classifications are somewhat subjective)

Figure 3. Radiation sources in the far- and vacuum uv.

The range of photon energies available from the above sources lies from about 0.2 to 1.2 electron volts. The dissociation energies of single bonds are of the order of 1 eV; hence in very broad terms the energy delivered to the molecule by the absorption of a uv photon is generally either only just, or not quite, sufficient to break one bond.

The precursors that have been used fall generally into three classes, as outlined in Fig.4. For the non-metallic elements the hydrides are available, with simple alkyl species as alternatives. For the main-period metals the simple alkyls are used. For the transition metals most work has concentrated on the carbonyls. All of these compounds contain the element either multiply bonded to the ligand, or bonded to several ligands, or both, and so the photolytic severance of a single bond is not in itself sufficient to liberate the central atom. Since the deposition is successful, it may be deduced that the photolytic bond scission is only the beginning of the story, and we should talk of "photo-initiated" deposition rather than photodeposition. That this is a recurrent feature of the process may be illustrated with a few examples.

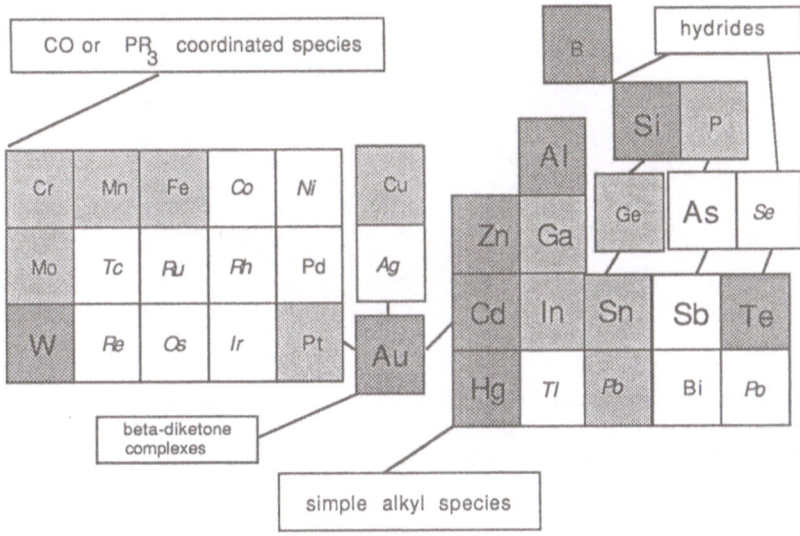

Figure 4. The precursors to the photodeposited elements.

The first example is the deposition of tungsten from $W(CO)_6$. Solanki and co-workers [22] showed some time ago that 248 nm irradiation of the vapour at room temperature produced a tungsten film of good electrical and mechanical properties but with a high residual oxygen level of up to 7 %. However using a broad-band lamp irradiation system combined with substrate heating to 300 - 350 degC, it is possible to obtain tungsten almost free of carbon and oxygen [23]. This temperature, although considerably below the temperatures at which tungsten is deposited from $W(CO)_6$ by pyrolysis alone, appears to be sufficient to bring about the decomposition of the relatively unstable species - probably $W(CO)_5$ - generated by the photolysis.

A converse example is provided by the complex trimethylgold - trimethylphosphine, $(CH_3)_3Au-P(CH_3)_3$. This is, against the general run of gold complexes, remarkably thermally stable [24], possibly because of electron donation from P to Au. However photolysis using the 248 nm KrF excimer laser line readily generates metallic gold without the need for significant substrate heating. It appears that this is because the photolytic scission of the Au-P bond leaves $Au(CH_3)_3$, which then breaks up spontaneously, being normally only stable below -50 degC. Figure 5 shows a patterned deposit obtained by the selective area photolysis of the gold complex [25].

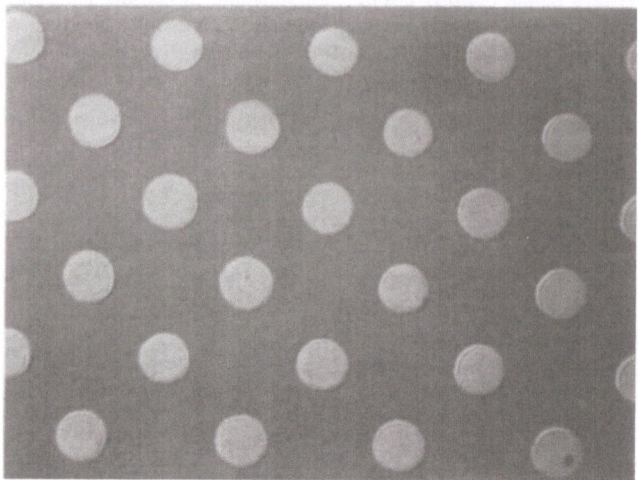

Figure 5. Patterned deposit of gold on an indium phosphide substrate, obtained by selective-area photolysis of trimethylgold-trimethylphosphine with 248 nm laser radiation. the pads are about 100 μm in diameter and 0.1 μm thick [25].

The technologically very important process for the production of amorphous silicon from disilane provides a further instructive example of the interplay of photon and thermal effects. Amorphous silicon may be produced by thermal or plasma-enhanced CVD from disilane as well as, more recently, photolytically. In all cases it has a residual content of hydrogen bound in the lattice, and in conventional CVD, as would be expected, the percentage of hydrogen depends inversely on the temperature of formation. Mishima and co-workers [26], in their work on the photolytic process, quote Pollock [27] as showing that the primary reaction is

$$Si_2H_6 \rightarrow Si_2H_5{}' + H'$$

and that the carbene-like species 'SiH$_2$' is then generated. This species polymerises to form the amorphous silicon matrix (SiH$_2$)$_n$, which then loses most of its hydrogen to scavenging H' radicals from the photo-reaction. The possibility of this H' scavenging reaction accounts for the generally lower H content - as measured by infrared spectroscopy - in the photo-produced material as compared with the thermal material. However it is notable that their data show some dependence of residual H content on the substrate temperature at the time of deposition. Free radical reactions are expected to be relatively temperature-independent, so a thermal component to the process may well still be significant.

A second type of problem arises when the organic species liberated during the photolysis is itself susceptible to photo-reaction. Cyclopentadienylindium I (CpIn) when photolysed with broad-band uv gives a deposit containing much carbon [28]. Since cyclopentadiene has a very strong broad absorption in the far uv, it appears likely that the cyclopentadienyl radical or other related species produced by the photolysis of CpIn will also be strongly absorbing, and hence susceptible to photopolymerisation or other undesirable side reactions producing eventually carbon.

LIKELY ROUTES TO NEW PRECURSORS.

As in all branches of exploration, one can envisage two distinct approaches to the development of new precursors for photodeposition. The first is the systematic approach, where one uses previous experience to predict where success is likely to be found. As indicated, this approach is currently proving successful with OMVPE. The second is to range much more widely and to try novel chemical routes with, it may seem, as little regard as possible for the guide of experience. It takes little imagination to guess which approach is likely to dominate the literature over the coming years! However to take the second approach is not automatically a bad thing since the other extreme, a fully systematic programme, may easily be a blinkered one. Nevertheless there seem to me to be some clear signposts pointing to where success may be found.

Firstly, close attention should be paid to the photochemistry of the liberated organic species. These should be stable to the wavelengths used, or at least if they are not they should generate stable species. This would seem to rule out aromatic or conjugated ligands unless their photochemistry is very well understood. Species such as alkylphosphines or amines may well prove valuable.

Secondly, as discussed above, a single photon will not usually supply enough energy to liberate the metal atom from all its organic ligands, but will generate an organometallic radical or electronically excited species. The aim of the chemistry will then be to allow this species to decompose cleanly to the metal atom, either spontaneously or with the minimal application of heat. To this end successful precursors are likely to contain the minimum number of metal-carbon bonds, and they should for preference be single bonds for ease of breaking. These requirements may be readily met in the organometallic chemistry of the main group elements, but will pose more problems among the transition metals, where the 18-electron rule favours highly coordinated species. However electron-donating ligands which are good leaving groups, such as PR_3, are again likely to find favour, since they can be removed in clean thermal dissociation reactions.

ACKNOWLEDGMENTS

I thank the Director of Research, British Telecommunications plc, for facilities granted during the preparation of this paper.

REFERENCES

1.   M.M.Faktor and J.Haigh, J. Crystal Growth 58, (1982), 291.
2.   D.H.Reep and S.K.Ghandhi, J. Electrochem. Soc. 130, (1983), 675.
3.   J.P.Duchemin, M.Bonnet, F.Koelsch, and D.Huyghe, J. Electrochem. Soc. 126, (1979), 1134.
4.   S.Takagishi and H.Mori, Jap. J. Applied Phys. 22, (1983), L795.

5.   H.Krautle, H.Roehle, A.Escobosa and H.Beneking, J. Electronic Materials 12, (1983), 215.

6.   Y.Pauleau, Thin Solid Films 122, (1984), 243.

7.   M.G.Jacko and S.J.W.Price, Can. J. Chem. 41, (1963), 1560.

8.   D.G.Tuck in Comprehensive Organometallic Chemistry Vol.1 ed. G. Wilkinson, F.G.A.Stone and E.W.Abel, Pergamon (1982).

9.   S.P.Denbaars, B.Y.Maa, P.D.Dapkus, A. Danner and H.C.Lee, reported at ICMOVPE III, April 1986.

10.   J.Haigh and S.O'Brien, J.Cryst. Growth 68, (1984), 550.

11.   M.G.Jacko and S.J.W.Price, Can. J. Chem. 42, (1964), 1198.

12.   M.R.Leys and H.Veenvliet, J. Crystal Growth 55, (1981), 145.

13.   J.Nishizawa and T.Kurabayashi, J. Electrochem. Soc. 130, (1983), 413.

14.   T.Fukui and Y.Horikoshi, Jap. J . Applied Phys. 14, (1980), L551.

15.   A.K.Chatterjee, M.M.Faktor, R.H.Moss and E.A.D.White, J. de Physique Colloque C5, 43, (1982),   491.

16.   J.Haigh   and S.O'Brien, J. Crystal Growth 67, (1984), 75.

17.   M.J.Ludowise, C.B.Cooper and R.R. Saxena, J. Electronic Materials 10 (1981), 1051.

18.   C.A.Larsen, C.H.Chen, G.B.Stringfellow and D.W.Brown; and D.C. Bradley, M.M.Faktor and S. Watts, reported at ICMOVPE III, 1986.

19.   R.Heckingbottom and K.Prior, to be published.

20.   a) D.J.Ehrlich, R.M.Osgood and T.F.Deutsch, IEEE Vol QE-16, (1983), 1233.
       b) P.K.Boyer, C.A.Moore, R.Solanki, W.K.Ritchie, G.A.Roche and G.J.Collins, Proc SPIE Int Soc Opt Eng 385, (1983), 120.

21.   D.J.Ehrlich and J.Y.Tsao, Applied Physics Letters 41, (1982), 297; and Applied Physics Letters 44, (1984), 267.

22.   R.Solanki, P.K.Boyer and G.J.Collins, Applied Physics Letters 41, (1983),1048.

23.   J.Haigh, Chemtronics 1 (1986) 134.

24.   G.E. Coates and C.Parkin, J. Chem. Soc. (1963), 421.

25.   M.R.Aylett and J.Haigh, Proceedings of MRS conference, Boston, 1985.

26.   Y.Mishima, M.Hirose, Y. Osaka, K.Nagamine, Y.Ashida, N.Kitagawa and K.Isogaya, Japanese Journal of Applied Physics 22, (1983), L46.

27.   T.L.Pollock, H.S.Sandhu, A.Jodhau and O.P.Strausz, J. Amer. Chem. Soc. 95, (1973), 1017.

28.   M.R.Aylett and J.Haigh in Laser Diagnostics and Photochemical Processing for Semiconductor Devices (MRS Symposium proceedings no. 17), ed. R.M. Osgood, S.R.J. Brueck and H.R. Schlossberg, North Holland, 1983.

III-V MOCVD GROWTH USING ADDUCTS AS SINGLE STARTING MATERIALS.

F. MAURY
UA-CNRS 445, ENSCT 118 route de Narbonne, 31077 Toulouse cedex France

ABSTRACT

New starting materials for the III-V MOCVD process have been investigated. Several adducts have been used successfully as single organometallic sources of III-V elements. A systematic study of the nature of the Lewis acid and base has revealed the importance of the various bond strengths. The covalent compounds are suitable candidates only if they are prepared in situ in the CVD reactor. A chemical mechanism where the RCl cooperative elimination is a prominent step is proposed for GaAs growth using $ClR'_2Ga.AsR_3$ and some rules for the choice of a precursor are drawn.

INTRODUCTION

There is increasing interest in using the MOCVD process to prepare III-V epitaxial layers. However, the chemistry of this process is poorly understood and some difficulties, generally related to the high reactivity of the usual precursors, limit its development. Consequently, new starting materials are sought. Because group III and V elements form Lewis acid base adducts, these were the first original precursors investigated. These compounds are less reactive and toxic than the alkyl and hydride reactants. Furthermore, they are probable intermediates in the classical MOCVD process and their use could facilitate the study of the chemical mechanisms.

Until now, adducts have been used for the preparation of III-V compounds following three basic ways:

1) As a single source of both group III and V elements [1,2];

2) By formation in situ from the component molecules to avoid unwanted chemical side reactions [3]; and,

3) As direct replacements for group III alkyls [4].

The choice of adducts depends on the desired use.

We have investigated the attractive idea of using adducts as a single source and we report in this paper some experimental results that have led us to propose a chemical mechanism as well as some general rules for the choice of a precursor useable in this CVD process.

RESULTS AND DISCUSSION

About fifteen compounds have been prepared and tried in a classical cold wall MOCVD reactor. Some give III-V epitaxial layers in a well defined temperature range and others give only polycrystalline layers or no deposit (table 1). Several comments can be made concerning these experimental results.

The first important point is to avoid the dissociation of the adduct prior to heterogeneous decomposition. Indeed, Manasevit[5] has shown that the ratio As/Ga in the gaseous phase must be greater than unity to obtain the proper stoichiometry and epitaxy of the layers. Consequently, adduct dissociation in the homogeneous phase would lead to an unfavourable composition. We think that it is this dissociation that explains why attempts to deposit GaAs from $ClMe_2Ga.AsMe_3$ fail and why excess In is

obtained with $Me_3In\text{-}PMe_3$ [4] . Furthermore this dissociation seems related

R. M. Laine (ed.), Transformation of Organometallics into Common and Exotic Materials: Design and Activation, 195–198.
© *1988 by Martinus Nijhoff Publishers.*

to the maximum temperature of deposition since this temperature increases as the donor-acceptor bond strength increases and would then limit the

Table 1:

Precursors used in this work and typical results obtained in a MOCVD reactor

Starting material		Substrate temperature	Comment
$ClEt_2Ga \cdot M^VEt_3$	M^V = N	500 - 700°C	no deposit
	P	500 - 700°C	polycrystalline
	As	450 - 650°C	epitaxy
$ClR'_2Ga \cdot AsR_3$	R'/R = Et/Et	450 - 650°C	epitaxy
	Et/Me	525 - 600°C	polycrystalline
	Me/Et	550 - 750°C	epitaxy
	Me/Me	450 - 750°C	no deposit
$(C_6F_5)_nMe_{3-n}Ga \cdot AsEt_3$	n = 1	600 - 725°C	epitaxy
	3	450 - 750°C	no deposit
$(ClR_2Ga \cdot AsEt_2)_2CH_2$	R = Me	550 - 700°C	polycrystalline
	Et	450 - 625°C	epitaxy
$(Et_2M^{III}\text{-}PEt_2)_3$	M^{III} = Ga	650 - 775°C	polycrystalline
	In	500 - 600°C	no deposit
$(ClMeGa\text{-}PEt_2)_n$	2 < n < 3	50 - 820°C	no deposit
$\left[InEt_3 + HPEt_2\right]$		500 - 580°C	epitaxy

temperature range of deposition [1,6] . All our preformed adducts contain a chlorine atom or a pseudo-halogen group (C_6F_5) bonded to the III element to

stabilize the coordination bond. It is chemically possible to increase the strength of the bond between the III and V elements but then unwanted side effects can appear. These effects include, for example, poor volatility and/or low surface mobility resulting from excessively bulky structure which occurs when we increase the number of halogen atom or pseudo-halogen groups on the III element (e.g.$(C_6F_5)_3Ga \cdot AsEt_3$) and also with the covalent compound

(e.g. $(Et_2M\text{-}PEt_2)_3$, M=Ga, In).

However, the in situ formation of covalent compounds may be obtained

with the system ($M^{III}R_3 + HM^VR_2$) and the use of a low pressure CVD reactor

allows us to reduce these problems and the formation of a covalent monomer intermediate such as $Et_2In\text{-}PEt_2$ leads to good InP epilayers [7].

The strength of the donor-acceptor bond is less than the covalent metal-carbon bond and a fortiori the metal-chlorine bond and is highly dependent on the organic groups attached to the group III and V elements. This can be seen with the series $ClR'_2Ga \cdot AsR_3$ (R', R = Me, Et) since the results are very different depending on the nature of the alkyl groups.

A high dissociation energy is not sufficient for an adduct to be a suitable precursor. Another important energy must be considered, it is the empirical cooperative elimination energy E_{el} that would correspond to the elimination of an XY molecule following the equations:

$$X\diagdown \underline{\underline{M^{III}}} \ \underline{M^V}\diagup^Y \longrightarrow \diagup \underline{\underline{M^{III}}} \ \underline{M^V}\diagdown + XY$$

$$E_{el} = D_{(M\underline{\underline{III}}X)} + D_{(M^V-Y)} - D_{(X-Y)}$$

where the various bond strengths are represented by D.

Thus the empirical determination of E_{el} shows, for example, that EtCl is eliminated more easily with GaAs (78 Kcal.mol^{-1}) than with GaN (105 Kcal mol^{-1}) and consequently could explain the results with ClEt$_2$Ga.NEt$_3$ (no deposit) and ClEt$_2$Ga.AsEt$_3$ (epitaxy). Similarly, in the case of ClEt$_2$Ga.AsR$_3$ (R=Me, Et) the EtCl elimination (78 Kcal.mol^{-1}) is easier than MeCl elimination (87 Kcal.mol^{-1}) and would explain why epitaxy is only obtained with the first adduct.

An Auger spectroscopy study of the chemisorption and the thermal heterogeneous decomposition of the molecule ClMe$_2$Ga.AsEt$_3$ has been realized.

These results would argue for a Langmuir-Hinshelwood type mechanism related to that of Schlyer and Ring [8], where the adduct is adsorbed on the surface of the substrate prior to reaction in such a way that its thermal decomposition may be initiated by EtCl elimination and followed by elimination of the other alkyl groups without dissociation of the GaAs bond [9].

The different chemical steps which lead to an epitaxial growth using an adduct as single source may be summarized as follows:

i)The coordination compound, having a sufficiently high vapor pressure, is transported in the vapor phase without dissociation,

ii)The molecules are adsorbed on the substrate in the deposition zone and,

iii)A cooperative elimination reaction takes place at the deposition temperature followed by the elimination of the other alkyl groups.

CONCLUSION

This work demonstrates that an organometallic compound may be used as a single source of both III and V elements to elaborate III-V semiconductors by the MOCVD process.

The adducts provide good precursors and a systematic investigation of the nature of the Lewis acid and base has revealed the importance of the various bond strengths. With adducts, the dissociation of the III-V bond does not occur before heterogeneous decomposition and consequently this bond must be sufficiently strong. However, the compounds characterized by a III-V covalent bond are not good candidates because of their bulky structure, excepted if they are prepared in situ in a low pressure CVD reactor.

A chemical mechanism where RCl elimination is a prominent step is proposed for the GaAs growth using ClR'$_2$Ga.AsR$_3$.

This study has shown that the adduct ClMe$_2$Ga.AsEt$_3$ is the best starting material for GaAs epitaxy and allows us to draw some rules for the choice of a precursor useable in this CVD process.

ACKNOWLEDGMENT

The author has prepared this paper in memory of Professor G. Constant.

REFERENCES

1 Zaouk, A., Salvetat, E., Sakaya, J., Maury, F., and Constant, G., J. Crystal Growth 55, 135 (1981)

2 Maury, F., El Hammadi, A., and Constant, G., J. Crystal Growth 68, 88 (1964)

3 Moss, R.H., and Evans, J.S., J. Crystal Growth 55, 129 (1981)

4 Benz, K.W., Renz, H., Weidlein, J., and Pilkuhn, M.H., J. Electron. Mater. 10, 185 (1981)

5 Manasevit, H.M., and Simpson, W.I., J. Electrochem. Soc. 116, 1725 (1969)

6 Zaouk, A., and Constant, G., J. Phys. Suppl. 12, 43, C5-421 (1982)

7 Maury, F., Combes, M., and Constant, G., Proceed. of EUROCVD IV, Eindhoven, Eds. Bloem, J., Verspui, G., and Wolff, L.R., 257 (1983)

8 Schlyer, D.J., and Ring, M.A., J. Electrochem. Soc. 124, 569 (1977)

9 Maury, F., Constant, G., Fontaine, P., and Biberian, J.P., J. Crystal Growth (in Press).

ORGANOMETALLIC GROWTH OF RuS$_2$

H.EZZAOUIA AND O.GOROCHOV
Laboratoire de Physique des Solides
C.N.R.S., 1 place Aristide Briand 92195 MEUDON PRINCIPAL

ABSTRACT

MOCVD (Metal Organic Chemical Vapor Deposition) is considered as an established method of growth of semiconductor materials and as a potentially promising method for the deposition of new inorganic materials. A polycrystalline or epitaxial layer of RuS$_2$ has been grown on various substrates such as silica (SiO$_2$), sapphire (Al$_2$O$_3$), graphite and GaAs. RuS$_2$ grows epitaxially when an oriented (100) face of a GaAs single crystal is used as a substrate. The obtained deposits are characterized by X-ray, microprobe and SIMS analyses, as well as, by electrical measurements. The kinetics of deposition is discussed.

INTRODUCTION

Metal Organic Chemical Vapour Deposition (MOCVD) and Metal Organic Vapour Phase Epitaxy (MOVPE) are important thin film growth techniques, well established in the field of semiconductor materials (1-3). These techniques were developed for the growth of III-V and II-VI semiconductor materials such as GaAs and CdS. MBE (Molecular Beam Epitaxy) and MOVPE are now considered as concurential methods. MBE leads to a deposition at a lower temperature than MOVPE, but in a similar range of temperatures MOVPE gives a typical growth rate of 10μm/hour while, with MBE, it is about 1-2μm/hour.

In semiconductor technology, the choice of precursor was generally restricted to metal alkyls (or tellurium) and hydrides of S, Se, As and P. Sulfides and selenides (i.e. ZnSe) were also grown by the use of thiophene and tetrahydrothiophene in order to avoid premature reaction of H$_2$S or H$_2$Se with metal alkyls. Obviously, the role of the substrate is crucial. For homoepitaxy, the quality of the single crystal substrate used and its pretreament are very important. It is known that substrate imperfections (as well as diffusion) not only induce a large number of dislocations in the nearby interface region but also far from the substrate (up to 1 μm or more). The choice of the substrate is limited by the lattice mismatch, by the reactivity and diffusion of the elements.

Both techniques are characterized by a series of complex reactions occuring at the proximity of the heated substrate. Under ideal growth conditions the reaction is surface controlled and prereaction is avoided as much possible, or is controlled concurren(voluntarily or not) to give the formation of suitable

R. M. Laine (ed.), Transformation of Organometallics into Common and Exotic Materials: Design and Activation, 199–203.
© *1988 by Martinus Nijhoff Publishers.*

radicals.

The design of reactors used in semiconductor technology by most laboratories is generally the same. A graphite susceptor heated by RF induction is centered in a vertical or horizontal quartz reactor with an inside diameter range of 2-12cm. The precursors are carried to the reactor in metal or glass tubes after bubbling 10-100 sccm of H through a metal organic liquid phase or passing H_2 through a solid powdered phase. Further dilutions with Ar or Ar/H_2 are generally needed in order to increase the total flow rate in the reactor to 100-1000 sccm. The temperature of the substrate is precisely monitored, but the exact surface temperature is rarely known.

The temperature of precursor sources, which controls the vapor pressure, is also an important parameter. The dependence of the growth rate on the partial pressure of the precursors is almost always non-linear, indicating that at high flow rates the reaction kinetics in the gas phase or at the surface play in important role.

The MOCVD techniques previously described for semiconductor technology can also be developed for the deposition of conductive or insulating inorganic films necessary for the fabrication of electronic devices (i.e. passivation of the semiconductor surface) or other purposes. Treatment or passivation of the surface in metallurgical technology is another application of MOCVD deposition when the temperature of the substrate must be limited to low values. The fabrication of photo sensitive thin films (e.g. solar cells), is another area where low cost technology is demanded.

RuS_2 is an interesting example of a material with a pyrite-type structure (isomorphous to FeS_2) whose CVT (Chemical Vapor Transport) growth with transporting agents such as I_2, Cl_2, HCl or others has been unsuccessful. The only example cited[2] in the literature of the crystal growth of RuS_2 by CVT appears not to be reproducible. On other hand, small single crystals of RuS_2 were litera grown with p or n-type conductivity by a flux method using Te, Sb or Bi as a solvent (4-5).

The crystallographic "a" parameter of RuS_2 is similar to that of GaAs, therefore, the expitaxial growth of RuS_2 on GaAs is possible.

RuS_2 is transparent in the visible region (E = 1.8eV) and has excellent mechanical (very hard material), chemical (stable in boiling aqua regia) and thermal (high melting point) properties. The infrared transparency of RuS_2 is also considered promising for its use as windows for IR detectors.

EXPERIMENTAL

An ambient pressure homemade MOCVD system was used for the described experiments. The reactor consisted of a 25mm by 500mm vertically fixed silica tube containing a graphite susceptor which is heated by a RF induction generator. At the top of the reactor four controlled gas flows (1, 1′, 2 and 2′) are introduced into the reactors.

    1 is composed of a 75% Ar/25% H_2 mixture used to carry the organometallic precursor into the reactor.

    1′ is the gas flow necessary to transport the reactants into the reactor.

    2 is the H_2S source

    2′ an Ar+H_2 mixture for the dilution of H_2S.

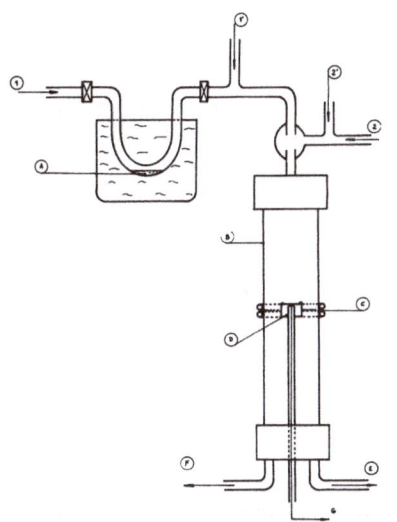

The first series of experiments were made using ruthenocene as the metal organic source of Ru. Prior to reaction, the reactor system is carefully outgassed under vacuum, followed by flusshing with Ar for several days. The flows of 1′ and 2′ are adjusted to the appropriate valves using flowmeters or mass flow controllers. The temperature of the metal organic source A is set and allowed to equilibrate for one hour. The substrate is then introduced, at which point valves 1 and 2 are opened, respectively. To obtain deposition, the temperature of the substrate is quickly increased to the desired value. Table 1 summarizes the conditions of a series of deposition experiments

Figure 1
- Schematic view of the MOCVD experiment

    The characterization of deposits was performed by x-ray analyses (powder diffractometer and Laüe technique), microprobe analysis and SIMS. The latter technique gives the possibility of determining the concentration profiles of RuS_2 epitaxially grown on GaAs. From these results it is possible to detect the interdiffusion of gallium into the RuS_2 layer and of sulfur into the GaAs substrate. By an appropriate choice of growth conditions the interdiffusion process can be minimized. Some transport properties of RuS_2 layers were made with semi-insulating substrates of GaAs.

Table 1

- Some experimental data on MOCVD deposition of RuS_2 in a 25mm diam. reactor.

EX	SUBSTRATE	TEMPERATURE OF SUBSTRATE	DURATION (MN)	TEMPERATURE (°C)(A)	FLOW 2 H$_2$S SCCM	FLOW 1 SCCM	FLOW 2' SCCM	FLOW 1' SCCM	THICKNESS (µm)
1	SiO$_2$	654	75	94	20	40	60	60	0.5*
2	SiO$_2$	704	90	94	20	40	60	60	1.2*
3	SiO$_2$	614	120	112	46	35	135	250	3.06*
4	SiO$_2$	614	15	112	46	35	135	250	1.9*
5	SiO$_2$	614	60	112	38	35	50	50	2.6*
6	SiO$_2$	614	40	112	38	35	50	50	1.8*
7	SiO$_2$	614	15	112	38	35	50	·50	1.25*
8	SiO$_2$	614	7	112	38	35	50	50	0.95*
9	GaAs	614	90	112	20	35	65	65	1.5**
10	GaAs	500	60	112	55	35	50	50	0.45**
11	GaAs	400	60	112	55	40	50	50	0.13**

* Polycrystalline layer   ** Epitaxial layer

DISCUSSION OF RESULTS

RuS_2 thin films have been grown on various substrates by a MOCVD technique. Powder diffraction patterns show the cubic pyrite structure with lattice parameters close to that of the literature value of pure RuS_2 (a = 5.61 Å). The calculated cell mismatch (0.6%) between GaAs and RuS_2 allowed for the formation of an epitaxially grown 100 layer of RuS_2 on the 100 face of GaAs (a = 5.65 Å), which was confirmed from both Laüe and powder diffraction patterns.

Microprobe analysis showed the deposit to be close to stoichiometry (RuS_2 ± 0.02). Electrical measurements indicated a large number of charge carriers and p-type conductivity. Depositions of RuS_2 were obtained at temperatures ranging from 300 to 800°C. Below 500°C the rate of growth decreases considerably.

The temperature of the organometallic source proved to be critical and was adjusted to approximately 110°C. At optimum conditions a growth rate of almost 10 µm/h was obtained. A non-linear growth was indicated by the fact that the growth rate decreased as a function of time. Short-term deposition on silica leads to deposits consisting of small polycrystalline domains. Upon further deposition on silica a microdendritic structure was observed.

CONCLUSION

     The classical MOCVD method, largely developed for the growth of III-V and II-VI semiconductors, has been successfully used for the deposition of a new material, RuS_2.

     The choice of rutherocene, as opposed to carbonyl derivatives (less stable at high temperature and air sensitive) commonly used in the literature, appears to be a good substitute as a precursor. Using GaAs as a substrate, the RuS_2 layer grows epitaxially. Under well-controlled conditions either p or n-type conductivity should be obtainable.

REFERENCES

1) H.M. MANASEVIT, J. of Crystal Growth, 22 (1974) 125-148.
2) J.B. MULLIN, S.J.C. IRVINE and D.J. ASHEN, J. of Crystal Growth, 55 (1981) 92-106.
3) J.B. MULLIN, S.J.C. IRVINE, J. GIESS and A. ROYLE, J. of Crystal Growth, 72 (1985) 1-12.
4) H. EZZAOUIA, J.W. FOISE and O.GOROCHOV, Mat. Res. Bull., 20 (1985) 1353-58.
5) Ibid., 20 (1985) 1421-1425.

SECTION D

SOL-GEL PROCESSING

DONALD R. ULRICH
Directorate of Chemical and Atmospheric Sciences
U.S. Air Force Office of Scientific Research, Bolling AFB,
Washington, D.C. 20332-6448, U.S.A.

ABSTRACT

The ultrastructural control of materials through sol-gel processes offers significant promise for the achievement of reliable performance in ceramics, glass and composites. Several examples of new structural, optical and electromagnetic materials with superior and unique properties are presented based on maximal homogeneity or heterogeneity attained through understanding of the fundamental chemistry. New concepts such as ceramic molecular composites and optically active gels have been derived through polymer, physical and synthetic chemistry. Scaling calculations based on molecular orbital calculations for prediction of the silanol polymerization mechanism, and hierarchical clustering predictions for the sol-gel derived ultrastructures are presented. The application of polymeric network theory to the design of gel ultrastructures is presented.

INTRODUCTION

Modern glass and ceramics have been the products of applied physics approaches, which have produced advances in ceramic science for the past four decades. The emphasis of these approaches has been on structural development at the microstructural level (>10,000A), property control through establishing the relationships between physical behavior and microstructure, and densification through high temperature and fabrication technologies. However, these approaches alone, based on high temperature and sophisticated fabrication technologies, still impose a severe limit on producing ceramics, ceramic composites and glasses of high reliability, particularly for use in severe environments. They do not control the variabilities arising from the physical chemistry of the materials which lead to unpredictable, catastrophic failures during use; and which inhibit the attainment of properties approaching the theoretical values in structural, electronic and optical ceramics, glass and ceramic composites.

As shown in Fig.1, the major advances in ceramics and glass required to attain the aforementioned objectives during the next 20 years will depend on an approach which emphasizes control through all aspects of the chemistry. This approach is called ultrastructure processing. Since 1978 the U.S. Air Force basic research program has focused on the chemical

R. M. Laine (ed.), Transformation of Organometallics into Common and Exotic Materials: Design and Activation, 207–235.
© *1988 by Martinus Nijhoff Publishers.*

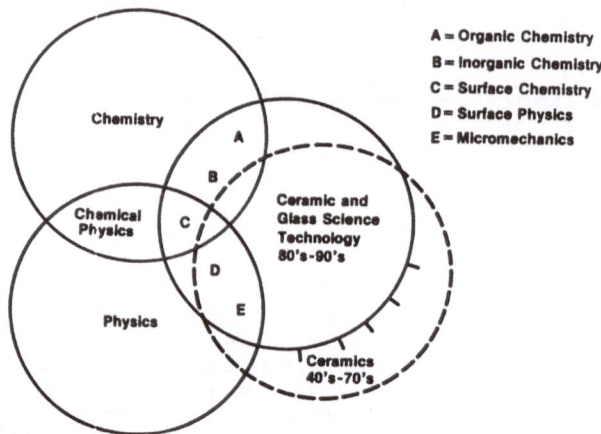

Fig. 1 Changes in the role of physics and chemistry as ceramics moves toward ultrastructure processing.

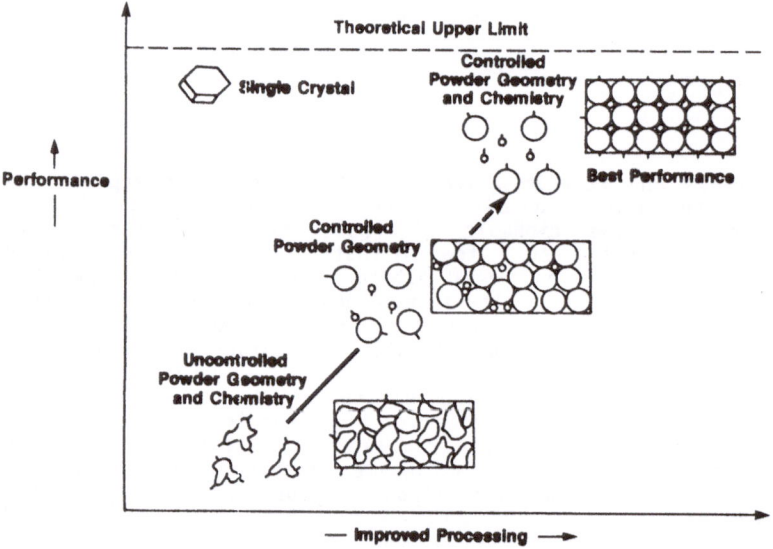

Fig. 2 The impact of ultrastructure processing on ceramic performance.

processing of ceramics and glass. It is therefore appropriate that the program falls within the Directorate of Chemical and Atmospheric Sciences at the Air Force Office of Scientific Research. Ultrastructure Processing has been the subject of three major international conferences co-chaired and supported by AFOSR in 1983, 1985, and 1987 (1)(2)(3) with support and sponsorship of a symposium entitled "Better Ceramics Through Chemistry" at the Materials Research Society spring meeting in alternate years - 1984, 1986, and 1988 (4)(5)(6).

Long term reliability of a material is usually limited by localized variations in the physical chemistry of the surfaces and interfaces within the material . Ultrastructure processing, therefore, refers to the molecular manipulation and control of uniquely homogeneous structures, extremely fine scale (100A) second phases, and controlled surface compositional gradients during the earliest stages of formation at scales of 50 to 1000 Angstroms (5 to 100 nanometers) through chemical processing. Chemical processing includes the manipulation of various types of sols, gels, and organometallic precursors at low temperature, as well as changes in the phase state of these substances at elevated temperatures. As will be illustrated by the results of AFOSR investigators in the following sections, this approach has resulted in crystalline ceramics, glass and composites with superior and unique properties, environmental insensitivity and reproducibility.

ULTRASTRUCTURE PROCESSING OF MULLITE

In ceramic processing, powder consolidation and sintering is the common approach for the processing of polycrystalline systems. In this approach, the three basic steps are: (1) processing and dispersion of powders, (2) consolidation of powders either with the use of liquid suspensions or dry processing techniques, and (3) densification of powder compacts with the application of heat and/or pressure. The powders employed are often characterized by uncontrolled geometry and chemistry or, at best, controlled powder geometry. As shown schematically in Figure 2 this results in microstructures (irregularly shaped particles or spheres) and ultrastructures (interphases, secondary phases, and pores between spheres) which produce levels of performance far below that of the theoretical limit.

Mullite ($3Al_2O_3 \cdot 2SiO_2$), a high temperature structural material, is a very good system to demonstrate this problem. In comparison to other high performance materials such as silicon nitride,, Si_3N_4, mullite is not considered as a high strength material at low temperatures. Its potential becomes apparent at elevated temperatures, while Si_3N_4 matrix ceramics start losing their strength above 1200°C. (Figure 3).

The attainment of properties which approach theoretical values in high temperature structural ceramics by the ultrastructure processing approach is schematically shown in Figure 2. The complete control of raw materials includes the design of molecules and chemistry for powders which upon densification provide the compositional stoichiometry necessary for glass free grain boundaries (ultrastructure).

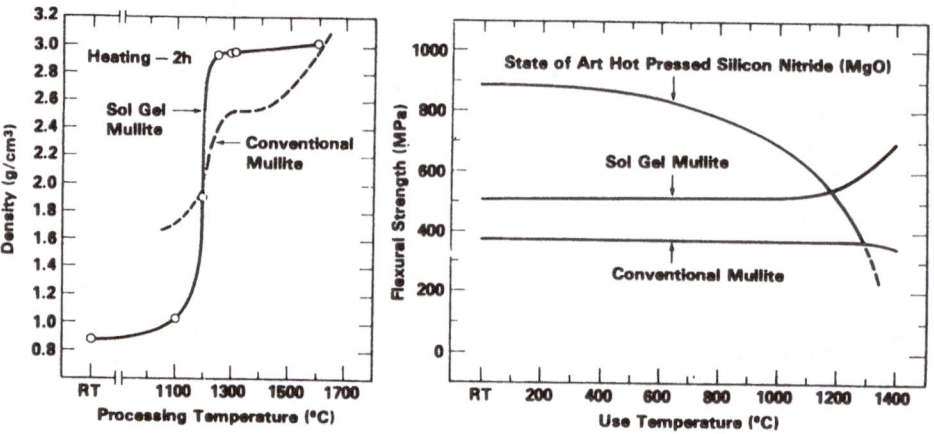

Fig. 3 Comparison of conventional and sol-gel mullite.

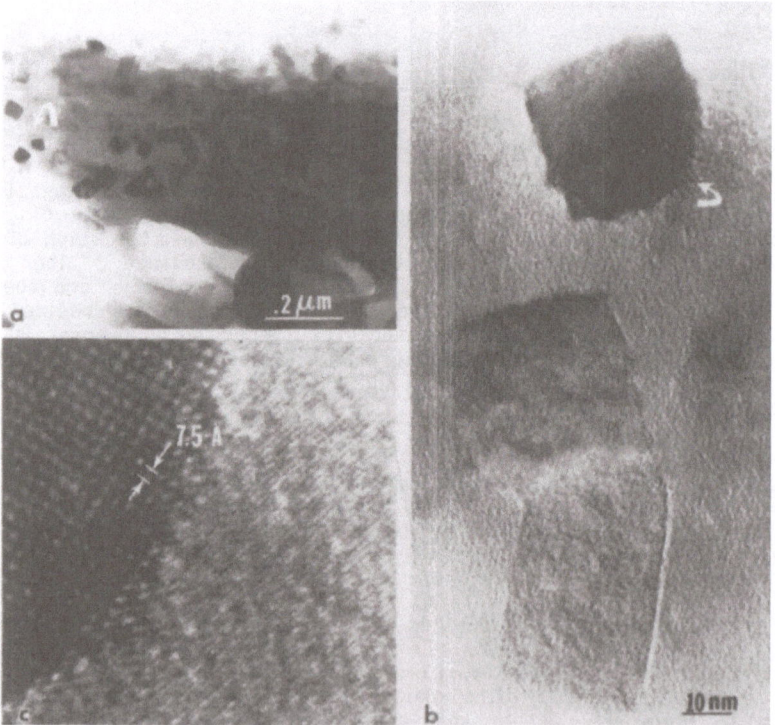

Fig. 4 Sol-gel mullite ultrastructure development.

The viability of the ultrastructure concept by chemical processing has been demonstrated by Professor Aksay of the University of Washington with the processing of mullite. As shown in Figure 4, Aksay prepared mullite by the molecular mixing techniques of sol-gel colloidal dispersion. Superior performance is demonstrated at elevated temperatures. It does not display strength decrease at least up to 1400°C because the scale of chemical uniformity approaches molecular dimensions; strength starts to increase at 1400°C. Under these condition it is possible to first sinter a mullite composition matrix to a highly dense state in an amorphous state. Next mullite crystals are nucleated and grown as shown in the bright field image of Figure 4(a). These crystals which are smaller than 100 nm are uniformly distributed in the matrix as shown in Figure 4(b); when these crystals grow into each other, glass free grain boundaries form although the rest of the matrix is still amorphous. By carefully observing the higher magnification image taken from a grain boundary region (Figure 4(c)), the fringes in either mullite grain meet at the boundary, and there is no separation between the two crystals. High resolution electron microscopy (HREM)(Figure 5) taken from these mullite crystals reveal structural details at the atomic scale. Mullite formed by sol-gel processing has grain boundaries free from any amorphous film and shows that the formation of these glass-free grain boundaries are responsible for the superior performance of mullite at high temperature.

The mullite forming gels were made from colloidal boehmite (AlOOH, aluminum monohydroxide) and tetraethoxysilane (TEOS). The sintering behavior of the mullite forming gels is compared with the sintering behavior of mullite prepared by conventional techniques (Figure 3). Aksay is able to form nearly fully dense (>99% of Theoretical Density) and translucent mullite at 1200°C. Conventional mullite requires 1600°C. The ability to sinter mullite at temperatures significantly lower than the usual 1600°C sintering temperature signifies a very important processing accomplishment in the elimination of process defects that otherwise force higher temperature sintering.

INFRARED TRANSPARENT MULLITE

Sintering of same sol-gel derived mullite at 1250° results in infrared transparent mullite, forming the basis for sol-gel passive infrared optics for the 3-5 micron region (Figure 6). The superior performance arising from sol-gel processing is compared to state-of-the-art hot pressed mullite which requires hot pressing at 1670°C followed by three hours of annealing at 1800 C.

HETEROGENEOUS NANOCOMPOSITES

The ultrastructured mullite is an example of making maximally homogeneous high performance ceramics from the sol-gel process. In 1982 under AFOSR support Professor Roy at Pennsylvania State University showed that the sol-gel route could be used to make maximally (i.e.,

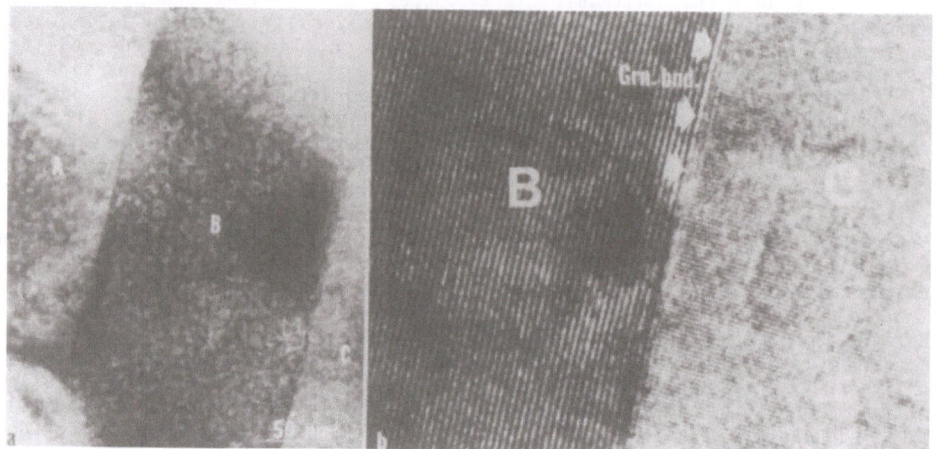

Fig. 5 Glass-free grain boundaries in mullite from sol-gels.

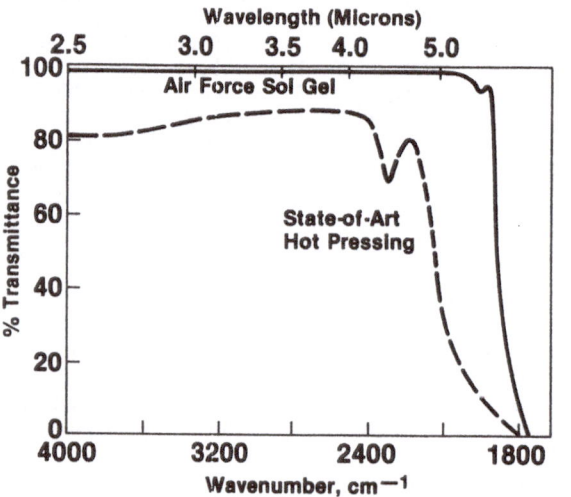

	HOT PRESSING	SOL GEL
Temperature	1670 °C	1250 °C
Time	10 Min	60 Min
Pressure	55 MPa	None
Annealing Temp/Time	1800 °C/ 3 Hrs	None
Size Limit	5.0 cm	No Limit
Micro-Cracking	Yes	None
Composition Control	No	Yes

Fig. 6 Infrared transparent mullite from sol-gels.

in the degree of interpenetration of phases) heterogeneous materials. He has demonstrated a wide range of high temperature ceramic uses of the sol-gel route for making this new class of materials which he calls nanoscale composites derived from multiphasic xerogels. (Figure 7) A nanoscale composite is a material which has two (or more) phases with the physical dimensions of the phases lying in the range 1-10 nm. The multiphases may differ in either composition or structure or both.

There are several examples of reliable and unique high temperature structural and electromagnetic ceramics made from this low temperature process. Figure 8 shows the design of high temperature calcium strontium zirconium phosphates with zero thermal expansion. As exemplified by $Ca_{0.5}Sr_{0.5}Zr_2P_6O_{24}$, this nanocomposite has zero thermal expansion from 0^oC to at least 500^oC. The end components of this ultrastructured ceramic are calcium zirconia phosphate (CZP) and strontium zirconia phosphate (SZP).

These materials can be designed to have electrically conducting or insulating properties, ferromagnetic properties or low frequency damping properties concurrently with the ultralow thermal expansion. One series based on sodium zirconium silicate and yttrium iron garnet provides high temperature, low thermal expansion magnetically absorbing nanocomposites (Figure 9).

A second example, also shown in Figure 9, are high temperature (i.e., melting point of 1700^oC) zero expansion engine ceramics based on $CaZr_{24}P_6O_{24}(CZP)$ and MgO. Of primary scientific importance is that phosphate containing ceramics are usually lower melting materials. However, the sol-gel nanocomposite route enables this material to demonstrate high temperature capabilities.

SOL-GEL OPTICS

Dr. Larry Hench at the University of Florida has pioneered the development of a low temperature sol-gel process for the rapid fabrication of large complex shapes of optical glass mirrors and components. Until this time the rapid, reliable production of gel monoliths has been difficult to realize because of the lack of control over drying stresses and cracking. Zarzycki shows that the drying stress is a function of pore size and rate of evaporation of the pore liquor, which depends on the liquid vapor pressure (7). Professor Hench has demonstrated the use of organic additions to alkoxide sols, termed drying control chemical additives (DCCA), to control the rate of hydrolysis and condensation, pore size distribution, pore liquor vapor pressure, and drying stresses. By use of the DCCAs, which include formamide (NH_2CHO), glycerol ($C_3H_8O_3$), and several organic acids, such as oxalic acid, ($C_2O_4H_2$), with alkoxide precursors, a wide range of sizes and shapes of optically transparent dried gel monoliths of SiO_2, SiO_2-TiO_2, $Li_2O-Al_2O_3-SiO_2-TiO_2$ and other ternary and quarternary glasses can be made within a few-day processing schedule.

Sol Gel Nanocomposites

- **Based on Ultra Heterogenity from Solution Sol Gels**

- **Mix Several Phases on a Scale of 10 Nanometers**

- **Low Temperature Process for Reliable Ceramic High Temperature Use**

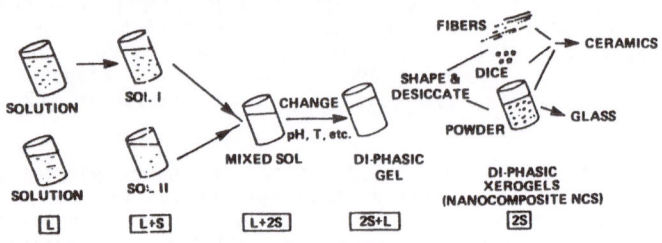

Fig. 7   Concept of sol-gel nanocomposites.

Nanocomposites

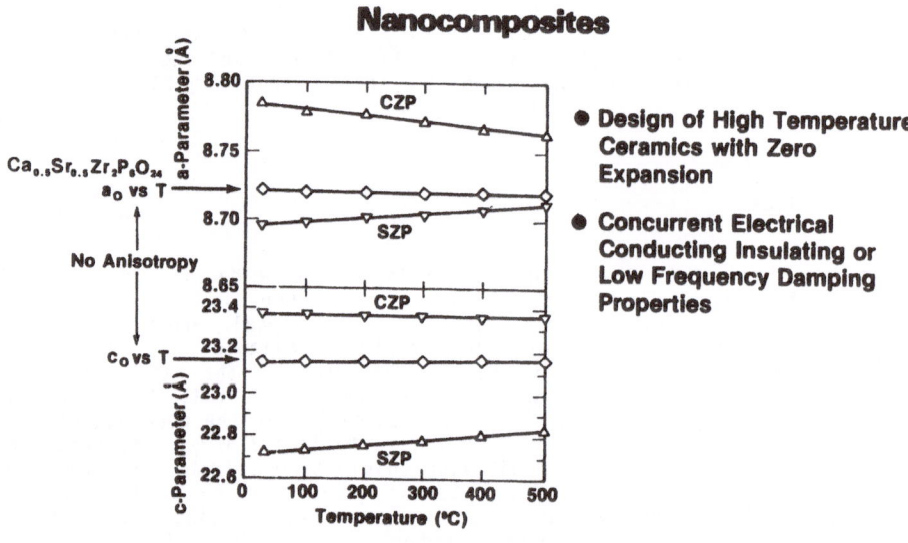

- **Design of High Temperature Ceramics with Zero Expansion**

- **Concurrent Electrical Conducting Insulating or Low Frequency Damping Properties**

Fig. 8   Sol-gel design of zero expansion nanocomposites.

● New Materials from an Ultrastructural Composite Approach

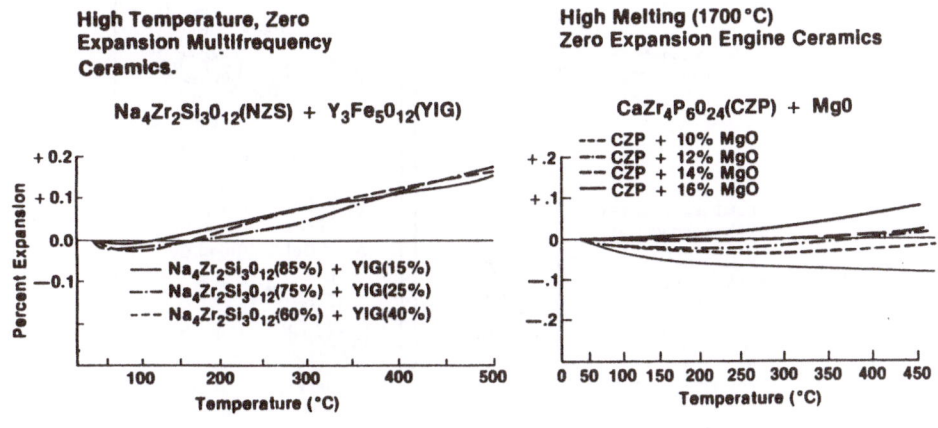

High Temperature, Zero Expansion Multifrequency Ceramics.

$Na_4Zr_2Si_3O_{12}(NZS) + Y_3Fe_5O_{12}(YIG)$

— $Na_4Zr_2Si_3O_{12}(85\%) + YIG(15\%)$
—·— $Na_4Zr_2Si_3O_{12}(75\%) + YIG(25\%)$
--- $Na_4Zr_2Si_3O_{12}(60\%) + YIG(40\%)$

Percent Expansion / Temperature (°C)

High Melting (1700 °C) Zero Expansion Engine Ceramics

$CaZr_4P_6O_{24}(CZP) + MgO$

--- CZP + 10% MgO
—·— CZP + 12% MgO
-- CZP + 14% MgO
— CZP + 16% MgO

Temperature (°C)

Fig. 9 Multifunctional sol-gel nanocomposites.

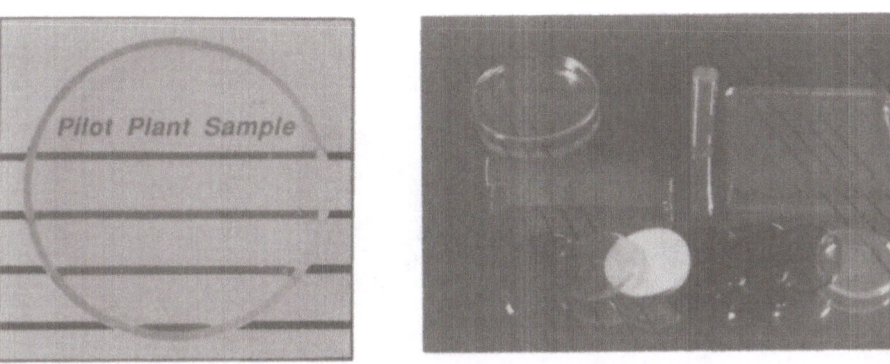

Fig. 10 Sol-gel/DCCA optical glass and filters.

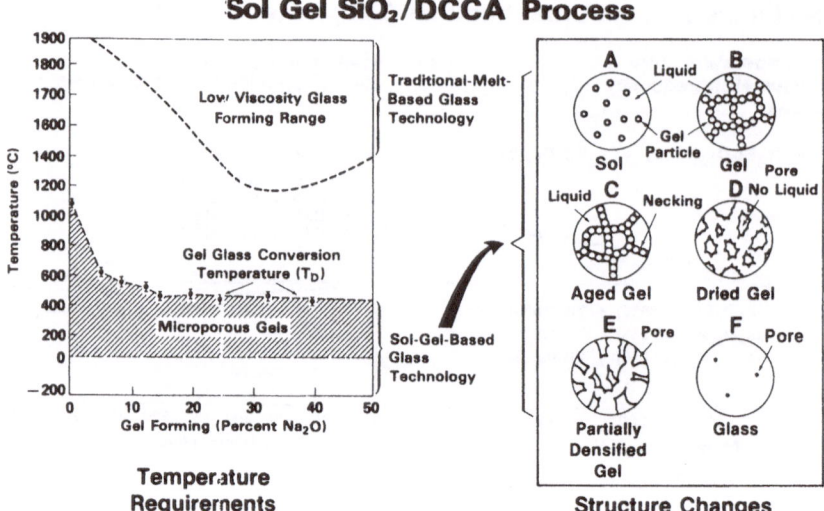

Fig. 11   Temperature requirements for sol-gel process compared to melt-derived glass. The sequence of structural changes associated with the sol-gel process is shown.

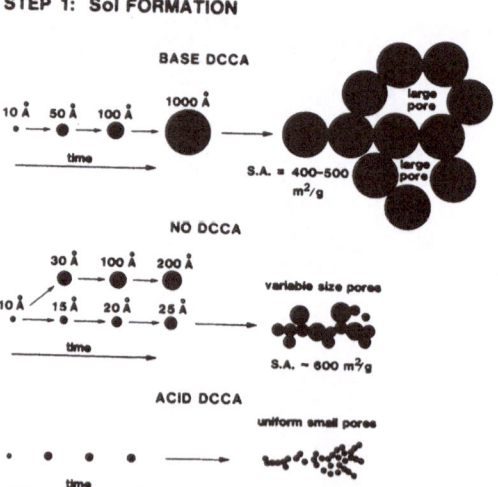

Fig. 12   Sol-gel pore size control with DCCAs.

STEP 2: GELATION

(A) Without DCCA

large pore distribution

(B) With DCCA

$SI(OR)_4$
+
$SI(OH)_4$
+
H_2O
in pores

small pore size distribution

STEP 3: AGING

(A) Without DCCA

(B) With DCCA

STEP 4: DRYING (Minimize Drying Stress Due to

$$p_i = \frac{2\gamma\cos\theta}{d_i} \quad \text{by}$$

Minimizing Δd_i; $\Delta p_i = 0$ when
$$\Delta d_i = 0)$$

(A) Without DCCA

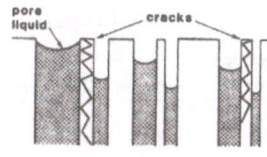

large differential evaporation
large stresses, large σ
distribution
→ CRACKING

(B) With DCCA

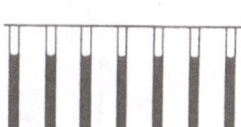

little differential evaporation
uniform σ
uniform stress distribution
→ NO CRACKS

STEP 5: DENSIFICATION

DCCA Eliminated as Vapor before Pore Closure

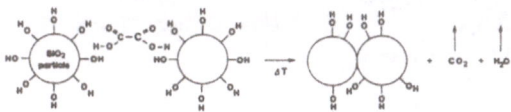

Fig. 12   Sol-gel pore size control with DCCAs (continued).

218

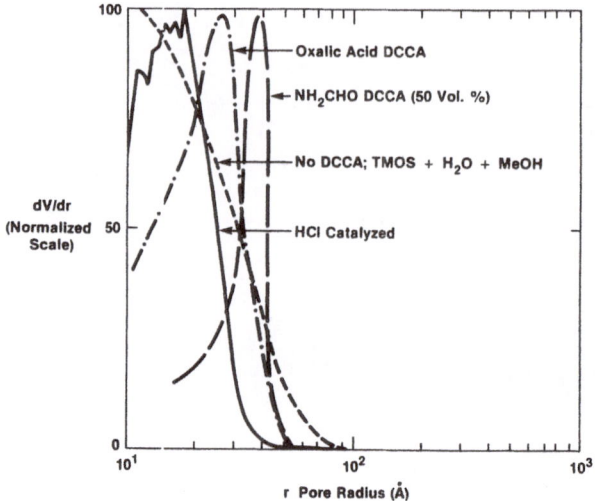

Fig. 13 Silica gel pore volume distributions from adsorption isotherms.

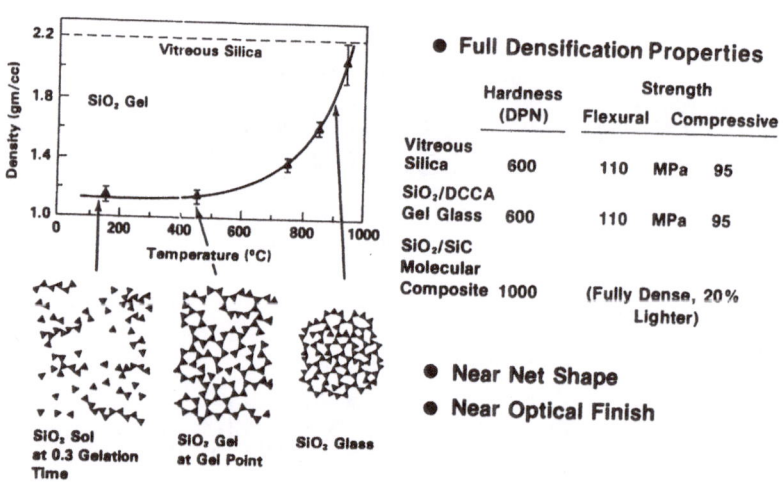

Fig. 14 Properties of fully densified SiO₂/DCCA sol-gel glass.

With the discovery of the DCCAs and a simple, unique thermal stabilzation process during densification, Professor Hench can now make silica (SiO$_2$) glass to complete densification with properties which equal or surpass those of melt-derived glasses. These are shown in Figure 10 where each spacing represents 2.54 cm. or one inch. Also shown are silica gel optical filters and laser host glasses produced by introducing transition or rare earth elements and sensitizers into the transparent and colorless silica gel matrix made by the DCCA - TMOS (tetramelthylorthosilicate) method. These advances have been protected by 23 patent applications filed jointly by the U.S. Force and the University of Florida.

The sol-gel SiO$_2$/ DCCA process is illustrated schematically in Figures 11 and 12. Figure 11 shows temperature requirements for the sol-gel process compared to traditional melt-based glass. The sequence of structural changes which accompany the gelation, polymerization and densification processes are shown. These must be controlled in order to produce monoliths with a range of densities.

Figure 12 shows the control of the sol-gel process with organic DCCAs. This has been determined from the work of Dr. Hench with Professor Jeri Jonas of the Chemistry Department at the University of Illinois. The role of DCCAs on hydrolysis and condensation mechanisms has been studied using high pressure, high resolution [29] Si NMR, high pressure Raman spectroscopy, and N$_2$ absorption - desorption. This interaction is an example of the fundamental understanding that can be achieved when the ceramic engineer and glass scientist collaborate with experts in chemical physics.

Without a DCCA a wide range of pore sizes and diameter of solids network are produced when gelation occurs. Differential growth of the silica network will thereby occur during aging due to local variations in solution-precipitation rates. The net effect is an aged gel structure such as depicted in Steps 2-5, with many regions susceptible to cracking during drying.

Addition of a basic DCCA such as formamide produces a large sol-gel network with uniformly large pores. An acidic DCCA, such as oxalic acid, in contrast results in a somewhat smaller scale network after gelation but also with a narrow distribution of pores (Figure 13). Thus, either basic or acidic DCCAs can minimize differential drying stresses by minimizing differential rates of evaporation and ensuring a uniform thickness of the solid network that must resist the drying stress. Achieving a uniform scale of structure at gelation also results in uniform growth of the network during aging which thereby increases the strength of the gel and its ability to resist drying stresses.

Hench's process enables all hydroxyls and DCCAs to be removed concurrent with complete pore collapse. The SiO$_2$ /DCCAs gel densification process is shown in Figure 14 with full densification properties.

Combining the SiO_2-DCCAs process with the thermal stabilization process has now produced the glass trademarked as Gelsil. The DPN hardness has been increased to 1100, nearly double that of melt derived SiO_2. The gels are dried and densified to near net shape with an optical surface finish of 1/10 to 1/20 λ .

The SiO_2 glass/ DCCA optical absorption spectra from the ultraviolet to the near infrared for Gelsil is compared to that of commercial vitreous silica. As can be seen the USAF SiO_2 gel glass (with DCCA and stabilization) shows complete transmission from 0.2 to 3.0 microns with no absorption bands. (Figure 15). Hydroxyls have been completely removed.

The complete control of hydrolysis, condensation, drying and densification is also reflected in the very low, temperature independent coefficient of thermal expansion (Figure 16). This indicated a network structure with very few non-bridging oxygens and non-saturated silicons. Hench recently has developed a process which yields the same results at temperatures which are 300° C lower.

PRECERAMIC POLYMER PRECURSORS FOR ULTRASTRUCTURES

Preceramic polymer precursors will play a major role in ceramic ultrastructures derived from sol gels. One example of this is discussed in the next section "Molecular Ceramic Composites".

There have been many pronouncements since 1981 in the polymer chemistry community by those who have claimed to "pioneer" the preceramic polymer area. In my opinion the credit for origination should go to Dr. Anthony Matuszko of AFOSR who had the vision to initiate the organometallic chemistry research for this area in 1957. Dr. Robert West of the University of Wisconsin was one of Matusko's original investigators in organosilicon chemistry. In 1957 very little was known about organosilicon chemistry.

During the years that followed, West made some surprising discoveries. Through an extensive study of the reactivity and properties of a number of organosilicon compounds, he found it was possible to control the chemistry of these compounds and thus affect their properties. Then in the late 60's and early 70's, West and his research group studied a series of silicon compounds called cyclic organopolysilanes.

During a lecture tour to Japan in 1976, West visited Yajima,who reported that he had discovered that these organopolysilanes could be pyrolyzed to form Beta-silicon carbide fibers after first being converted to polymers. Their process (Figure 17) for making this material begins with permethylpolysilane (I), which is then pyrolyzed above 400° C to produce a polymeric carbosilane (II). From the hexane-soluble, a nonvolatile portion of polymeric carbosilane fibers can be melt-spun. These fibers are crosslinked by surface oxidation in air, then further pyrolyzed at 1300° C for 1 hour in a nitrogen atmosphere to make black Beta-silicon carbide fibers (III).

221

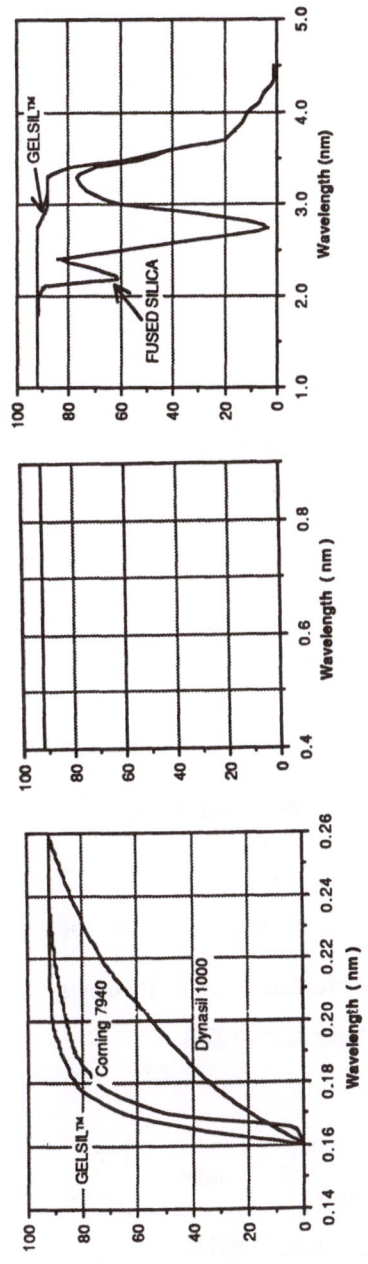

Fig. 15 Optical absorption spectra of fully dense SiO2 sol-gel glass.

222

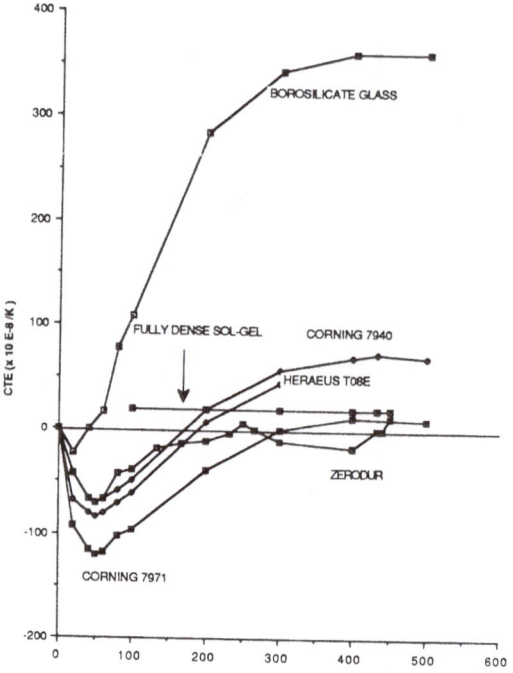

Fig. 16 Profile of coefficient of thermal expansion – temperature for fully
dense SiO_2 sol-gel glass.

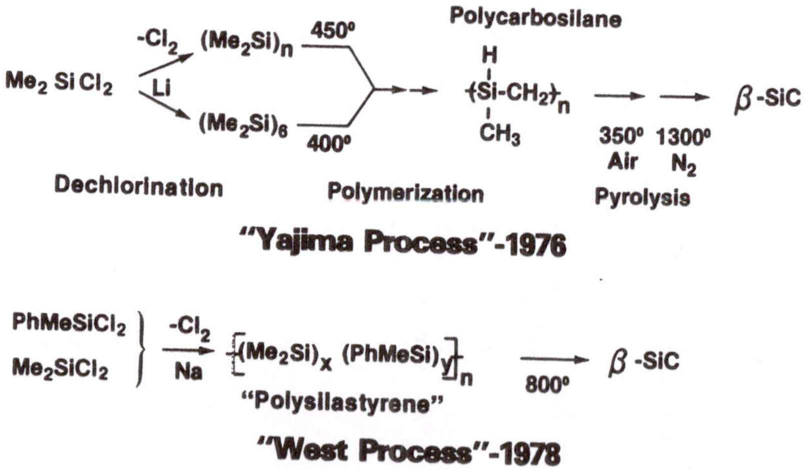

Me_2SiCl_2 $\xrightarrow[\text{Li}]{-Cl_2}$ $(Me_2Si)_n$ $\xrightarrow{450°}$

$(Me_2Si)_6$ $\xrightarrow{400°}$

Polycarbosilane

$\{Si-CH_2\}_n$ (with H above Si and CH_3 below) $\longrightarrow \longrightarrow$ β-SiC

350° 1300°
Air N_2

Dechlorination   **Polymerization**   **Pyrolysis**

"Yajima Process"-1976

$\left.\begin{array}{l} PhMeSiCl_2 \\ Me_2SiCl_2 \end{array}\right\}$ $\xrightarrow[\text{Na}]{-Cl_2}$ $\{(Me_2Si)_x (PhMeSi)_y\}_n$ $\xrightarrow{800°}$ β-SiC

"Polysilastyrene"

"West Process"-1978

Fig. 17 Preceramic polymer pyrolysis to silicon carbide for molecular
composite ultrastructures.

However, the technological potential of the Japanese-made polysilanes as ceramic precursors was limited by the necessity of first converting the polysilanes to polycarbosilanes, then fractionating them to transform them into tractable materials. Permethylpolysilane is a crystalline, white incoherent powder that decomposes before melting and is insoluble in common solvents.

In 1978, Professor R. West of the Department of Chemistry, University of Wisconsin, decided to make an improved ceramic precursor. He found that polymers consisting of mixtures of phenylmethylsilylene (PhMeSi) and dimethylsilylene (Me$_2$Si) with PhMeSi/Me$_2$Si ratios varying from 3:1 to 20:1 were less crystalline and could be softened to viscous liquids at high temperatures. In a series of experiments at the Air Force Materials Laboratory, these materials (called polysilanes) were used to strengthen silicon nitride ceramic parts. Silicon nitride ceramic bodies were soaked in a melted polysilane to fill the pores. Then, the bodies were heated to 1400°C to convert the polymer to silicon carbide, which filled the pores and made the ceramic stronger.

This discovery led to the synthesis of copolymers of dimethylsilylene and phenyl-methylsilylene with the PhMeSi/Me$_2$Si ratio near 1. Mixtures of dimethyldichlorosilane and phenylmethyldichlorosilane were condensed in toulene with sodium. After 36 hours, the reaction was cooled and carefully treated with water to produce a meltable, gummy polymer. Since its structure resembles polystyrene, it was named polysilastyrene.

Polysilastyrene is quite soluble in organic solvents, such as benzene, toluene, tetrahydrofuran, or acetone, and therefore can be characterized by the usual methods of polymer chemistry. It can be molded, cast into films from solution, or drawn into fibers, Significantly, its fibers can be converted directly to silicon carbide without first being converted to polycarbosilane (Figure 17).

MOLECULAR CERAMIC COMPOSITES

Molecular level ceramic composites of SiO$_2$/SiC have been achieved at the University of Florida as a result of interinstitutional collaboration between the University of Florida, the University of Wisconsin, and the 3M Co. As stated in the previous section, pioneering efforts of Professor R. West of the University of Wisconsin, under AFOSR sponsorship, has led to the development of a range of polysilane organometallic precursors for SiC. Dr. B. Lee in Professor Hench's sol-gel ceramic research group at the University of Florida has discovered a means of using chemical free radical initiators to crosslink silane precursors in situ. The crosslinking process is equally applicable to silane oligomers as well as polymers and greatly increases the yield of SiC ceramic after pyrolysis.

224

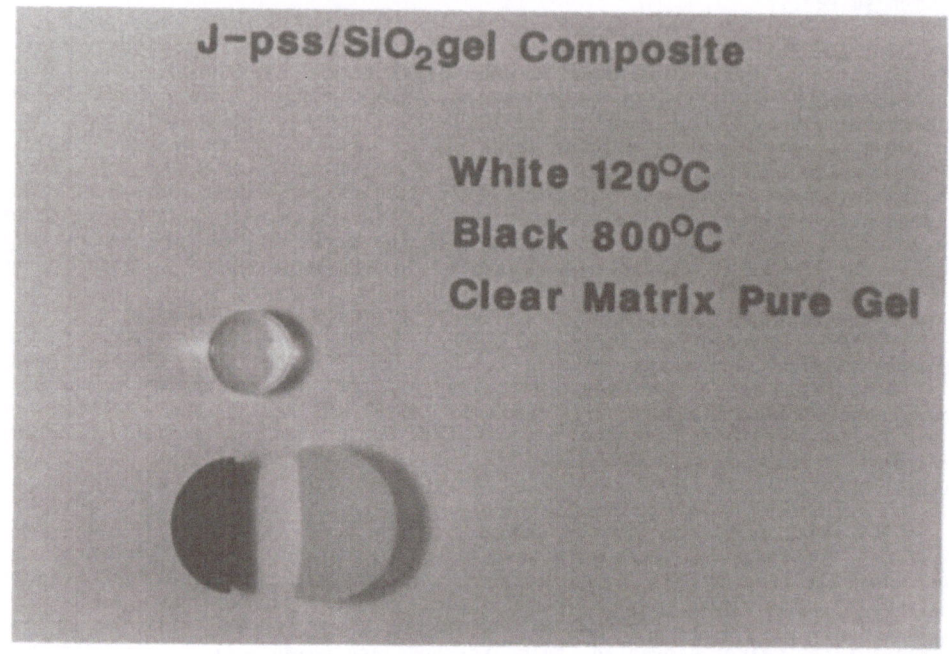

Fig. 18   SiC/SiO$_2$ Molecular Ceramic Composites.

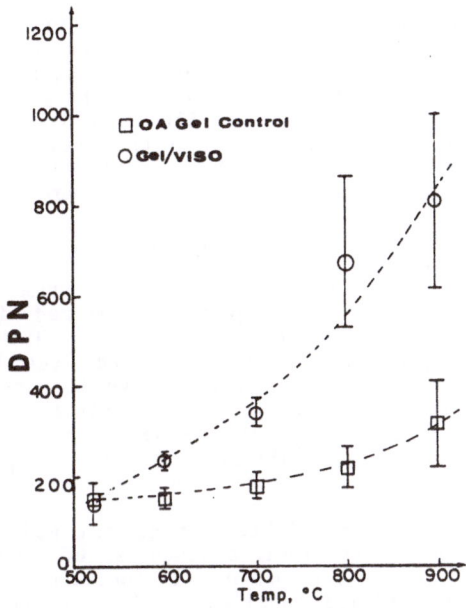

Fig. 19 Diamond Pyramidal Hardness as a function of pyrolysis temperature for microporous gel (OA) and impregnated with silane oligomer.

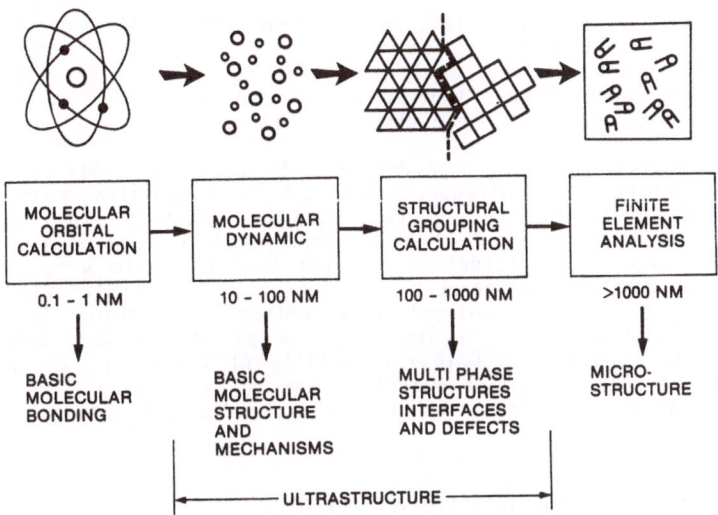

Fig. 20 Scaling calculations for ultrastructural ceramics.

This crosslinking process makes it possible to combine the polysilane SiC precursor and crosslinking agent with the SiO_2 sol and drying control chemical additives (DCCA's) developed by Professor Hench's team. A cogellation of the crosslinked silane and silica sol occurs. This forms a monolithic silica structure containing silane, Pyrolysis in an inert atmosphere converts the crosslinked silane to SiC which is dispersed with the SiO_2 matrix on a molecular level of ultrastructural homogeneity. Figure 18 compares the ceramic molecular composites densified at 120°C (white) and 800°C with the densified clear gel-derived SiO matrix.

An exciting feature of this process is that the silane/silica sol is fluid and can be easily pumped at or near room temperature to fill molds of nearly any size or geometry with great uniformity. During cogellation an object is formed that precisely replicates the shape and surface features of the mold. The shrinkage that occurs during drying and pyrolysis is uniform and can be accurately predicted. Consequently this process should make it possible to produce either very complex shapes or very large objects rapidly and inexpensively. Extrusion of the sol at the point of cogellation into large structures is also a possibility with this unique molecular composite system.

The physical properties of the lightweight ceramic molecular composites are also impressive. Hardness and fracture toughness of the low density composites with only 3 vol. % carbide reinforcement is 200 to 300% higher than microporous pure silica monoliths. For example, as shown in Figure 19, the molecular composite pyrolyzed at 800°C has a low density of 1.32 ± 0.1 g/cc. The K_{IC} to ratio is 1.3, which is in the vicinity of that of hot pressed 20 v/o SiC/borosilicate composites (8). Recently lightweight composite structures have been developed with a hardness and toughness equivalent or superior to fully dense silica (Figure 13). Consequently, the strength or fracture toughness to density ratios, which are especially important for design of large scale structures, are very much superior for these SiO_2/SiC molecular composites as compared with pure vitreous silica.

A related process has also been developed by the University of Florida scientists for impregnating ultraporous matrices or composites with silane precursors and crosslinking agents. During pyrolysis the SiC phase forms within the matrix on a molecular scale. The SiC greatly toughens and hardens the structure without increasing the weight. Since this silane/SiC impregnation process can be used in conjunction with the University of Florida sol-gel DCCA method for making large scale gel-glass monoliths, and controlling pore size and pore size distribution, it provides a unique capability of tailoring the physical properties of ceramic and glass materials for a wide range of defense applications without increasing weight or decreasing payload.

SCALING CALCULATIONS FOR SOLUTION DERIVED CERAMICS

Figure 20 shows the calculation scheme being pursued for the ultrastructure control of ceramics through chemical processing. At the ultrastructure level molecular orbital calculations and molecular dynamics are being applied for the basic molecular design and reaction mechanism prediction which will lead to the required ultrastructure in glass and ceramics. These are addressing, respectively, basic molecular bonding and basic structure and mechanisms. Micromechanics, which models microstructure, interfaces with ultrastructure via structural grouping calculations.

MOLECULAR ORBITAL CALCULATIONS OF THE SILANOL POLYMERIZATION MECHANISM

Major Larry Davis and Lieutenant Colonel Larry Burggraf of AFOSR are using the MNDO (Modified Neglect of Diatomic Overlap) semi-empirical molecular orbital method developed by M.J.S. Dewar of the University of Texas under AFOSR support to investigate the detailed chemical reaction mechanism of silanol polymerization to silica (9). In particular they are addressing the sequence of steps that leads from silicic acid in aqueous solution to silica, under conditions that promote anionic polymerization.

There have been a number of theoretical studies involving silicon-containing molecules using these and other molecular orbital methods. All of these have involved fairly small molecules. Davis and Burgraff compared MNDO calculations and concluded that MNDO should be a useful tool to study larger silicon-containing molecular systems. This conclusion led to this current study of silanol polymerization.

The results indicate that the mechanism strongly favors five-coordinate silicon anion intermediates, and is not simply a nucleophilic displacement of hydroxide ion by the $Si(OH)_3O^-$ anion at the silicon of silicic acid as commonly assumed. Davis and Burggraf have discovered a sequence of steps that begin with addition of silicon-containing anions to silicic acid, followed by rearrangement of these large anions that are produced, and that finally result in another silicon-containing anion being eliminated to serve as a catalytic species for the polymerization. The predicted activation energy for this sequence of steps is in agreement with the experimental activation energy for silanol polymerization.

Pentacoordinate silicon anions appear to be quite important in systems which contain small anions because of the ease of addition of these anions to tetrahedral silicon compounds. The calculations of Davis and Burggraf support these siliconates as possible chain-carriers in anionic silanol polymerization. One of the major keys to understanding the polymerization process is how and when water can be eliminated as the polymerization proceeds. Their results show that water is more easily eliminated when the coordination at the silicon can change from hexavalent to pentavalent rather than pentavalent to tetravalent. After water elimination, a series of addition and rearrangement steps can result in the desired siloxane bond in a neutral product. Some of the intermediates in this series of rearrangements are cyclic structures characterized by trivalent bridging oxygen atoms.

Thus a consistent mechanism has been produced for the anionic polymerization of silicic acid to silica. Whether the rate-determining step is water elimination or addition and rearrangement of the siliconates must await more accurate calculations on the water elimination activation enthalpies. In either case, however, the importance of hypervalent siliconates in anionic polymerization has been established.

Future work will focus on evaluating cationic silanol polymerization mechanisms and delineating theoretically the role and mechanisms of Professor Hench's DCCAs.

POLYMER NETWORK THEORY AND GELATION

The late Paul Flory, Nobel Laureate in Polymer Chemistry, had been supported by AFOSR from 1959 until his death in September 1985. At that time Flory had been working with Professor Uhlmann of the University of Arizona, also under AFOSR sponsorship , on applying his network and kinetic theories of polymeric gelation to silica sol gels.

The most extensively investigated sol-gel material has been SiO_2 prepared from TEOS. It has been shown that pronounced differences in structure result depending whether processing is carried out under acid or basic conditions. It has also been demonstrated that the water-to-TEOS ratio is an important parameter. It has been sometimes suggested that under high pH and higher water content colloidal particles are developed, while under conditions of low pH and low water content linear polymers form.

It must be recognized, however, that these views represent rather considerable approximations to reality and are undoubtedly incorrect in detail. While fine colloidal size particles are indeed formed at high pH and high water content, these are by no means the dense particulates familiar to colloidal chemists. Indeed, the best evidence suggests that these should be thought of as dense, lacey particulates. Further, under conditions of low pH while more polymeric gels are formed, considering the functionality of TEOS and its hydrolysis products (functionality=4), it is unlikely that long linear chains will be formed. Insight obtained from the study of many organic systems with functionalities greater than 2 have indicated that branching takes place early in polymerization and structures with appreciable three-dimensional cross-linking are obtained. As such reactions continue, gelation is observed. On this basis, the polymeric structures produced with TEOS under low pH conditions should probably be expected to display appreciable branching and cross-linking.

This work of Flory and Uhlmann is summarized schematically in Figure 21 (10). They said that the convensional wisdom is incorrect. Silica gel formation proceeds through chemical reactions between functional groups by a polycondensation and not by flocculation of particles. The elastic characteristics predicted and experimentally verified are those for a molecular network and not for an aggregation of a dispersion of globular particulates.

Ceramist Interpretation

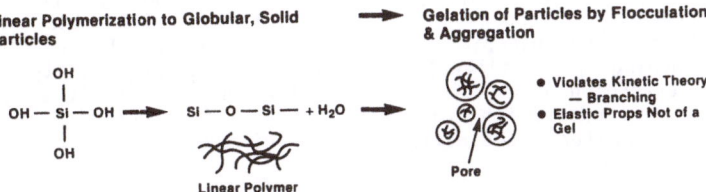

Flory—Network Theory + Kinetic Theory into Gelation Theory

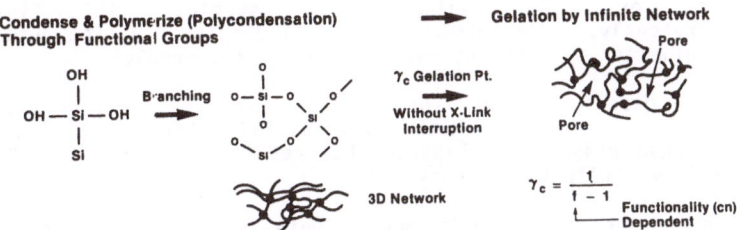

Fig. 21   Flory polymer network theory and gelation applied to silica sol-gels.

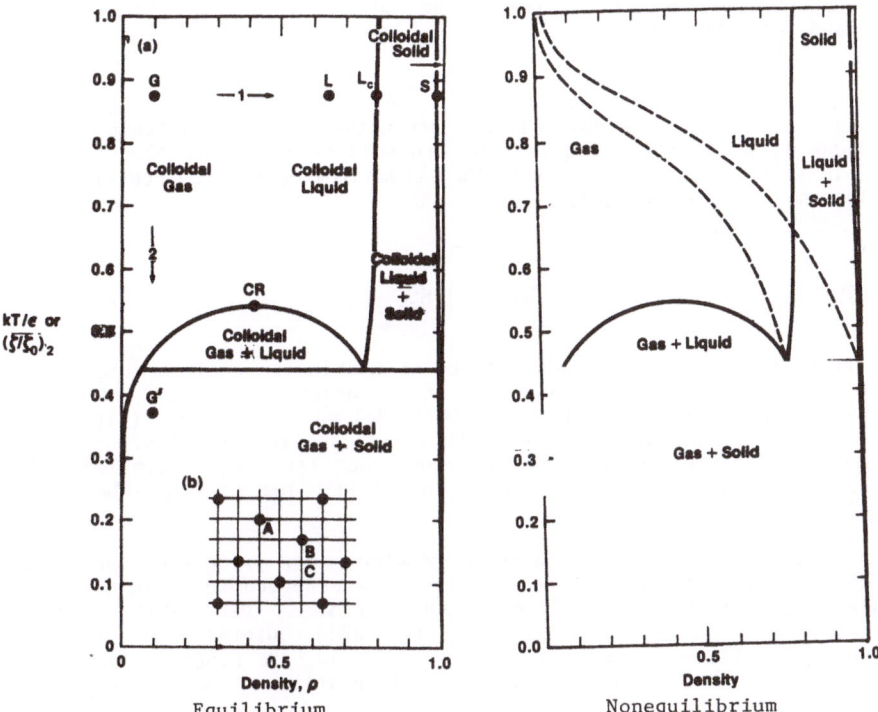

Equilibrium   Nonequilibrium

Fig. 22   Theoretically calculated colloidal phase diagrams for predicting the
ultrastructure of chemically derived ceramics.

The assumption of linear polymerization to globular, solid particles violates the kinetic theory of branching. The gel point observed in a poly-condensation process may be identified with the incipience of a network of a macroscopic size. The occurrence of gelation requires that the branching probability γ , defined as the probability that a molecular chain extending from one branch point in the structure leads to another branch, must exceed its critical value γ_c. As shown in Figure 21, f is the functionality (or coordination number) of the branch point, which is 4 for polymeric silicates.

The network, or gel, makes its appearance abruptly in the course of the polymerization which, however, continues without interruption at and beyond the gel point. The proportion of gel increases continuously at the expense of the sol-comprising species of finite size. Simultaneously, the density of reticulation in the network increases, as is reflected by the increase in the elastic modulus.

This description of the gelling process is abundantly confirmed by comparisons of theory with experimental observations on branched and cross-linked polymers. Experimental results on molten silicates appear to be consistent with this theory.

One can obtain gels under both high pH and low pH conditions, but the character of these gels differs appreciably -- most notably in the pore size distribution of the dried gels. Such differences have important consequences for subsequent processing of the dried gels to produce dense glass and ceramic bodies.

The detailed characteristics of the chemistry and structure remain to be properly elucidated even in the case of TEOS. In case of other alkoxide or other organometallic precursors the available state of knowledge is much poorer and it can properly be said that almost nothing is known in satisfactory detail about the chemistry and sequence of structural development in such systems.

HIERARCHICAL CLUSTERING IN SOL-GEL SYSTEMS

Professor Aksay of the University of Washington collaborated with Flory relative to the network theory of polymer gelation as applied to Aksay's theoretically calculated equilibrium colloidal phase diagram (Figure 22) for predicting the ultrastructure of chemically derived ceramics. The example of mullite described earlier was derived from an extension of this collaborative work to a calculated nonequilibrium, but preliminary phase diagram.

In the first step of ceramic processing, the processing and dispersion of powders, colloidal dispersion techniques are needed to eliminate particle clusters that form uncontrollably due to van der Waals attractive interactions. This approach is especially useful when working with multiphase particle systems in the 10 to 10 m range. In the second step, consolidation of powders, transitions from dispersed to consolidated state start as a nucleation and growth process of first generation particle clusters. These particle clusters form when either the interparticle binding energy, E, or the particle number density, , in the suspension exceeded a critical value.

Aksay outlined these transitions with a theoretically calculated E vs. p colloidal phase diagram (Figure 22). The tentative conclusion was that the low particle number density was described by Flory's theory and that the high particle number density was described by the traditional flocculation of hard particle theory. Preliminary evidence indicates that both cases may contribute to hierarchically clustered microstructures.

Hierarchically clustered microstructures are due to multiple clustering of particle clusters. Consequently, in such hierarchically clustered microstructures, the classification of the void space follows a similar trend as first, second, and third generation voids. Unlike the particle clusters that we wanted to eliminate in the dispersion stage, the size and the spatial arrangement of the particle clusters that form can be controlled by simply changing the colloidal consolidation path on the phase diagram. In the third step, the evolution of the microstructure follows a path influenced by the hierarchy of particle clustering. First generation particle clusters sinter faster and at significantly lower temperatures than the higher order void regions. This evolutionary path may be characterized by the methods familiar to us in statistical mechanics.

NONLINEAR OPTICALLY ACTIVE SOL-GELS

In preliminary AFOSR research supported with Dr. Alan Buckley at the Celanese Research Company, nonlinear optical sol-gels were derived by the processing of organics and polymers with sols and gels, such as 2- methyl -4- nitroaniline with TEOS (Figure 23). In this work, high phase matcheable second harmonic generation (SHG) was observed for frequency doubling from $\lambda = 1.064$ μm (near IR) to $\lambda = 0.532$ μm (green). (Figure 24). Patents have been filed for a wide range of sol-gel: organic and sol-gel; polymeric nonlinear optical materials.

OTHER USAF BASIC RESEARCH PROGRAMS IN SOL-GEL MATERIALS

Other AFOSR basic research programs in sol-gel glass, ceramics and composites are listed in Table 1.

SUMMARY

In this paper the processing of ceramics, glass and composites with superior and unique properties has been described. Examples have been cited based on either the homogeneity or heterogeneity

232

NONLINEAR OPTICAL SOL-GEL GLASS: MNA

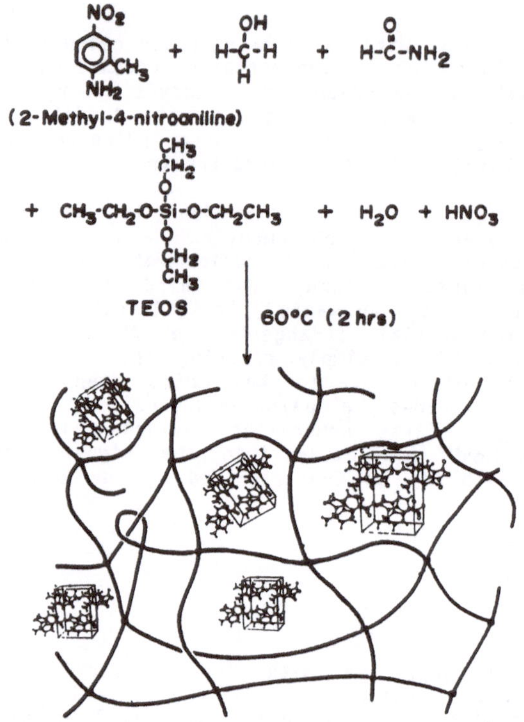

Fig. 23 Nonlinear
Optically
Active
Gels.

(2-Methyl-4-nitroaniline)

TEOS   60°C (2 hrs)

NONLINEAR OPTICAL SOL-GEL GLASS: MNA

- No chemical degradation of nonlinear
  optical organic sites during sol-gel
  process.

- Specific polar orientation of organic
  sites same as in single crystal.

Fig. 24 High Second
Harmonic Generation
in Sol-Gel Organic/
Polymer:SiO_2 Glass

- High phase matchable SHG as
  observed for distribution of 100μ
  size crystals.

  (nir) $\omega + \omega \rightarrow 2\omega$ (green)
  $(\lambda : 1.064\mu m)$ $(\lambda : 0.532\mu m)$

TABLE 1

OTHER USAF BASIC RESEARCH PROGRAMS IN SOL-GEL MATERIALS

o   DIAMOND - GEL COMPOSITES
              - Hard
              - High Thermal Conductivity
o   NONLINEAR OPTICAL AND ELECTROOPTICAL CERAMICS
              - Lithium Niobate   Square Hysteresis
              - Potassium Sodium Niobate Loops
              - Lead Zirconate Titanate   x (2) Second Order Nonlinearities
o   INORGANIC-ORGANIC GELS
              - Long Relaxation Times
              - Self-Healing
              - Polishable
o   BORON CARBIDE-ALUMINUM ARMOR
o   ULTRASTRUCTURED COMPOSITES
              - Pore Interconnection and Orientation
                -- Active or Passive Polymer Impregnation
                -- Gallium Aluminum Arsenide Impregnation-
                   Transverse Optical Bistability
                -- Mechanical Ductility
                -- Passive or Active Ceramic Impregnation
              - Electroactive and Nonlinear Optical Copolymerized
                 Sol-Gel - Polymer Composites
              - Preceramic Polymer Sol-Gel Nanocomposites
o   SINGLE MODE SOL-GEL WAVEGUIDES
              - Electrolytic Exchange
o   NOISE FIGURES OF MERIT   > Polytypic Boundaries
     NOISE EQUIVALENT CIRCUITS   > In Gel Multilayers
o   SILOXANE SILICA GELS
              - Enhanced Toughness
o   ORDERED GELS
              - Liquid Crystallinity in Gels
o   RIGID ROD, LIQUID CRYSTALLINE POLYMER (POLYBENZOTHIAZOLE) - SILICA
     GEL INTERPENETRATING NETWORKS
o HfO; Al_2O_3; SiC - ZrO_2 COATINGS FOR CARBON - CARBON COMPOSITES.
o   NONOXIDE GELS
              - SiC
              - Chalcohalides
              - Halides
o   SUPERCRITICAL FLUID PROCESSING OF GELS
o   SiC GEL INTERNAL MATRIX PROTECTION OF CARBON - CARBON MATRICES.
o   COLLOID DERIVED FIBER-REINFORCED SOL-GEL COMPOSITES

derived from the sol-gel approach. Examples of advanced composites such as molecular ceramic composites or NLO sol-gel glass-organic composites processed at the ultrastructure level via sol-gel processing have been presented.

These unique materials have come about, however, only through a critical mass, long term basic research investment in understanding the chemistry which is essential to the molecular design and processing of reliable high performance ceramics, glass and nanocomposites. Also essential to achieving these goals has been catalyzing the interaction and collaboration of ceramists and glass scientists with inorganic and organometallic synthetic chemists, polymer chemists and physicists, theoretical chemists, and chemical physicists. Several examples have been given where the successful collaboration is leading to the understanding of the fundamental process chemistry and new materials.

As shown by the breadth of the Air Force basic research program, there are many promising areas for the investigation of ceramics derived through chemical processing.

Not included in this presentation are the other chemically based AFOSR programs in transformation processing based on organometallic, polymeric and microemulsion precursors or changes in phase state, and micromorphology processing. The latter describes the chemical and solution synthesis and special surface - chemical manipulation of ultrafine, submicron particulates as well as hierarchically clustered or ordered assemblages.

What is fundamental to each of these approaches is to be able to design the molecule and predict the reaction route which upon processing will yield the ultrastructural grouping essential to achieving predictable properties and performance in ceramics, glass, nanocomposite, molecular composites, or even fiber-reinforced composites.

ACKNOWLEDGEMENT

Special appreciation is expressed to my wife, Mrs. Eleanor Ulrich and to Mrs. Cathy Faustmann of the AFOSR staff for their help in the preparation of this paper as well as to Dr. Donald L. Ball for his encouragement and support. We are indebted to the dedication, achievement and enthusiastic inspiration of all the investigators sponsored by AFOSR in this area of research, many of whom could not be listed here.

REFERENCES

1   Ultrastructure Processing of Ceramics, Glasses and Composites, edited by L. L. Hench and D. R. Ulrich, (Wiley Interscience, New York, 1984).

2. Science of Ceramic Chemical Processing, edited by L. L. and D. R. Ulrich (J. Wiley, NY 1986).

3. Proceedings of the Third International Conference on the Ultrastructure Processing of Ceramic, Glasses and Composites, edited by J. D. Mackenzie and D. R. Ulrich, (Wiley Interscience, New York, (1987), in press.

4. Better Ceramics Through Chemistry, edited by C. J. Brinker,, D. E. Clark, and D. R. Ulrich (Elsevier - North Holland, 1984).

5. Better Ceramics Through Chemistry II, edited by C. J. Brinker, D. E. Clark, and D. R. Ulrich (Mat. Res. Soc., Pittsburgh, PA, 1986).

6. Better Ceramics Through Chemistry III, Materials Research Society Spring Meeting, April 1988, Reno, Nevada.

7. J. Zarzycki, in Ultrastructure Processing of Ceramics, Glasses and Composites, edited by L. L. Hench and D. R. Ulrich, Wiley Interscience, New York (1987).

8. B. Lee, Chemically Derived Ceramic Composites, Ph.D. Dissertation, University of Florida, 1985.

9. L. P. Davis and L. D. Burggraf, to be published in Proceedings of the Third International Conference of the Ultrastructure Processing of Ceramics, Glass, and Composites, edited by J. D. Mackenzie and D. R. Ulrich, Wiley Interscience, New York, (1987).

10. P. Flory, in Science of Ceramic Chemical Processing, edited by L. L. Hench and D. R. Ulrich (Wiley Interscience, New York, 1986).

TECHNICAL CONVERSION OF ALKOXIDES TO OXIDES, YESTERDAY, TODAY AND TOMORROW
- illustrated by examples taken from the sol-gel dip coatings on glass

HELMUT DISLICH
Schott Glaswerke, Hattenbergstr. 10, 6500 Mainz, Federal Republic of Germany

ABSTRACT

Coatings are the most important sol-gel products. These can be single oxides or well defined multicomponent oxides produced from alkoxides. Some present and future sol-gel dip coatings are described.

INTRODUCTION

This lecture will cover a period of about forty years research and development in the SCHOTT glassworks, recalling the pioneer work of W. Geffcken and H. Schroeder and, more recently, the work of N. Arfsten, P. Hinz, E. Hussmann and their colleagues.

$$\text{Alkoxide} - \text{Alk} \rightarrow \text{Oxide} \qquad (1)$$

(1) describes the subject of this lecture in a schematic and simplified manner. Any chemist not familiar with the technical background is directly tempted to ask: ... surely there is some mistake? Oxides are plentiful raw materials and alkoxides must be prepared by a complicated method, often from the oxides ...
The question is a natural one, and the reasons are shown in Fig. 1.

- The resulting oxides are extremely pure
- They can be prepared directly in the desired forms,
  i. e as:
  solid mass
  fibres
  films
  extremely fine powder
  hollow spheres
  thin coatings

FIGURE 1

This is not only true for single oxides like SiO_2, TiO_2, Al_2O_3 and so on, but also for defined multicomponent oxides used as glass, ceramics, and glass-ceramics [1], since the equation of the following is also valid, in the same simplified sense as (1).

$$\text{Alkoxide}^{I} + \text{Alkoxide}^{II} + \text{Alkoxide}^{III} \rightarrow \text{Multicomponent Alkoxide}^{I, II, III}$$
$$- \text{Alk}^{I}, \text{Alk}^{II}, \text{Alk}^{III} \rightarrow \text{Multicomponent Oxide} \qquad (2)$$

A few practical examples will probably illustrate this schematic representation and keep your interest awake. The following products can be made without any necessity for going through the customary intermediate melted phase [2]:

236

R. M. Laine (ed.), Transformation of Organometallics into Common and Exotic Materials: Design and Activation, 236–249.
© 1988 by Martinus Nijhoff Publishers.

- Ge or B doped SiO_2-preform for telecommunication fibres

- Borosilicate glass hollow spheres, filled with deuterium and tritium for nuclear fusion by laser

- Caesium aluminum silicates etc., unknown in nature

- Alkali-resistant fibres with a high content of ZrO_2 for reinforcement of cement

- Unsupported TiO_2-SiO_2 films of 1 - 100 μm thickness

- Micropellets with encapsulated radioactive waste

- Extremely fine powder (< 1 μm) of $BaTiO_3$, $SrZrO_3$ etc.

- Transparent thin coatings on glass

FIGURE 2

One must add that everything on Fig. 2 has only been carried out in the development laboratories, that is to say in small units as yet. The exception is the production of thin films on glass, which have been in mass-production for many years; we can look back with pleasure on a series of small celebrations in the glassworks as yet another product in this range reached the magic million square meter production mark. The reason why this well-established products should be mentioned today is that the technical development started at SCHOTT under the initiative of W. Geffcken and H. Schroeder. First beginning was at the end of the thirties, and continued in the present department in the last three decades by broadening the chemical basis all over the Periodic System. A well-known slogan is "Better Ceramics through Chemistry".

The interest in the chemistry of the sol-gel process (known internally at SCHOTT as "glass from the bottle" [3]) is now world-wide, and several of the above-mentioned applications could become interesting in a mass-production sense. The development of thin films continues at SCHOTT and elsewhere, and we anticipate further practical applications.

We will now concentrate on thin films on glass to illustrate the principles of the sol-gel process.

THE PRESENT TECHNICAL AND ECONOMIC IMPORTANCE OF OXIDE FILMS ON GLASS

The conventional industrial measure of tons per year is meaningless for thin films, since the usual application is in the field of optical interference phenomena, i. e. the layers are of the order of 0.1 to 0.5 μm thick. The producers of alkoxides cannot therefore reckon with vastly increased markets as far as tonnage is concerned. However, the picture can be quite different for certain special products. Thus the firm Stauffer offers defined multicomponent alkoxides for the purpose of producing multi-component oxides, e. g. mullite or spinel.

More impressive than the tonnage are the areas of oxide films produced each year; these run into millions of square meters. Since very large areas can be coated with very few tons of alkoxide, the degree of refinement of the product is extremely high, and this justifies the participation of the SCHOTT glassworks in this market.

The oxides are produced at 500 °C and they consequently posses a high degree of environmental stability, are wipe-resistant and have an excellent large-aera optical homogeneity, and many other good properties which the author will deal with later in this lecture.

Since this is primarily a scientific and technical meeting, where the economic aspects should not be unduly emphasised, no tables of Dmarks per year are shown, but the areas where these films on glass are especially important [4, 5, 6] are listed.

Vehicles	– rear view mirrors, glare reduced, heatable and aspheric (because of the blind spot)
Window glass	– IROX®, sun-reflecting, neutral grey, reduces cooling costs
Oil paintings and glass showcases	– non-reflecting coatings, better view with good protective properties.

FIGURE 3

These are all established markets. At the present time the market is expanding in the vehicle branch (Daimler, BMW, Audi, Volvo among others), is sluggish in the building market, and in the field of oil paintings in museums a great many extremely valuable works deserve an appropriately high quality glass.

Figs. 4 - 7 show some rear view mirrors in the sequence plane, spherical, aspherical and heatable; Fig. 8 shows a building in Frankfurt with IROX® window glass (Frankfurt is a stronghold for IROX®).

FIGURE 4 Rear view mirror (plane)

FIGURE 5 Rear view mirror (spherical)

FIGURE 6 Rear view mirror (aspherical)

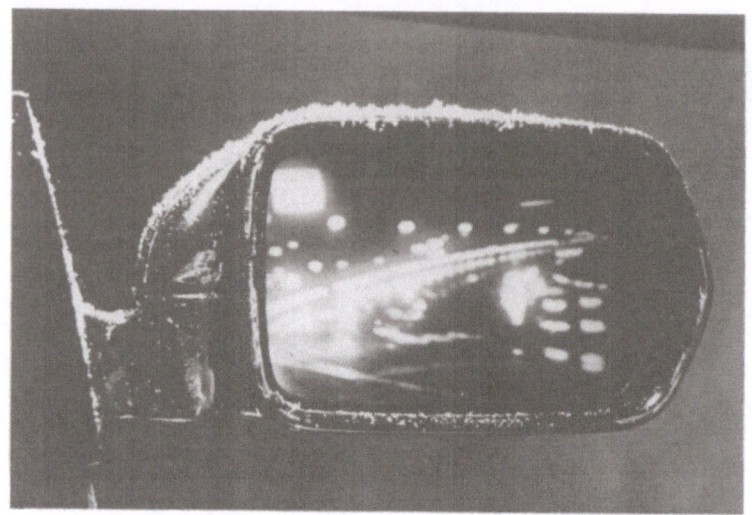

Fig. 7 Rear view mirror (heatable)

Fig. 8 Euro-House in Frankfurt

Fig. 9 Geno-House in Stuttgart

The coated glass can be thermally hardened, which is not automatically the case for other types of coating. Thus the glass can also be used as panels in the spandrel area of the building, as shown in Fig. 9.

Furthermore, the coated glass can be curved after the heating process, as is demonstrated by the spherical and aspherical rear view mirrors.

The applications mentioned up to now are the major ones, from the point of view of turnover; however, it should also be mentioned at this point that other applications such as heat-reflecting glass, cold-light mirrors, selectively reflecting mirrors, achromatic optics, conversion filters, anti-glare filters, glass with transparent conducting coatings, and color effect filters have all been produced by SCHOTT using the same chemistry and the same process [7].

THE CHEMISTRY AND THE PRODUCTION PROCESS FOR OXIDE COATINGS FROM ALKOXIDES

The chemical principles

The process is based on the hydrolysis and condensation scheme shown in (3).

$$Si(OCH_3)_4 + 4H_2O \rightarrow Si(OH)_4 + 4CH_3OH\uparrow$$

$$Si(OH)_4 \xrightarrow{Temp.} SiO_2 + 2H_2O\uparrow \tag{3}$$

Synthesis of SiO_2 (schematic)

The silicic acid ester is dissolved in alcohol. Temperatures of 400 - 500 °C are necessary for the second step. The reaction actually runs through a variety of intermediate stages and can be applied for many other elements.

The principle of the production process

Fig. 10 shows how the plates of glass of up to 12 m² area are dipped into the alcohol solution and evenly pulled out in an atmosphere which is carefully adjusted with respect to temperature, humidity and air circulation. The alcohol solvent and the hydrolyzed alcohol evaporate and the coating becomes solid. Then the firing takes place at 400 - 500 °C and the coating is stable and clear as glass.

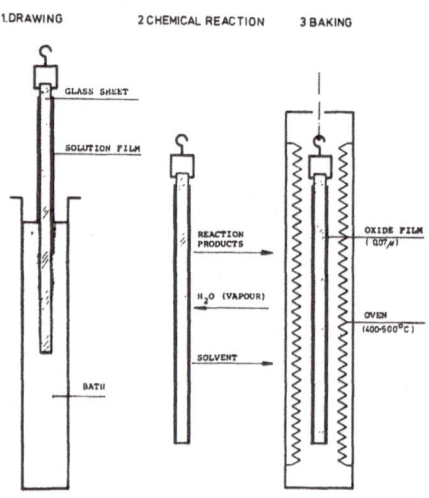

Fig. 10   Production in the dipping process

The composition of various coatings

The rear view mirrors consist of three layers of three distinct oxides, one on top of the other: TiO_2-SiO_2-TiO_2. The color and the reflection properties can be adjusted via the thickness of the layers.

The IROX® layer is made of TiO_2 with embedded palladium, i. e. it is a cermet layer (ceramic and metal). Different absorption can be achieved by adjusting the palladium content. The non-reflecting coating consist of TiO_2/SiO_2-TiO_2-SiO_2, i. e. it is composed of three individual layers of moderate, high and low refractive index, respectively.

The ability of the layers to be combined with each other (up to 12 may be combined) and the ability to mix the starting solutions to achieve any desired intermediate value of the refractive index accounts for most of the above-mentioned products. Silicon alkoxides give the low-refracting SiO_2, and titanium alkoxides yield the highly refracting TiO_2, and these are therefore the most important, but in no way the only alkoxide to be used in this way.

Fig. 11 shows the refraction of SiO_2/TiO_2 layers as a function of the molar proportion of TiO_2 ($\lambda = 550$ nm).

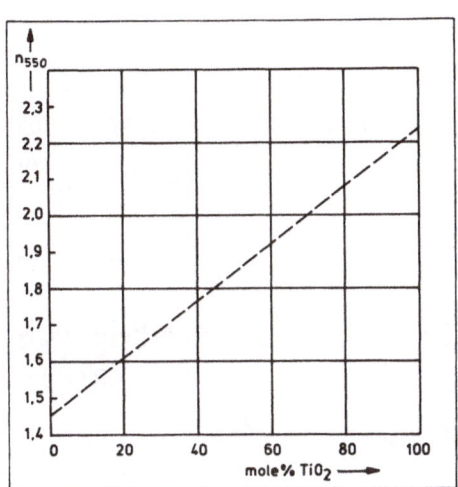

Fig. 11 Refractive index of SiO_2/TiO_2 films as a function of the molar proportion of TiO_2 ($\lambda = 550$ nm)

Fig. 12 shows which elements have been used to date in the production of oxides from alkoxides.

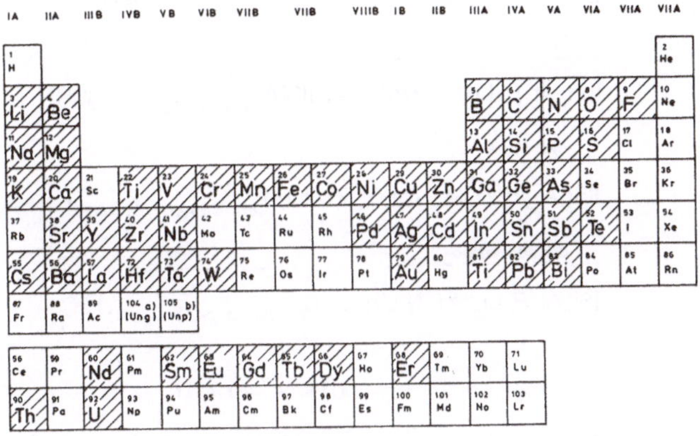

Fig. 12 Elements used to date in the sol-gel process

As mentioned at the start, there is the further possibility of making defined multicomponent layers. These can be glassy, crystalline or glass-ceramic. The basis for these products is the high mutual reactivity of the alkoxides (or their partially hydrolyzed intermediates), as shown in (4) for the example of a spinel synthesis.

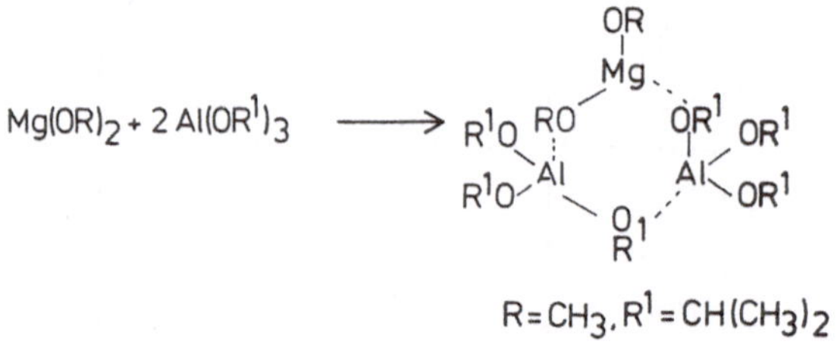

$$Mg(OR)_2 + 2 Al(OR^1)_3 \longrightarrow$$

$$R = CH_3, R^1 = CH(CH_3)_2$$

$$Mg[Al(OR)(OR^1)_3]_2 + 8 H_2O \longrightarrow Mg[Al(OH)_4]_2 + 2 ROH + 6 R^1OH$$

$$Mg[Al(OH)_4]_2 \longrightarrow Mg[AlO_2]_2 + 4 H_2O \qquad (4)$$

Spinel synthesis

The most complicated coating which has been produced by us to date is a glass ceramic with eight different elements, as shown in (5).

$$m\, Si(OMe)_4 + n\, Al(Osec.Bu)_3 + o\, P_2O_5 + p\, LiOEt +$$
$$q\, Mg(OMe)_2 + r\, NaOMe + s\, Ti(OBu)_4 + t\, Zr(OPr)_4$$

$$\underline{complexation}$$

$$[(Si_mAl_nP_oLi_pMg_qNa_rTi_sZr_t)(OR)_{4m+3n+p+2q+r+4s+4t}(OH)_{3o}]$$

$$\underline{hydrolysis}$$

$$[(Si_mAl_nP_oLi_pMg_qNa_rTi_sZr_t)(OH)_{4m+3n+p+2q+r+4s+4t+3o}]$$

$$\underline{condensation}$$

$$[Si_mAl_nP_oLi_pMg_qNa_rTi_sZr_t\, O_{4m+3n+3o+p+2q+r+4s+4t}] \qquad (5)$$

Synthesis of a glass ceramic

The varied and flexible chemistry of the alkoxides enables the design of tailor-made coatings. The basic chemistry is described elsewhere [8], and we will shortly publish a description of the production process and the properties of the layers so obtained [9].

The special importance of alkoxide chemistry for the sol-gel process in general and for layers in particular can only be hinted at in this short presentation; for instance, the choice of the group R in an alkoxide $Me(OR)_x$ will have a decided influence on the rate of hydrolysis, as will also the value of the pH. Furthermore, the precondensation in solution is also important, as are the temperature, reaction time, humidity of the working atmosphere after the formation of the layer, and so on. The producers of alkoxides offer a constantly growing variety of interesting starting materials, which can be further modified chemically by the researcher on the spot. Here are two good examples:

If it is necessary to reduce the rate of hydrolysis of a particularly sensitive alkoxide, this can be easily achieved by the use of a chelating agent, for example acetyl-acetone; the alkoxide group is partially replaced by a chelate group.

If two alkoxides of drastically differing rates of hydrolysis are to be reacted with one another in solution (which is one considerable advantage of the sol-gel process in the production of homogeneous final products), then the slowly hydrolyzing component is hydrolyzed first in isolation and then added to the quicker component. (6) shows that one also profits from the added advantage gained by the possibility of reaction between a group Me^I-OH and a second group Me^{II}-OR.

$$Si(OR)_4 + H_2O \rightarrow (OR)_3Si-OH + ROH\uparrow$$

$$(OR)_3Si-OH + ROTi(OR)_3 \rightarrow (OR)_3Si-O-Ti(OR)_3 + ROH\uparrow \qquad (6)$$

Many such examples could be given, simply using classical alkoxide chemistry. Actually, the real novelty is the application of this knowledge to the preparation of oxides with desirable properties, as is admirably expressed by the title of the book [10] "Better Ceramics Through Chemistry". Thus the "mesh" of the grid of an oxide layer can be controlled by the reaction course, giving various refraction values at a constant chemical composition. The temperature can be used to control the crystal structure, as is shown by anatase, brookite and rutile for TiO_2.

THE FUTURE OF SOL-GEL DIP COATINGS

Now that we have dealt with the undoubted advantages of the process, it is time to turn to the question of the economic future in this field. One must be careful here, since unknown economic and market factors will dictate the answer to this question, in combination with technical innovation. It will be attempted to illustrate the opinion, based on a few examples, whereby it cannot be taken into account the future possibilities of competing processes such as the sputter, CVD, or spray methods etc. Chemical variability plays a major role in favour of the sol-gel dip coating process, and it will be expected to see further developments in the areas of optoelectronic magnetic and electrochrcmic oxide layers, protective layers (including anti-corrosion coatings), and photosensitive sulfide films, but one must be careful in predicting too much. On the other hand, the following two fields seem to be more concrete in their development potential:

Contrast enhancement filter for monitors and terminals

SCHOTT has developed detachable screens with layers on both surfaces, resulting in a higher contrast and reduced reflection on terminals and personal computers [11], as shown in Fig. 13.

Fig. 13

This is achieved by a sequence of TiO_2/SiO_2-, TiO_2- and SiO_2-interference layers applied to large areas of glass using a dip coating process with alkoxide solutions. The screens can be cut to any shape. A variety of float glasses of differing strength of absorption can be used as supporting glass, in order that a variety of different types of terminal with differing light intensities can be accommodated. The stronger the absorption, the more favorable is the ratio of useful light to instrusive light, i. e. the higher the contrast. Fig. 15 demonstrates the very small residual reflection obtained using a layered screen of this sort.

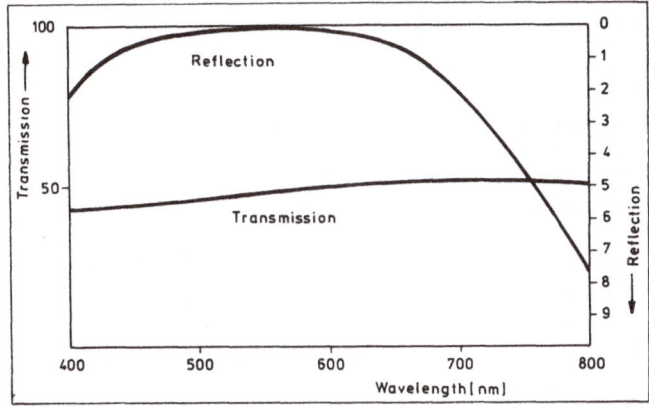

Fig. 14   Reflection and transmission curves of an anti-reflec-
ting contrast enhancement filter, coated on both sides

The layers are scratch-proof (as measured by MIL-C-675 A), resistant to
the common chemical agents, and easily cleaned.

Electrically conducting transparent indium oxide layers

Various highly conducting layers can be produced by doping with certain
oxides; the indium oxide layers are best known in this respect, and these may
be produced by either the vacuum or the sol-gel process [12]. The transmission and reflection values of these glass products are important points of
quality: Fig. 15 demonstrates these for a good conducting layer of 25 $\Omega/\square$
resistance [13] produced by the sol-gel method.

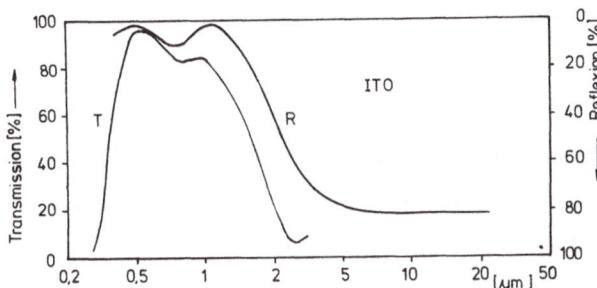

Fig. 15   Reflection and transmission of an ITO-coated glass

The layers can be produced to tailor-made specifications, and are therefore useful for a great variety of purposes, such as conducting layers in
displays, or as IR-reflecting layers of high transmission for visible light,
as shown by the selection of basic data [13] of Table I.

248

Table I

Properties of ITO-films optimized for application in displays (a, c) and
for window-system (b)

	(a)	(c)	(b)
Film thickness [nm]	20 – 30	∿ 80 – 100	270
Charge carrier density [cm^{-3}]	4×10^{20}	$5 - 6 \times 10^{20}$	$5 - 6 \times 10^{20}$
Mobility [$cm^2 \, V^{-1} \, s^{-1}$]	15 – 20	60 – 70	35 – 40
Conductivity [$\Omega^{-1} \, cm^{-1}$]	800 – 1 200	5 000 – 6 000	3 000 – 3 500
Sheet resistance [$\Omega/\square$]	500	∿ 25	10

Doping with oxides from the 2nd to the 5th main group of the periodic
table (or with oxides from subgroups 1 to 5, or subgroup 8) can be used to
vary the electrical conductance over several orders of magnitude. The same
applies for the sensitivity to etching by acids. Display layers can be
structured by rapid etching, whereas for certain other purposes attack by
acids cannot be tolerated at all [14].

SUMMARY

The technical conversion of alkoxides to oxides ("sol-gel process") is
the subject of much interest world-wide; in particular the underlying chemistry and the resulting (or potential) special products are the focus of attention. This article has described some of these sol-gel dip coating products from SCHOTT which are already on the market, and furthermore some products still in development, which could have good prospects. The general
characteristics of the process is the chemical variation which can be
achieved, and this leads to a variety of opportunities for special products.
Thin layers on supports gain many distinct advantages on using the sol-gel
manufacturing procedure, and for this reason the technique probably has
especially good prospects in this particular application.

REFERENCES

[1] Dislich, H., Angew. Chem. Int. Ed. Engl. No. 10 (6), 363 (1971)

[2] Dislich, H., J. Non-Crystalline Solids 57, 371 (1983)

[3] Dislich, H., and Hinz, P., Schott Information (English) 1, 17 (1981)

[4] Prospect Rückspiegel auf interferenzoptischer Basis. Deutsche Uhrglas-
    fabrik, Grünenplan, FRG

[5] Prospect IROX® solar-reflective glass. Function and design in modern
    architecture. The complete colour-neutral glass from Schott. Schott
    Glaswerke, Mainz, FRG (1981)

[6] Prospect Non-reflective glass. Deutsche Spezialglas AG, Grünenplan,
    FRG (1981)

[7] Prospect Produktionsinformation "Gläser mit Oberflächenschichten", Schott Glaswerke, Mainz, FRG (1981)

[8] Dislich, H., Hinz, P., and Kaufmann, R., US-Patent 3,847,583, Jenaer Glaswerke Schott & Gen., Mainz, FRG (1969)

[9] Dislich, H., in: Sol-Gel Technology (Klein, L., ed), Thin Films from the Sol-Gel Process, to be published by Noyes Publication, Park Ridge, N. J., USA

[10] Brinker, C. J., Clark, D. E., and Ulrich, D. R. (eds) Materials Research Society Symposia Proceedings, Vol. 32., New York. Elsevier Science Publishing Company, Inc. (1984)

[11] Arfsten, N. J., and Hussmann, E., not published, Schott Glaswerke, Mainz, FRG

[12] Arfsten, N. J., Kaufmann, R., and Dislich, H., in: Ultrastructure Processing of Ceramics and Composites (Hench, L. L., and Ulrich, D. R., eds) 189-196, John Wiley & Sons, New York (1984)

[13] Arfsten, N. J., J. Non-Crystalline Solids 63, 243 (1984)

[14] Arfsten, N. J., not published, Schott Glaswerke, Mainz, FRG

SOL-GEL PROCESSING OF TRANSITION METAL OXIDES

J. LIVAGE

Spectrochimie du Solide, Université Pierre et Marie Curie,
4 Place Jussieu, 75230 Paris, Cedex 05, France

1. CHEMICAL ENGINEERING OF THE MOLECULAR PRECURSORS

Sol-gel processing involves the use of molecular precursors, mainly alkoxides, as starting materials. A macro molecular network is then obtained through hydrolysis and condensation reactions. Major advances in ultra-structure processing will require an emphasis that relates chemical process to gel formation and powder morphology.

Oxide gels are almost always obtained according to the same procedure. Hydrolysis can be performed in the presence of an acid or a base catalyst. These catalysts are supposed to favor either electrophilic or nucleophilic substitutions leading to linear or branched polymers (2).

More recently, chemical additives, such as formamide or oxalic acid, have been used as Drying Control Chemical Agents (DCCA), leading to monolithic SiO_2 glasses (3). In this paper, we would like to point out that such additives may chemically react with the precursor at a molecular level and therefore modify the whole hydrolysis condensation process. This has clearly been shown when acetic acid is added to pure $Ti(OBu^n)_4$. An exothermic reaction takes place leading to a clear solution. An infrared study of this solution shows a set of two bands around 1500 cm^{-1}, typical of stretching vibrations of the $(COO)^-$ group. The frequency separation ($\Delta v = 130$ cm^{-1}) between the symmetric and the anti-symmetric component suggests that CH_3COO^- acts as a bidentate ligand according to the reaction:

$$Ti(OR)_4 \xrightarrow{\quad CH_3COOH \quad} Ti(OR)_x(OAc)_y + ROH$$

A new molecular precursor is obtained in which titanium is now six-fold coordinated. A stoichiometric compound $Ti(OBu^n)_2(OAc)_2$ is obtained when $(OAc)/(Ti) = 2$ (4). It can be isolated as single crystals and x-ray diffraction studies show that acetate bridges link titanium atoms.

Chelating acetates are less easily hydrolyzed than OR groups, increasing the time of gelation. This favors the polycondensation process and leads to monolithic gels in which high polymers, rather than small colloidal particles, are present. The chelating effect of acetic acid is not unique. It can be observed with other carboxylic acids leading to polynuclear carboxylates (5).

R. M. Laine (ed.), Transformation of Organometallics into Common and Exotic Materials: Design and Activation, 250–254.
© 1988 by Martinus Nijhoff Publishers.

2.   A PREDICTIVE MODEL FOR HYDROLYSIS-CONDENSATION REACTIONS
      Organic chemists often use simple models to predict the
reactivity of their compounds. Hydrolysis of $Si(OR)_4$ in the
presence of an acid or base catalyst can thus be des cribed in
terms of electrophilic or nucleophilic substitutions (2).
Such models would be also very convenient in the field on
inorganic polymerization, but they are not yet widespread.
      Recently, M. Henry suggested that Sanderson's ideas could
be extended in order to describe the chemical reactions
involved in the sol-gel process (6). His model is based on
the principle of electronegativity equalization that may be
stated: when two or more atoms initially different in electronegativity combine, they adjust to the same intermediate
electronegativity in the compound. This intermediate electronegativity is defined as the geometric mean of the stability
ratios of all the atoms before combination. It then becomes
possible to work out a distribution of partial charges on combined atoms in order to describe their chemical reactivity.
The chemical behavior in inorganic and metal-organic molecular
precursors can be easily deduced from this model and compared
with experiments. Depending on the partial charge of the OH
group in an M-OH bond, different reactions may occur, such as
acid or base ionization and oxolation or olation condensation.

3.   ORDERED AGGREGATION IN GELS
      Aggregation of colloidal particles is a very important
phenomenon in colloid science. It leads to the formation of
gels but must be avoided when monodispersed powders are
desired. Because of the Brownian motion, aggregation is usually a random process and is therefore quite difficult to
study. A major advance in this field has been made during the
last few years by using small-angle X-ray scattering and fractal theory. The fractal dimension of colloidal aggregates can
be measured from the power-law decay of the static structure
factor in the Porod region (8). A model can then be suggested
describing the process as diffusion-limited aggregation or
cluster aggregation (9).
      Sometimes, however, ordered aggregates can be observed.
They are formed under special conditions and lead to the
so-called "colloidal crystals" (10).
      Ordered colloidal systems were observed a long time ago
in transition metal oxide colloids. They are responsible for
the irridescent aspect of tungstic acid and β FeOOH sols. The
symmetric arrangement of monodispersed plate-like particles
gives rise to the so-called "schiller layers" (11). The equilibrium distances between the associated particles can be as
large as a few thousand Angstroms. Anisotropic-ordered, colloidal phases have also been observed with V_2O_5 gels. These
gels are made of fiber-like particles that look like flat ribbons (12). Concentrated colloidal solutions separate under
suitable conditions into a concentrated and a dilute phase.
In the concentrated phase, the colloidal particles are arrang-
ed with their major axis parallel to each other, giving rise

to a neumatic anisotropic structure called "tactoid" (13).
These tactoids are dispersed in an optically isotropic-dilute
sol phase called "atactosol."

Ordered, reversible aggregates change to ordered, irreversible aggregates if the distances between the individual
particles are carefully reduced. This can be done by suitable
changes in or of the dispersion medium or by its gradual
removal. This leads to ordered, solid aggregates called
"crystalloids" in which all particles are mutually oriented.
These crystal-like bodies share with classical crystals the
ability to "dissolve" on addition of a proper swelling agent;
the product of dissolution being, in this case, colloidal particles instead of ions.

In the concentrated, ordered tactoid phase, V_2O_5 ribbons
are separated by water layers. On careful addition of coagu-
lating electrolyte V_2O_5 tactoids gradually shrink to a
fraction of their original size maintaining, and even enhanc-
ing, their internal anisotropy (13). Such a shrinkage can
also be observed when a V_2O_5 gel is deposited onto a glass
substrate. After drying, a strongly anisotropic coating is
obtained. X-ray diffraction diagrams of these V_2O_5,nH_2O
xerogels exhibit a series of 00ℓ peaks corresponding to a one-
dimensional interlayer spacing arising from the stacking of
the V_2O_5 ribbons along a direction perpendicular to the
substrate (14). The basal spacing then depends on the water
content, and therefore on the partial water pressure above the
sample. At low water content (n < 5), the basal spacing d
varies by steps of about 2.8 Å each, corresponding to the
intercalation of 1 water layer. This stepwise process is
observed up to 3 water layers; beyond this value (for n > 5),
a continuous swelling occurs leading to a more or less viscous
gel.

The formation of ordered aggregates in transition metal
oxide gels opens new possibilities for making ceramics. A
careful shrinkage of such aggregates, before sintering, would
lead to a material in which all individual particles are mutually oriented. This could be very interesting for magnetic
ceramics or anisotropic semiconducting coatings.

4. ALL-GEL DEVICES

Gels are usually considered as intermediates in the processing of glasses and ceramics. Therefore, many studies are
made in order to get a better understanding on how a gel can
transform into a monodisperse powder or a monolithic glass.
We would like to show that gels could also be considered as
materials and used as such for making micro-ionic devices.

Gels are basically biphasic systems containing a solvent
trapped into a solid network. We would then expect properties
arising from both phases and, moreover, the interface between
the liquid and solid being very large, unexpected interface
properties could also be observed.

Transition metal-oxide gels therefore exhibit specific properties (15):

~ Electronic properties ariging from the hopping of unpaired electrons between metal ions in different valence states, in the solid particles.

~ Ionic properties arising from proton diffusion through the liquid phase when it is water.

Several applications of thesegels have already been described. They take advantage of the special viscosity of the gel state. Large coatings can therefore be easily deposited at room temperature on any kind of substrate.

Antistatic coatings have been obtained from V_2O_5 gels. These gels are mixed conductors where both electronic and ionic conductivity is observed depending mostly on the water pressure above the sample. A rather high conductivity of about 10^{-2} Ω^{-1} cm^{-1} is observed under ambient conditions.

Electrochromic layers have also been made from colloidal WO_3. They reversibly switch from white to blue upon applying a voltage of about ± 2 volts.

Many other properties and applications could be described. It is even possible to make an "all-gel" device where all active compounds are gel layers deposited one on top of the other. An electrochromic display device, for instance, could be obtained by using:

- An I.T.O., or doped, SnO_2 gel for making a transparent conductive layer

- An electrochromic WO_3 layer

- An electroytic gel made, for instance, from an hydrous oxide.

Such devices have already been made in the laboratory. They open a wide range of new possibilities in the field of microionics.

REFERENCES

1. H. Dislich and P. Hinz, J. Non-Crystalline Solids, 48, 11 (1982).

2. L. C. Klein, Ann. Rev. Mater. Sci., 15, 227 (1985).

3. L. L. Hench, in "Ultra structure processing of ceramics, glasses, and composites," ed. L. L. Hench and D. R. Ulrich (Wiley-Interscience), 1984.

4. S. Doeuff, M. Henry, C. Sanchez, and J. Livage, J. Non-Crystalline Solids, 89, 206 (1987).

5. J. Catterick and P. Thornton, Adv. Inorg. Chem. and Radiochem., 20, 291 (1977).

6. J. Livage and M. Henry, in "Ultrastructure Processing of Advanced Ceramics", J.D. Mackenzie and D.R. Ulrich, Eds. John Wiley, (in the press).

7. R. T. Sanderson, J. Chem. Ed., 36, 507 (1959).

8. D. W. Schaefer and K. D. Keefer, Phys. Rev. Lett., 53, 1383 (1984).

9. D. W. Schaefer and K. D. Keefer, Phys. Rev. Lett., 53, 1383 (1984).

10. P. Pieranski, Contemp. Phys., 24, 25 (1983).

11. H. Zocher and W. Heller, Zeit. Anorg. Allgem. Chemie, 186, 75 (1930).

12. J. J. Legendre and J. Livage, J. of Colloids and Interface Science, 94, 75 (1983).

13. W. Heller, in Polymer Colloids II, ed. Fitch (Plenum Press, New York), 353 (1980).

14. J. J. Legendre, P. Aldebert, N. Baffier, and J. Livage, J. of Colloids and Interface Science, 94, 84 (1983).

15. J. Livage and J. Lemerle, Ann. Rev. Mater. Sci., 12, 103 (1982).

THE SOL-GEL PROCESS: PRESENT AND FUTURE

J. LIVAGE
Spectrochimie du Solide, Université Pierre et Marie Curie
4, place Jussieu, 75230 Paris, Cedex 05, France

ABSTRACT

The revival of the sol-gel process began about 15 years ago. Many results have been published since then, but the future of this technology depends mainly on whether it will be able to make better materials or even completely new materials. Therefore, a real process mastery is required from both a scientific and technologic point of view. The chemistry of molecular precursors and inorganic polymerization reactions has to be studied in detail. Accurate information will not be obtained until many characterization techniques are really adapted to the specific problems of gels and colloids. The physical and chemical parameters governing aggregation and gelation have to be studied in order to understand the sol-to-gel-to-materials conversion Despite some remarkable results, sol-gel science is still in its infancy. Its development requires the collaboration of scientists coming from different fields: ceramists, chemists, physicists, etc.

INTRODUCTION

The sol-gel synthesis of glasses and ceramics is a two-step process. The first one is based on hydrolysis and condensation of molecular precursors. The second one requires drying and densification of gels in order to obtain a bulk material. Both steps have been known for a long time. Synthesis of oxide gels from alkoxides were already reported during the last century. The ancient art of pottery was based on firing of gels made from clay and water. The uniqueness of the socalled "sol-gel" process is to go all the way from the molecular precursor to the product, allowing better control of the whole process and the synthesis of "tailor-made" materials. In this discussion, we shall discuss the state-of-the-art and then look forward in order to define what progress is necessary so that the sol-gel process can compete with others.

THE SOL-GEL PROCESS: STATE OF THE ART

The first sol-gel synthesis of SiO_2 from silicic acid ester was reported more than a century ago [1]. Sol-gel dip-coatings were patented for the first time in 1939 [2]. However, the real revival of the sol-gel process started around 1970 with the work of R. Roy [3] and H. Dislich [4]. Nowadays an increasing interest, both from academic and industrial laboratories, is devoted to this new technology, and several products have already appeared on the market: thin-film coatings on glasses [5] and

R. M. Laine (ed.), Transformation of Organometallics into Common and Exotic Materials: Design and Activation, 255–260.
© *1988 by Martinus Nijhoff Publishers.*

fibers [7]. Let us first try to summarize the most significant results obtained during the last decade.

Most work was devoted to monolithic glasses made through hydrolysis-condensation of silicon alkoxides [7] or destabilization of colloidal silica [8]. The main advantages of these processes are the high purity of the raw materials, especially useful in preparing fiber-optic preforms, and the fact that melting can be avoided, which leads to unusual noncrystalline solids [9]. The main drawback comes from the formation of cracks on drying. Two methods seem to be able to overcome this drawback: one is hpercritical drying [10], and the other, drying control chemical additives (D.C.C.A.) [11].

Ceramics have an especially good prognosis in the production of submicronic powders of controlled size and morphology [12,13]. Many interesting results can be found in the literature describing the synthesis of monodispersed colloids either from inorganic [14] or metal-organic [15] precursors. Porous materials in which the pore size, pore volume, and surface area can be controlled, have also been successfully obtained via the gel route. Current examples include selective membranes for ultrafiltration [16], catalytic surfaces for coal liquefaction [17], and AR coatings [18].

Coatings appear to be one of the most promising applications of the sol-gel process. Optical films deposited on glass windows by dip-coating are already commercially available from Schott [5]. Potential applications may be in the area of electro-active coatings. Today possibilities include semiconducting V_2O_5 [19], ferroelectric P.L.Z.T. [20], barium titanate [21], and electrochromic WO_3 [22].

Fibers can be spun directly from gels at room temperature without adding a viscosity-increasing agent. Al_2O_3 and $SiO_2-Al_2O_3$ fibers are already commercially available from 3M and Sumitomo. Fibers from other systems (SiO_2, SiO_2-TiO_2, SiO_2-ZrO_2) have also been reported in the literature [23].

This brief survey of the literatuyre shows us the wide range of applications based on the sol-gel process. Let us now look toward the future.

THE SOL-GEL PROCESS; PROBLEMS AND PROGRESS

Many significant results have been obtained during the last decade. Several products, such as fibers or coatings, are already on the market. However, the sol-gel science is still in its infancy, and a lot of progress has to be achieved before the sol-gel process becomes really competitive.

The Chemistry of the Sol-Gel Process

The sol-gel process is based on inorganic polymerization chemistry. The literature contains numerous descriptions on how to make monolithic gels or monodispersed colloids. However, the lack of scientific understanding is still real.

Most molecular precursors are alkoxides. Their chemistry has already been studied in detail [24]. However, even in solution, these alkoxides exit as oligomers rather than monomers. We really need to know their exact molecular structure as a function of the alkoxy group, the concentration,

and the solvent. Very little information is available, and the structure
of the molecular precursors still remains a matter of controversy.

Chemical additives are often used in order to improve the process.
They can be catalysts [7] or D.C.C.A. [11]. They play a very important
role, and presumabley react chemically with the precursor at a molecular
level, leading to new precursors [25]. The whole hydrolysis-condensation
process is therefore completely modified. Almost nothing is known today
about these chemical reactions and the nature of the chemically modified
precursors.

Hydrolysis and condensation are involved in the formation of polymeric
species, but a real knowledge of their mechanisms has yet to be reached.
Simple systems used as models should be studied in more detail. This would
be very difficult because both hydrolysis and condensation occur at the
same time. They can hardly be separated experimentally.

A real science of inorganic polymerization and inorganic polymers has
to be developed. However, experimental techniques that allow unambiguous
characterization of condensed species in a solution are not yet available.
A great deal of work has to be done in order to adjust the existing
techniques to this specific problem. The ideal technique should be able to
give "in situ" information of condensed species in the solvent. It should
also be valid all the way, from small molecular species to colloids and
solids. It should at least be able to give information on the dynamics of
the system. We usually have to deal with complicated systems in which
different species are in constant evolution. Even once a gel is formed,
ageing continues and drastic chemical and structural modifications are still
observed. An increasing number of publications appear in the literature,
reporting F.T.I.R., Raman, NMR, E.X.A.F.S... experiments. Such work has to
be encouraged and diversified.

At last, most studies are performed on oxides obtained from alkoxides.
New precursors and new materials should be studied. Progress has already
been made in the direction of sulfides and nitrides [26][27]. Hydrolysis can
be replaced by sulfidolysis or ammonolysis. The field of halide glasses
could also benefit from development of halogen chemistry leading synthesis
of fluoride gels [28].

The technology of the sol-gel process

The future of sol-gel technology depends mostly on the answer to the
two following questions [29].
. Is it possible to make existing products better and more cheaply ?
. Is it possible to imagine new products accessible only through this
  technology ?

. Better ceramics through chemistry

This is the name given to a symposium of the Materials Research
Society. It gives the right answer to the first question. Sol-gel technology
should, at least in principle, allow complete control of the process all the
way, from the molecular precursor to the ceramic or the glass. This would
lead to "tailor-made" materials in which, not only the stoichiometry, but
moreover the "ultrastructure" can be controlled. This of course requires a
real process mastery specially during the sol to gel and gel to glass or
ceramic conversion [30-32].

Aggregation and gelation occur that dictate the ultrastructure of the resulting material. At the present time, it is not clear if any of the available models adequately describe concentration induced aggregation and gelation [33]. Again, experimental techniques have to be developped in order to study these processes and determine the main physical and chemical parameters. Recently, Small Angle X-Ray Scattering (S.A.X.S.) and fractal analysis have been used with success at Sandia [34].
Such studies should be encouraged. It is now known that chemical additives [11] or hypercritical drying [10] lead to monolithic glasses. The role of the pore size and pore distribution has to be elucidated, together with a deeper understanding of the differential stresses created by the departure of the solvent [28].

. New products

It would be impossible to predict today what new products will be obtained via the gel route. Most of them will be completely unexpected. A survey of the literature could, however, give us some useful guidelines.

The sol-gel process avoids melting. Unusual non-crystalline solids can therefore be obtained either by extending the normal glass formation limits or by obviating problems of phase separation. Examples of these are given in a recent paper of J.D. Mackenzie |9|.

Unusual microstructure can also be obtained leading to new composites. Diphasic xerogels in which small metal particles are dispersed into a ceramic matrix have recently been synthetized by R. Roy [35]. Porous silica derived from gels can be impregnated with PNMA. The resultant composites have interesting mechanical properties. Porous silica containing SiC as a dispersed phase and then impregnated with organic polymers have even more interesting properties [9].

Mixed organic-inorganic gels, so-called Organically Modified Silicates (ORMOSIL) have been synthetized by H. Schmidt [36]. They are made from organo-metallic silicon derivatives exhibiting non-hydrolyzable Si-C bonds. Such materials open a completely new field for organic-inorganic plastics that could exhibit interesting mechanical and physical properties. Some of then have already been patented as soft abrasive powders, anti-scratch coatings or contact lenses [37].

Many completely unexpected products will appear within the next years taking advantages of the uniqueness of the sol-gel process. We can already predict that at least some of them will have real commercial applications.

REFERENCES

1. Ebelmen, Ann. 57, 319 (1846).

2. W. Geffcken and E. Berger, Dtsch. Reichspatent 736 411 (1939), Jenaer Glaswerk Schott and Gen, Jena.

3. R. Roy, J. Am. Ceram. Soc., 52, 344 (1969).

4. H. Dislich, Angew. Chem. Int. Ed. Engl. 10, 363 (1971).

5. H. Dislich and P. Hinz, J. Non-Cryst. Solids, <u>48</u>, 11 (1982).

6. 3M Co. Catalog, "Nextel Ceramic Fiber".

7. L.C. Klein, Ann. Rev. Mater. Sci., <u>15</u>, 227 (1985).

8. E.M. Rabinovich, D.W. Johnson, J.B. MacChesney and E.M. Vogel, J. Non-Cryst. Solids, <u>47</u>, 435 (1982).

9. J.D. Mackenzie, J. Non-Cryst. Solids, <u>73</u>, 631 (1985).

10. J. Zarzycki, M. Prassas and J. Phalippon, J. Mater. Sci., <u>17</u>, 3371 (1982).

11. L.L. Hench, in Science of Ceramic Chemical Processing, p.52, L.L. Hench and D.R. Ulrich Ed., J. Wiley (1986).

12. Int. Conf. on Ultrastructure Processing of Ceramics, Glasses and Composites (1983), Gainescille, Floride, USA, Ed. L.L. Hench and D.R. Ulrich (Wiley, New-York 1984).

13. Symposium "Better Ceramics Through Chemistry" Materials Research Society 1984, Spring Meating, Albuquerque, New Mexico, USA.

14. E. Matijevic, Acc. Chem. Res. <u>14</u>, 22 (1981).

15. E.A. Barringer and H.K. Bowen, J. Am. Ceram. Soc., <u>65</u>, C-199 (1982).

16. A. Larbot, J.A. Alary, J.P. Fabre, C. Guizard and L. Cot, in "Better Ceramics through Chemistry II", Ed. C.J. Brinker, D.E. Clark and D.R. Ulrich (Mat. Res. Soc. Pittsburgh, P.A. 1986).

17. H.P. Stephens, R.G. Dosch and F.V. Stohl, Ind. I and D. Eng. Chem. Prod. Res. and Dev., <u>24</u>, 15 (1985).

18. R.B. Pettit, C.S. Ashley, S.T. Reed and C.J. Brinker, to be published in "sol-gel technology" Ed. L.C. Klein, Noyes, Park Ridge, NJ, (1987).

19. C. Sanchez, F. Babonneau, R. Morineau, J. Livage and J. Bullot, Phil. Mag. B., <u>47</u>, 279 (1983).

20. K.D. Budd, S.K. Day and D.A. Payne, Brit. Ceram. Soc. Proc., <u>36</u>, 107 (1985).

21. R.G. Dosch, in "Better ceramics through Chemistry" p.157, Ed. C.J. Brinker, D.E. Clark, D.R. Ulrich, (North-Holland 1984).

22. A. Chemseddine, R. Morineau and J. Livage, Solid State Ionics, <u>9-10</u>, 357 (1983).

23. S. Sakka, Am. Ceram. Soc. Bull., <u>64</u>, 1463 (1985).

24. D.C. Bradley, R.C. Mehrotra and D.P. Gaur, Metal Alkoxides, Academic Press, New-York, (1978).

25. S. Doeuff, M. Henry, C. Sanchez and J. Livage, J. Non-Cryst. Solids, <u>89</u>, 206 (1987).

26. R.R. Chianelli and M.B. Dines, Inorg. Chem., 17, 2758 (1978).

27. C.J. Brinker, Comm. Am. Ceram. Soc., C-4 (1982).

28. D.R. Ulrich, Ceramic Bull., 64, 1444 (1985).

29. J. Wenzel, J. Non-Cryst. Solids, 73, 693 (1985).

30. C.J. Brinker and C.W. Scherer, J. Non-Cryst. Solids, 70, 301 (1985).

31. C.J. Brinker, E.P. Roth, G.W. Scherer and D.R. Tallant, J. Non-Cryst. Solids, 71, 171 (1985).

32. C.J. Brinker, G.W. Scherer and E.P. Roth, J. Non-Cryst. Solids, 72, 345 (1985).

33. J. Zarzycki, in "Science of Ceramic Chemical Processing", p.21, Ed. L.L. Hench and D.R. Ulrich, (Wiley, New-York, 1986).

34. K.D. Keefer, in "Science of Ceramic Chemical Processing", p.14, Ed. L.L. Hench and D.R. Ulrich, (Wiley, New-York, 1986).

35. R.A. Roy and R. Roy, Mat. Res. Bull., 19, 169 (1984).

36. H. Schmidt, J. Non-Cryst. Solids, 73, 681 (1985).

SOL-GEL DERIVED THIN FILMS: CRITICAL ISSUES[*]

C. J. BRINKER
Sandia National Laboratories, Albuquerque, New Mexico 87185-5800

ABSTRACT

The critical physical and chemical issues concerning sol-gel-derived thin films are reviewed. The fundamental physical issues which are currently poorly understood are concentration-induced aggregation and gelation and the physics of drying porous media. These phenomena depend in turn on chemical issues such as the relative rates of condensation and evaporation during film deposition. Preliminary experiments demonstrate the tailoring of thin film structure by controlling precursor polymer structure prior to deposition and the relative rates of condensation and evaporation during deposition.

I. INTRODUCTION

The sol-gel process for making glass is now an intensely studied topic in materials science. Within just the last five years seven symposia have been devoted largely to this subject [1-7]. Presumably, this interest reflects the numerous, diverse technological applications perceived for gel-derived glass products. Although thin films and coatings comprise the largest potential source of gel-derived products, the physics and chemistry of film formation from solution has been vitually ignored. This report outlines the current level of understanding of gel-derived thin films, enumerates the most important physical and chemical issues yet to be resolved and summarizes the results of studies currently in progress designed to elucidate the physics of film formation.

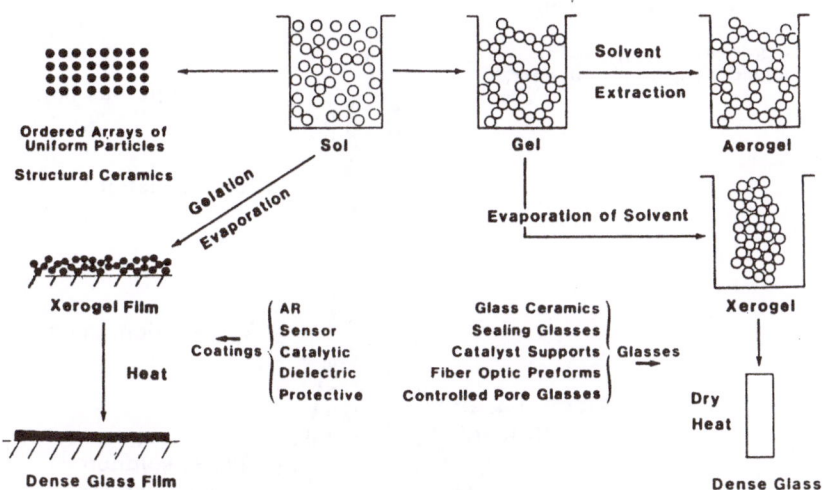

Fig. 1   Schematic of the sol-gel glass forming process

*This work performed at Sandia National Laboratories supported by the U.S. Department of Energy under contract number DE-AC04-76DP00789.

R. M. Laine (ed.), Transformation of Organometallics into Common and Exotic Materials: Design and Activation, 261–278.
© 1988 by Martinus Nijhoff Publishers.

I.1 Sol-Gel Processing

Sol-gel processing of glass and ceramics uses metal alkoxides of network forming cations (e.g. $M(OR)_x$ where M is Si, B, Ti, Al, etc., and R is often an alkyl group) as soluble, ceramic precursors. In alcohol/water solutions the alkoxide groups are removed stepwise by acid or base catalyzed hydrolysis reactions and replaced with hydroxyl groups (Eq. 1). Subsequent condensation reactions involving the hydroxyl groups yield macromolecular solution species composed of inorganic oxide linkages (Eqs. 2 and 3).

$$Si(OR)_4 + H_2O \rightarrow (RO)_3Si-OH + ROH \qquad (1)$$

$$2 (RO)_3 Si-OH \rightarrow (RO)_3Si-O-Si(OR)_3 + H_2O \qquad (2)$$

$$(RO)_3Si-OH + Si(OR)_4 \rightarrow (RO)_3Si-O-Si(OR)_3 + ROH \qquad (3)$$

Gelation typically occurs by a secondary process in which these primary species link together to form a network extending throughout the solution. At the gel point, physical properties of the system such as viscosity and elastic moduli diverge. Because virtually all elements of interest to ceramists are available as metal alkoxides [8], metal organics and/or soluble salts, sol-gel processing is potentially adaptable to any ceramic system of interest.

Bulk gels (commonly called monoliths) are made by casting a solution in a mold followed by gelation and drying (see Fig. 1). If drying occurs by evaporation of the solvent, the network collapses somewhat and the resulting product is termed a xerogel. If the solvent is removed under supercritical conditions, shrinkage is minimized and the resulting expanded gel is termed an aerogel.

Fig. 2 Schematic of the
        dip coating process.
        Regions corresponding
        to concentration,
        aggregation, gelation
        and drying are indicated.

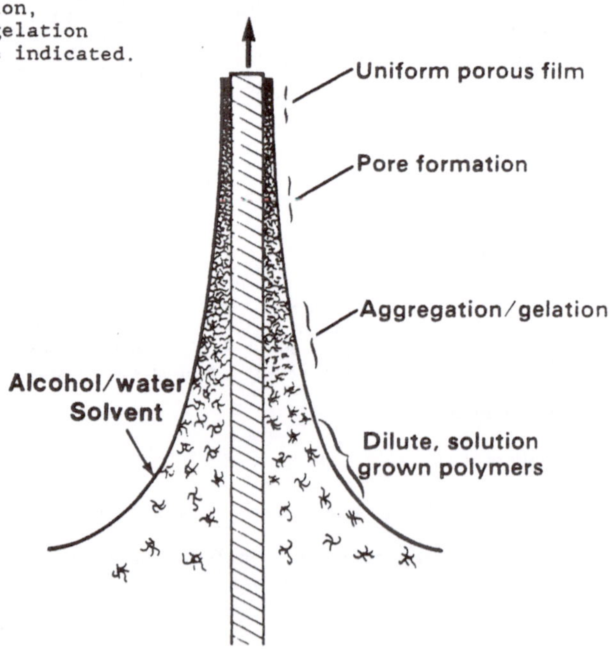

Uniform porous film

Pore formation

Aggregation/gelation

Alcohol/water Solvent

Dilute, solution grown polymers

Prior to gelation the solution (often in a more dilute form) can be deposited on ceramic, metal, semiconductor, and/or plastic surfaces by a wide range of methods including: dipping, spinning, spraying, electrophoresis, and thermophoresis. The dipping process is represented schematically in Figure 2. Briefly stated, the major difference between gelation in bulk and in thin film systems is that for films, gelation is induced by rapid evaporation of solvent. In seconds, rather than days, the solution structure is transformed to a gel and porous xerogel. Thus the relative rates of hydrolysis, condensation, esterification, AND EVAPORATION dictate the resulting film structure. This will be discussed in more detail below.

II. CRITICAL ISSUES

II.1 Applications

Critical issues in sol-gel processing of films are a consequence of their current and potential uses. Thus, a brief summary of film applications will serve as a preface to this section. Applications for gel-derived films may be loosely divided into four categories: protective, optical, electrical and porous. Protective films provide corrosion or abrasion resistance to the underlying substrate. Current examples include oxidation protection for black chrome solar selective absorbers [9], and silicon [10], anti-scratch coatings for acrylics [11], and corrosion-resistant overlayers and interlayers for flexible solar mirrors [12]. Potential future applications are limitless.

Optical coatings alter the reflectance, transmittance, or absorption of the substrate. IROX (TiO_2/Pd) coated architectural glass is the best example of the present state of the art [13]. Additional examples include antireflective (AR) coatings on glass [14-16], silicon [17], and plastics [18], and highly reflective multi-layers. Potential future applications include thin film waveguides, photochromics [19], and oriented birefringent films.

Electrical films may be further subdivided into electrically passive or active categories. Passive films include spin on glasses(SOG) which have been developed as interlayer [20], planarizing [21], or low temperature alternatives to native thermal SiO_2 in MOS technologies. A future application, which could dramatically impact the microelectronics industry, is passivation of compound semiconductors, e.g., GaAs or InP, which have poor, mixed native oxides [22]. Perhaps the largest potential application of sol-gel derived thin films is in the area of electro-active films. Today, applications include semiconducting vanadium pentoxide [23], ferroelectric PLZT [24], barium titanate [25], and electrochromic tungsten oxide [19] films. Future applications are expected to include electro-optic materials for shutters, windows, and switches in integrated optic devices, and electrostrictive materials for thin film transducers useful, for example, in surface acoustic wave devices (SAWS).

Porous films constitute those in which the pore size, pore volume, surface area and/or surface reactivity are tailored to achieve specific goals. Current examples include selective membranes for ultrafiltration [26], catalytic surfaces for coal liquifaction [27], and controlled pore AR coatings [18]. Future applications are envisioned in ultra-low dielectric constant films for MOS devices and in thin film sensor devices in which, by controlled accessibilty or reactivity, the resistivity, capacitance, modulus, or weight of the porous film is changed by selective absorption and/or reaction with specific environmental molecules.

II.2 Physical Issues

We have large gaps in our knowledge concerning the pertinent physics which describes deposition and subsequent consolidation of polymeric solutions species on substrate surfaces. The physics of film formation from solution underlies all applications of gel-derived thin films, because, by analogy to bulk systems [28], it is expected to dictate the film microstructure (pore size, surface area, pore volume, extent of crosslinking, etc.) and consequently the film densification behavior.
Some insight into the film formation problem may be gained by re-consideration of Fig. 2. We first must consider the structure of the polymeric species which, together with their solvent, comprise the coating solution. Small angle x-ray scattering (SAXS) results obtained by Schaefer and co-workers [29-31] indicate that, depending on the conditions of hydrolysis and condensation, polysilicate species formed in solution can vary from weakly branched linear polymers to discrete porous clusters to fully dense colloidal particles. It is possible to distinguish between some of these various structures by power law analyses of the scattered intensity I(K) as a function of the scattering vector K

$$I(K) \propto K^{-X} \tag{4}$$

where for uniform Euclidian objects (e.g., colloidal particles) $X=4$ and for mass fractal objects, $X=D$, the fractal dimension, which in three dimensions is less than 3 [30]. Fractal objects differ from Euclidian objects in that they have dilation symmetry, i.e., if a portion of the object is magnified, the resulting magnified structure looks identical to the original structure. D relates the molecular weight, M, to the radius, r:

$$M \propto r^{D} \tag{5}$$

Thus, for fractal objects density decreases radially according to:

$$\rho = 1/r^{3-D} \tag{6}$$

Although D quantifies the random object according to Eqs. 5 and 6, D unfortunately is not unique to a particular growth process or topology. For example, ideal linear polymers, swollen branched polymers, and lattice animals have fractal dimensions of 2, while both percolation clusters and diffusion-limited aggregates have fractal dimensions of 2.5 [30].
Regardless of the form of the initial solution species (e.g., chains, porous clusters, or colloids), during film deposition, the initial liquid-like structure is transformed to a highly condensed, porous xerogel. This film forming process certainly involves at least four physical phenomena: 1) Upon initial evaporation of solvent the solution species are concentrated. By analogy to organic polymer systems this is similar to changing from a good to a theta solvent [32]. For polymeric species (characterized by D less than 3) this is expected to result in a collapse or compaction of the original solution structure (increased D) [33].
2) Further solvent removal induces rapid aggregation as solution species are forced into close proximity. At present it is not clear if any of the available aggregation models adequately describe concentration-induced aggregation. Conventional wisdom suggests that this process should result in structures similar to cluster-cluster aggregates [34] (in three dimensions D=1.75-1.8 [35]), however, Kolb and Hermann dispute this idea [36]. For any of the possible aggregation models, the ultimate aggregate structure depends on the "sticking coefficients", e.g., the probability, P, that an encounter will result in bonding. If the sticking coefficient is high (P approaches 1), individual clusters are prevented from interpenetrating, and very open aggregate networks result (reduced D). Conversely, if the probability of sticking is low, porous clusters might be

expected to completely interpenetrate resulting in more compact structures (increased D). Colloidal silica aggregates with varying sticking coefficients (induced by varying pH) have been investigated by light scattering [37]. The light scattering results show that the resulting fractal dimension in solution varies from 1.77 (high sticking coefficient) to 2.08 (low sticking coefficient). A pertinent question regarding film formation is: Can interpenetration occur in the time scale of the coating process (several seconds)? Preliminary experiments conducted in our laboratory provide an answer to this question. These experiments are described in section III.1.

3) A third phenomenon induced by evaporation during the deposition process is gelation. According to theories proposed by Stauffer [38] and de Gennes [39], gelation might be expected to occur by percolation of individual aggregates. If so, the fractal dimension of the system at the gel point would be the fractal dimension of percolation clusters at the percolation threshold. However, recent two-dimensional simulations of the sol-gel transition modelled by irreversible aggregation of clusters yield D_{gel}=1.75 [40]. Kolb and Herrmann state that their cluster aggregation model results in structures noticeably distinct from both flocculation clusters and percolation clusters; however, for the case of irreversible aggregation (sticking probability = 1), the concentration of clusters cannot exceed the percolation threshold p_c. Therefore, with regard to film formation, the following questions arise: As evaporation of solvent proceeds, what is the critical concentration at which gelation occurs? And, is concentration-induced gelation adequately described by any of the existing models? The answer to the first question must certainly depend on the relative rates of sticking and evaporation. With regard to the second question, to my knowledge there have been no theoretical or experimental studies of concentration-induced gelation.

Both the aggregation and gelation processes are further perturbed by shear stresses which accompany both dip- and spin-deposition processes. Because the solution viscosity is shear-rate dependent near the gel point, some restructuring of the aggregate occurs during deposition depending on the extent of shearing. This might provide the system with more opportunities to break and reform; however, it is unclear whether this process would enable the system to "anneal" toward a more thermodynamically favored structure (increased D).

4) At the gel point the viscosity and elastic modulus of the film increase abruptly. Further removal of solvent requires the remaining liquid to flow through channels within the gel network. This drying process is initially accompanied by compaction of the solid phase which prevents the formation of high energy, liquid-gas interfaces within the gel (cavitation). However, because the skeleton can continue to condense as reactive groups confront one another, the viscosity and modulus of the network continue to increase. After a critical extent of crosslinking, the gel network can support the hydrodynamic compressive force created by liquid-gas menisci. At this point porosity invades the film. Scherer has recently modeled the drying of a gel film attached to a non-deformable substrate [41]. His model can account for both the change in the viscosity of the network with continued condensation and the permeability of the network to the evaporating solvent; unfortunately, these properties have not as yet been determined experimentally. A major finding of the model is that if the film is rigidly attached to the substrate, i.e., there is no strain in the plane of the substrate, the dried film is always under tensile stress. To begin to understand the structural manifestations of the drying process, we need to know both the relative rates of condensation and evaporation (see Section II.3), and the manner in which crosslinking increases the viscosity and modulus. From a theoretical standpoint there are no models which describe the compaction of fractal aggregates, percolation clusters, etc. In addition, there are no published reports of experiments designed to follow the structural evolution of gel-derived

films during drying. Finally, it is unclear at what point the film becomes rigidly attached to the substrate.

The above discussion outlines our meager understanding of the physical processes underlying film deposition from solution. However, this is only the tip of the iceberg. Based on extensive studies of the densification of bulk gels [28,42,43], we may rationally extrapolate to films and conclude that the densification behavior will depend markedly on the microstructure, extent of crosslinking, excess free volume, etc., which in turn depend on the deposition process. Film densification occurs one-dimensionally normal to the substrate which increases the level of tensile stresses in the plane of the film [44]. Tensile stress is in turn expected to retard the densification kinetics [44], further complicating the densification problem. Structure and stress is also expected to influence the crystallization behavior, etc. Thus we have erected a house of cards: All subsequent steps depend on the previous step about which we know little.

II.3 Chemical Issues

Many of the critical chemical issues surrounding film formation are inseparable from the physical issues outlined above. Clearly, the structures of polymeric species comprising the coating solution as well as the structure of the deposited film depend largely on the sticking coefficient which, defined in chemical terms, is proportional to rates of Eqs. 2 and 3. From countless experiments conducted on silica systems it is apparent that the condensation rate (as reflected in the gelation time) can be varied over several orders of magnitude by variation of the pH alone [45]. In addition, Keefer [46] and Schaefer, et al., [45,47] have shown that by varying the relative rates of condensation and hydrolysis and/or the functionality of the precursor molecules, a wide spectrum of solution species ranging from dense colloids to mass fractal clusters to surface fractals can be generated. For any particular polymeric precursor structure (chains, clusters or colloids), the relative rates of condensation and evaporation during deposition largely define the resulting film structure. For example, if during the deposition process the condensation rate is large with respect to the evaporation rate, the aggregation process will be analogous to diffusion-limited aggregation except that Brownian diffusion is augmented significantly by solvent evaporation. The result will be open structures far from equilibrium. If the converse is true, the aggregation process will be reaction-limited, and compact, ordered structures result. For example, Budd, et al. [48], found the deposition of lead titanate films under reaction-limited conditions resulted in partially crystalline films compared to completely amorphous films formed under conditions where evaporation was expected to be rate limiting. Schmidt and co-workers found that by partially replacing alkoxide groups with non-hydrolyzable and non-condensable alkyl groups, the resulting organically-modified films were so compact that they exhibited no measurable surface area [49]. Nelson, et al., deposited colloidal silica films under conditions in which aggregation was diffusion-limited and obtained porous films useful as catalytic surfaces, while in the opposite case much denser films were obtained useful for protective coatings [50]. All of these results are explained on the basis of the relative rates of evaporation and condensation. In general, as the condensation rate is reduced with respect to the rate of evaporation, the reactant species are allowed more opportunities to attain thermodynamically favored condensation sites. Under these conditions, we expect to reduce the surface area and increase the density and extent of order in the film.

Not only are the relative rates of condensation and evaporation important with respect to the structure of the resulting film but also the relative rates of bond breakage (hydrolysis and alcoholysis reactions which occur by the reverse of Eqs. 2 and 3) and evaporation. Bond breakage allows restructuring to take place concurrently with deposition and is enhanced preferentially at strained or chemically unstable sites. Thus bond breakage and reformation accompanying film deposition are expected to anneal imperfections and move the system toward thermodynamic equilibrium. Chemical restructuring has been treated in several theoretical and experimental studies [51,52]. It was concluded that preferred sites for restructuring exist due to differing reactivity or differing accessibility to reactant molecules. Sharp features at perimeter sites are most likely to undergo dissolution and reformation leading to a general smoothing trend.

We should also not neglect the effects of chemical equilibrium implied in Eqs. 2 and 3. If the condensation reaction produces alcohol as a product, the evaporation of alcohol during deposition will shift the chemical equilibrium to the right. However, if condensation involves primarily hydroxyl groups, excess water in solution or in the coating environment is expected to retard the condensation process. An example of this is the study by Glaser and Pantano [53] who deposited silica films with varying H_2O/OR molar ratios (10:1 to 1:1). As the ratio increased, the density of the corresponding film increased, which is contrary to the effect seen in bulk gels. This pheneomenon appears to be explained by the fact that under the short time scale of the deposition experiment, increased water concentrations reduced the rate of condensation with respect to the rate of evaporation resulting in more compact structures.

From the above discussion it is apparent that chemistry strongly influences the physical processes accompanying film deposition. Clearly the question of paramount importance concerning film deposition and resulting structure is: What are the relative rates of condensation, hydrolysis, and evaporation which occur during film deposition? To my knowledge there have been no experiments designed to address this question.

A second chemical issue, which is even more elusive, involves the interaction of the coating solution with the substrate. Perhaps, because under most conditions, sol-gel coating solutions readily wet and adhere to a variety of substrates, there are no perceived problems concerning the coating-substrate interface. Yet, the mere fact that gel-derived films adhere to both oxide and non-oxide surfaces should provoke questions concerning the adhesion mechanism. The rate of attachment of the solution species to the substrate relative to the evaporation rate will determine the level of tensile stress in the dried film. More localized stresses manifested as altered M-O-M bond angles and bond distances result at the immediate interface due to the discontinuity in chemical composition. Presumably the interface is also characterized by additional defects, for example, wrong, missing, or dangling bonds.

A third issue concerns the reaction of the film constituents with environmental molecules during film deposition or during subsequent thermal treatments. Schott's large scale IROX deposition facility relies on the deposition ambient to promote hydrolysis of titanium alkoxide solutions during the coating process [13]. According to Eqs. 1-3, hydrolysis must precede condensation. Therefore, the rate of condensation during film formation depends on the rate of hydrolysis which in turn depends on the diffusion of water from the coating ambient through the applied film. Although IROX is a well established product, there appears to be no fundamental understanding of how the rate of hydrolysis affects the physical processes involved in film formation.

Reactions with environmental molecules at elevated temperatures have been used to chemically modify both bulk gels and thin films. Most notably, NH_3 has been reacted with porous, SiO_2 and multicomponent silicates to synthesize oxynitride compositions [53-56]. The rate of nitridation increases with increasing Lewis acidity and basicity of the metal and oxygen atoms, respectively [54]. In silicate systems, the Lewis acidity is of course increased by replacement of Si with group III elements (e.g., B, Al, etc.). In a pure SiO_2 system, the Lewis acidity of Si and the Lewis basicity of O increase as the Si-O-Si bond angle is reduced [57].

Cyclic trisiloxanes with reduced Si-O-Si bond angles have been reported for porous silicate gels and high surface area powders dehydrated above 250°C [58]. The formation of these species in silicate films is expected to enhance the nitridation rate. Brow, for example, observed that nitridation of gel-derived silicates increased greatly above 700°C [56]. He concluded that both the high surface area and the high reactivity of the dehydroxylated surface caused gel-derived silica films to incorporate nitrogen more rapidly and at lower temperatures than thermal silica. With regard to films (strongly adhering to a non-deformable substrate), an important question is how tensile stresses resulting from drying and consolidation affect the chemical reactivity. According to Brow, the rate of nitridation of films exceeds that of bulk gels, but he explains that this could also be a consequence of the reduced diffusion distances in films [56].

A final issue which has received little attention is how the coating solution alters the substrate surface. For example, do acidic or basic alcohol/water solutions alter the chemical composition or phase assemblage of the substrate surface? If so, do chemical species originating on the substrate surface redistribute themselves throughout the thickness of the film (generally a few hundred nanometers thick)? The answers to these questions are crucial to the development of gel-derived dielectrics useful as replacements for thermal SiO_2 or passivation layers for III-V compound semiconductors.

There are few published papers which provide insight with regard to the chemical issues summarized above. Smith, et al., used SIMS to depth profile sol-gel layers deposited on silver in a protective application [59]. After application and thermal processing, there was clear evidence for interdiffusion of silver and the coating constituents (Si, B, Al). However, the origin of the interdiffusion, for example, the deposition process or the subsequent heat treatment, was not clearly identified. Pantano and co-workers [60] have used SIMS and Auger spectroscopy to investigate the chemistry occuring during densification and nitridation of silica films deposited on silicon. Although the precise interface chemistry was not clearly resolved, it was apparent that during densification at temperatures up to 1150°C the silicon interface was significantly oxidized. Presumably in a multi-component system the thermal oxide would incorporate components from the deposited film, e.g., B, P, Al, etc. This is the concept underlying spin-on dopants. It has been used in an advantageous manner by Hayashi, et al., [61] to modify the interface of silica films deposited on GaAs substrates.

III. ON-GOING EXPERIMENTS

The preceeding sections presented a brief overview of the critical physical and chemical issues surrounding solution processing of thin films. In this section, I present results of some recent experiments designed to elucidate the roles of solution structure and chemistry in the morphological development of sol-gel-derived thin films. The goal of this study is the tailoring of thin film properties (e.g., porosity, refractive index, pore size, densification temperature, etc.) by controlling polymer growth in solution prior to film deposition.

To date three polysilicate systems have been investigated: I, highly interpenetrating, weakly branched polymers formed near the isoelectric point; II, weakly interpenetrating porous clusters; and III, non-interpenetrating, dense colloidal particles.

III.1 System I

System I silicates are synthesized near the isoelectric point of SiO_2 (pH≃2) where the condensation rate is minimized [62]. This system has been studied extensively both in solution prior to gelation and during densification [28-31]. SAXS investigations indicate that the polysilicate species near the gel point have fractal dimensions, D=2, characteristic of mass fractal objects [31]. Because under the synthesis conditions employed all potential crosslinking points are expected to be equally probable, Schaefer and Keefer interpret D = 2 as indicative of structures called lattice animals [31]. In a percolation process lattice animals are the largest clusters which occur below the percolation threshold [31]. A computer simulated lattice animal is shown in Fig. 3.

Fig. 3   2-D lattice-animal
           percolation-cluster
           simulation after
           Ref. 31.

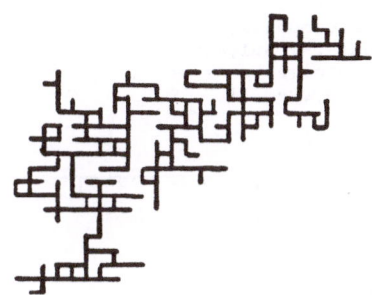

Analyses of the Guinier region of the scattering curves give information concerning the size of the scattering entities. An important conclusion derived from the Guinier scattering regime concerning system I silicates is that the solution species are highly interpenetrating prior to gelation [30]. This result is based on the concentration dependence of the Guinier radius shown in Fig. 4. Dilution of the system (10:1 with ethanol) prior to measurement disentangles the polymers resulting in an apparent increase in size, i.e., the Guinier radius approaches the true radius of gyration when the system is dilute. Conversely, as the system becomes more concentrated, the polymers become highly overlapped and entangled. Interpenetration is the distinguishing feature of system I silicates.

Fig. 4   Guinier radius as
a function of time
from gelation
t_{gel} - t for dilu-
ted and undiluted
system I silicates
[30].

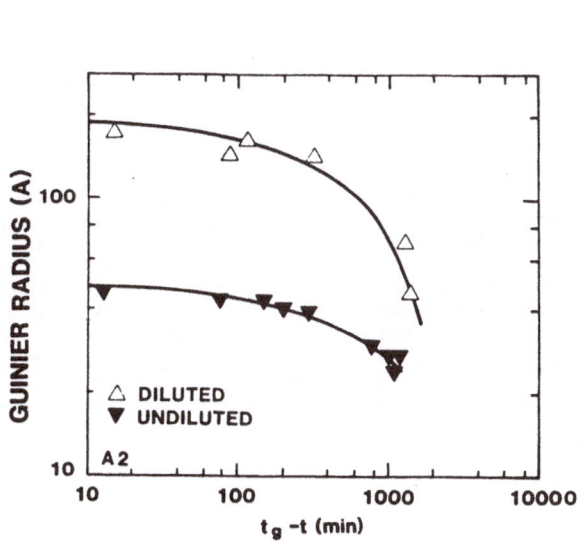

270

With regard to film formation an important consideration is whether or not polysilicates in dilute solution can interpenetrate appreciably during the film formation process. To answer this question, film forming solutions were prepared by growing polymers under concentrated conditions for normalized times $((t_{gel} - t)/t_{gel}$ in Fig. 4) ranging from 0.01 to 0.95 followed by dilution to 0.05 wt%. The solutions were immediately coated on single crystal silicon substrates and the refractive index and thickness were measured ellipsometrically. The results are shown in Fig. 5. Regardless of the initial size of the precursor polysilicate species, the refractive index (and density) of the deposited films are identical and near the theoretical value for fully dense v-SiO$_2$ viz. 1.45. This shows that if the condensation rate is low with respect to the rate of evaporation, the polymers can fully interpenetrate during the course of the deposition process. As we will see, this is contrary to the effect observed in system II.

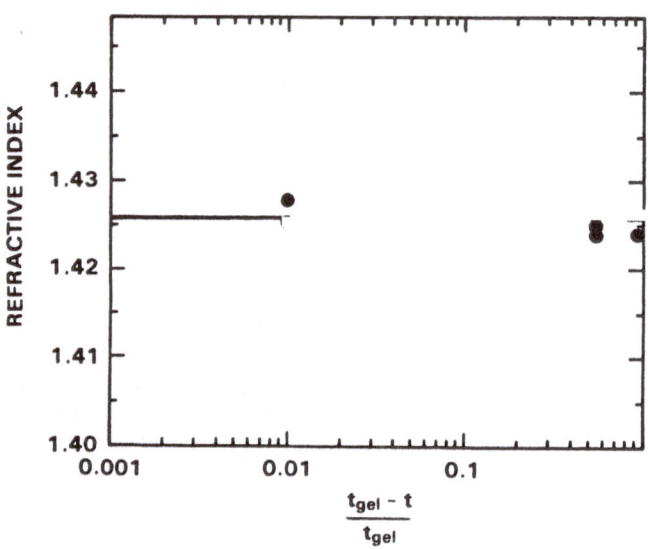

Fig. 5 Refractive index of system I films as a function of the normalized time from gelation ($t_{gel} - t/t_{gel}$).

III.2 System II

System II silicates are formed at intermediate pH where the rate of condensation is significantly increased [63]. Although solution species prior to gelation are again characterized as mass fractal objects (D = 2.0 to 2.5), a distinguishing feature of system II polysilicates is that in solution they apparently do not interpenetrate significantly, and therefore show no dependence the Guinier radius on concentration (Fig. 6), i.e., both the diluted and undiluted solutions exhibit the same Guinier radius. Again the important question regarding film formation is: Do system II silicates interpenetrate significantly during film deposition? To answer this question an experiment was performed similar to the one described above.

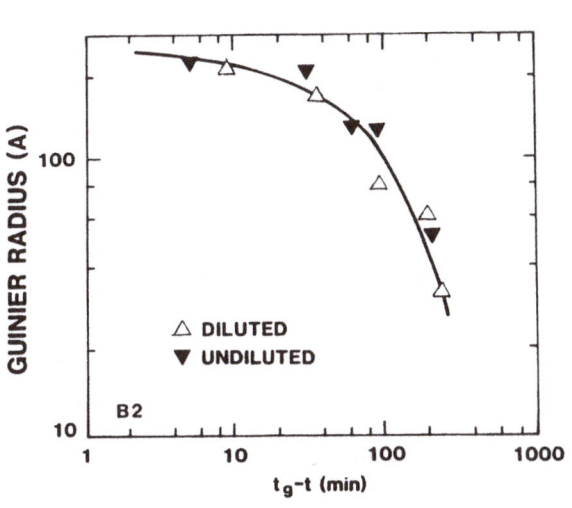

Fig. 6 Guinier radius as a
function of $t_{gel} - t$
for diluted and
undiluted system II
silicates [30].

Multicomponent silicate species were grown in dilute solution at 50°C and
pH 3.5. After various stages of growth, the polysilicate solutions were
deposited on silicon substrates and n and t were measured
ellipsometrically. The results, plotted in Fig. 7 as a function of the
aging time, show an inverse dependence of refractive index on polymer size
prior to deposition: the refractive index decreases (pore volume
increases) with original polymer size. This result is consistent with non-
interpenetrating polymers. Because the original solution species are mass
fractals, the density of each polymer decreases with distance from the
center of mass according to Eq. 6. Thus, if interpenetration does not
occur, the density of an assemblage of polymers should decrease with

Fig. 7 Size of system II
silicates measured by
light scattering prior
to deposition [70] and
refractive index, n, of
the deposited film as a
function of "aging" time
at pH 3.5 and 50°C.

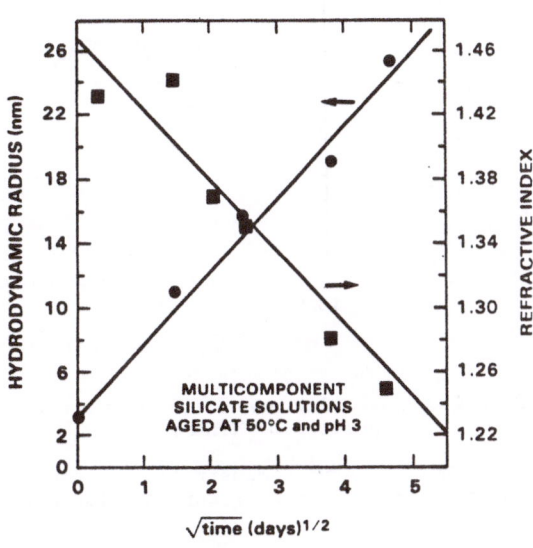

272

polymer size. To confirm this hypothesis SAXS was performed on the coating solution and the resulting film deposited on a weakly x-ray absorbing MYLAR substrate. Solution species are characterized by $D = 2.4$ (Fig. 8). The film is characterized by $D = 2.8$ and $D = 1.7$ for dimensions less than and greater than 14 Å, respectively (Fig. 9). This suggests that on a small dimensional scale (<14 Å) the original structure has become more compact due either to increased concentration or interpenetration. However, on a larger dimension scale the structure has become more ramified. The dimension 1.7 is consistent with cluster-cluster aggregation when the sticking coefficient equals one (see section II.1).

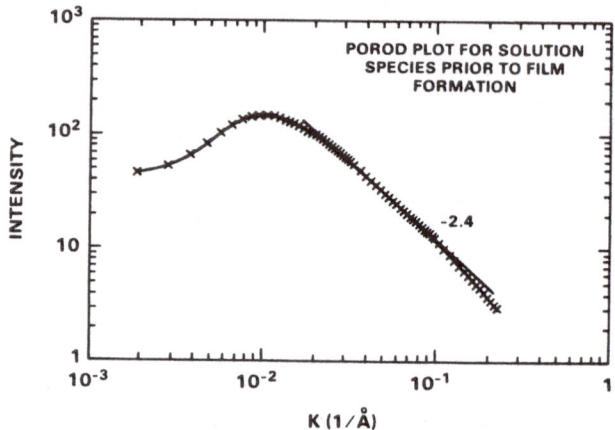

Fig. 8   Power law scattering curve for multicomponent system II silicate solution prior to film deposition.

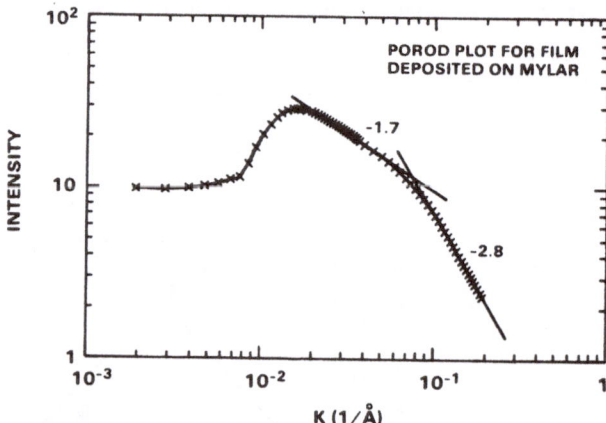

Fig. 9   Power law scattering curve for multicomponent system II film deposited on MYLAR.

Thus, we have demonstrated two completely different effects depending on the relative rates of condensation and evaporation during film deposition. When the condensation rate is reduced with respect to the evaporation rate (System I) compact structures result; conversely, as the relative rate of condensation is increased (System II) more open structures result. As suggested by Fig. 7, this latter behavior can be used to precisely tailor the refractive index.

Fig. 10 demonstrates the effect of film structure on subsequent densification behavior and suggests how to tailor structure for specific applications. Multicomponent solutions were deposited on silicon substrates from both unaged solutions and solutions aged at 50°C for 1 week. Following deposition the substrates were heated to 500°C, cooled, immersed in a fluorosilicic acid etch, dried, and measured ellipsometrically. Prior to etching the "unaged" film shows a significantly higher refractive index than the "aged" film, and during etching the unaged film shows no change in refractive index compared to a monotonically decreasing index measured for the aged film. These results indicate that after the 500°C heat treatment the unaged film contained no open porosity and according to the refractive index was relatively dense. The aged film was initially much less dense and during immersion the etchant solution penetrated the film causing its density to decrease further. The differences in behavior are explained by the dependence of viscous sintering on the pore dimension. During viscous sintering, the shrinkage rate is expected to be proportional to the surface energy divided by the product of viscosity and pore diameter [43]. Unaged films have smaller characteristic pore dimensions (and therefore higher surface areas) both of which enhance the densification rate. Conversely, the porosity in aged films is thermally stable to high temperatures. Thus, depending on the intended use, for example, protective or catalytic coatings, it is possible to tailor the film properties by controlling polymer growth in solution prior to film deposition. Protective versus porous properties are demonstrated in Fig. 10.

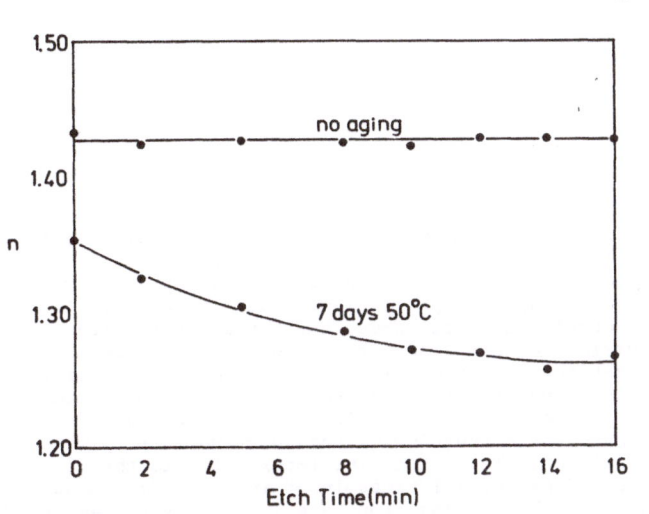

Fig. 10 Refractive index, n, versus immersion time in fluorosilicic acid for a system II film deposited from an "unaged" solution and a solution aged 7 days at 50°C prior to deposition.

III.3 System III

System III solutions are composed of dense, colloidal silica particles prepared by a modification of the Stöber process [63]. A system III film

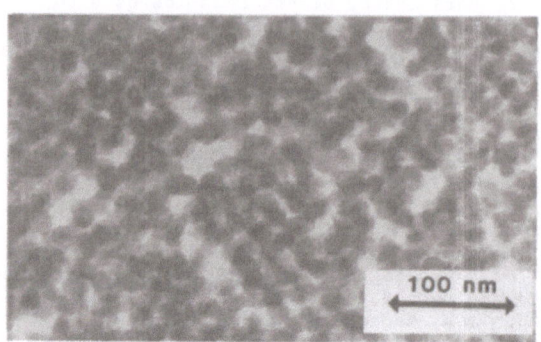

Fig. 11 TEM micrograph for a) system III silicates deposited on a carbon grid and b) multicomponent system II films deposited on silicon. Both films have equivalent pore volumes, (n = 1.25); however, the different scales of structure are clearly evident.

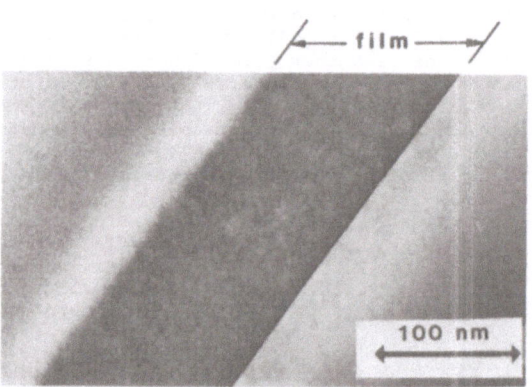

deposited on a TEM substrate is shown in Fig. 11 where it is compared to a system II film (n = 1.25) deposited on silicon in order to demonstrate the extremeley different dimensional scales of microstructure possible by gel processing. The nearly monodisperse spherical particles which comprise the system III films cannot completely fill space, and hence their refractive index is limited to a maximum of n = 1.32 for hexagonal close packing according to the Lorentz-Lorenz relationship [64]. Aggregation prior to film deposition or during deposition is expected to decrease the refractive index further. This is demonstrated in Fig. 12 where the refractive index of system III films is shown as a function of the pH at which they were aggregated prior to deposition. The refractive index reflects the condensation rate (indicated by the gel times inset in Fig. 12). Lower refractive indices result as the condensation rate is increased, because the aggregate structures formed both prior to and during deposition are more open as a consequence of the high "sticking" coefficients. System III should serve as a reasonable model system to investigate the effects of the relative rates of condensation and evaporation during film deposition on the structure of the resulting film. Along with Hurd an in situ light scattering apparatus has been fabricated for this purpose [65].

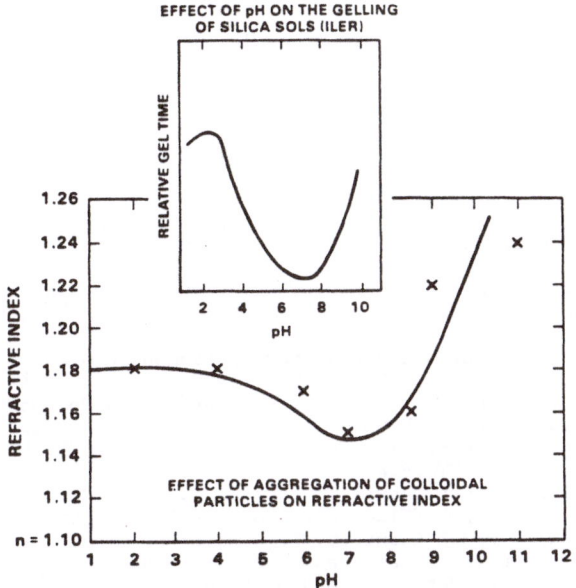

EFFECT OF pH ON THE GELLING OF SILICA SOLS (ILER)

EFFECT OF AGGREGATION OF COLLOIDAL PARTICLES ON REFRACTIVE INDEX

REFRACTIVE INDEX OF DEPOSITED FILMS AFTER DIFFERENT EXTENTS OF AGGREGATION.

Fig. 12   Refractive index, n, for system III films as a function of the pH of aggregation prior to film deposition (1 hour aggregation times). Inset: Gel times for colloidal silica as a function of pH [after Ref. 62].

This brief experimental section has summarized current experiments designed to elucidate the dependences of film structure on the precursor structure and the deposition process. Additional experiments are under way which investigate pore structure [66], interface chemistry [67], condensation kinetics [68], and densification kinetics [69]. Clearly, many more experiments must be conducted to address all the critical issues involved in sol-gel processing of thin films.

IV. CONCLUSIONS

This paper summarizes the critical physical and chemical issues surrounding film formation from solution. A key physical issue is the understanding of the concentration-induced aggregation and gelation processes which constitute film formation. These phenomena depend in turn on chemical issues, for example, the relative rates of condensation and evaporation. Thus, the physics and chemistry of film formation from solution are highly interdependent issues. Initial experiments have demonstrated the potential of tailoring film structure by controlling polymer growth prior to deposition.

ACKNOWLEDGEMENTS

Discussions with R. K. Brow, C. S. Ashley, A. J. Hurd, C. Pantano, K. Budd, D. W. Schaefer, K. D. Keefer, and J. E. Martin are gratefully acknowledged. The technical assistance of C. S. Ashley, S. T. Reed, E. Vernon, and C. R. Hills is greatly appreciated.

REFERENCES

1. Proceedings of the International Workshop on Glasses and Glasss Ceramics from Gels, Oct. 8-9, 1981, J. Non-Cryst. Solids 48, 1-230 (1982).
2. Proceedings of the Second International Workshop on Glasses and Glass Ceramics from Gels, July 1-2, 1983, J. Non. Cryst. Solids 63, 1-300 (1984).
3. Proceedings of the Third International Workshop on Glasses and Glass Ceramics from Gels, Sept. 12-14, 1985, J. Non. Cryst. Solids.
4. Ultrastructure Processing of Ceramics, Glasses and Composites, edited by L. L. Hench and D. R. Ulrich, (Wiley Interscience, New York, 1984).
5. Science of Ceramic Chemical Processing, edited by L. L. Hench and D. R. Ulrich (J. Wiley, NY 1986).
6. Better Ceramics Through Chemistry, edited by C. J. Brinker, D. E. Clark, and D. R. Ulrich (Elsevier - North Holland, 1984).
7. Better Ceramics Through Chemistry II, edited by C. J. Brinker, D. E. Clark, and D. R. Ulrich (Mat. Res. Soc., Pittsburgh, PA, 1986).
8. D. C. Bradley, R. C. Mehrotra, and D. P. Gawr, Metal Alkoxides, (Academic Press, London, 1978).
9. R. B. Pettit an C. J. Brinker, SPIE - Optical Coatings for Energy Efficiency and Solar Appl. 324, 176 (1982).
10. R. K. Brow and C. G. Pantano, Appl. Phys. Letters 48, 27 (1986).
11. H. Schmidt and B. Seiferling in Better Ceramics Through Chemistry II, edited by C. J. Brinker, D. E. Clark, and D. R. Ulrich (Mat. Res. Soc., Pittsburgh, PA 1986).
12. R. B. Pettit and C. J. Brinker, SPIE - Optical MAterials for Energy Efficiency and Solar Energy Conversion IV, 562, 256 (1985).
13. H. Dislich, J. Non-Cryst. Solids, 63, 237 (1984).
14. H. M. McCollister and N. L. Boling, U.S. Patent 4,273,826, June 1981.
15. B. E. Yoldas and D. P. Partlow, Appl. Optics, 23, 1418 (1984).
16. C. J. Brinker and R. B. Pettit, Sandia National Labs Report SAND 83-0137, 68 (1983).
17. R. B. Pettit, C. J. Brinker, and C. S. Ashley, Solar Cells, 15, 267 (1985).
18. R. B. Pettit, C. S. Ashley, S. T. Reed, and C. J. Brinker, to be published in Sol-Gel Technology, edited by L. C. Klein, Noyes, Park Ridge, NJ, 1987).
19. J. Livage, Solid State Chemistry 1982, Proceedings of the Second European Conference, Veldhoven, The Netherlands, 7-9 June 1982, edited by R. Metselaar, H. J. M. Heijligers and J. Schoonman (Elsevier, The Netherlands (1983).
20. Y. W. Lam and H. C. Lam, J. Phys. D: Appl. Phys. 9, 1477 (1976).
21. S. K. Gupta and C. G. Audain, Proceedings of SPIE, 469, 179 (1984).
22. T. P. Ma and K. Miyauchi, Appl. Phys. Letters, 34, 88 (1979).
23. C. Sanchez, F. Babonneau, R. Morineau, J. Livage, and J. Bullot, Phil. B, 47, 279 (1983).
24. K. D. Budd, S. K. Dey, and D. A. Payne, Brit. Ceram. Soc. Proc. 36, 107 (1985).
25. J. Fukushima, Yogyo Kyoaishi, 83, 204 (1975).
26. A. Larbot, J. A. Alary, J. P. Fabre, C. Guizard, and L. Cot, in Better Ceramics Through Chemistry II, edited by C. J. Brinker, D. E. Clark, and D. R. Ulrich (Mat. Res. Soc., Pittsburgh, PA, 1986).

27. H. P. Stephens, R. G. Dosch, and F. V. Stohl, Ind. I and D. Eng. Chem. Prod. Res. and Dev., 24, 15 (1985).
28. C. J. Brinker, W. D. Drotning, and G. W. Scherer in Better Ceramics Through Chemistry, edited by C. J. Brinker, D. E. Clark, and D. R. Ulrich (Elsevier - North Holland, 1984).
29. D. W. Schaefer, K. D. Keefer, and C. J. Brinker, Polym. Preprints. Am. Chem. Soc., Div. of Polym. Chem., 24, 239 (1983).
30. D. W. Schaefer and K. D. Keefer in Better Ceramics Through Chemistry, edited by C. J. Brinker, D. E. Clark, and D. R. Ulrich (Elsevier - North Holland, 1984).
31. D. W. Schaefer and K. D. Keefer in Better Ceramics Through Chemistry II, edited by C. J. Brinker, D. E. Clark, and D. R. Ulrich (Mat. Res. Soc., Pittsburgh, PA 1986).
32. P. J. Flory, Principles of Polymer Chemistry, Cornell University Press, London (1953).
33. J. E. Martin and A. J. Hurd, submitted, J. Appl. Cryst.
34. P. Meakin, Phys. Rev. Lett., 51, 1119 (1983).
35. P. Meakin and Z. R. Wasserman, Phys. Lett. 103A, 337 (1984).
36. M. Kolb and H. J. Hermann, to be published.
37. C. Aubert and D. S. Cannell, Phys. Rev. Lett. 56, 738 (1986).
38. D. Stauffer, J. Chem. Soc. Faraday Trans. II, 72, 1354 (1976).
39. P. G. de Gennes, Scaling Concepts in Polymer Physics, (Cornell, Ithaca, 1979).
40. M. Kolb and H. J. Hermann, J. Phys. A., 18, L435-441 (1985).
41. G. W. Scherer, submitted, J. Non-Cryst. Solids.
42. T. A. Gallo, C. J. Brinker, L. C. Klein, and G. W. Scherer in Better Ceramics Through Chemistry, edited by C. J. Brinker, D. E. Clark, D. R. Ulrich, (Elsevier - North Holland, 1984).
43. G. W. Scherer, C. J. Brinker and E. P. Roth, J. Non-Cryst. Solids, 72, 369 (1985).
44. G. W. Scherer and T. Garino, J. Am. Ceram. Soc., 68, 216 (1985).
45. C. J. Brinker, K. D. Keefer, D. W. Schaefer, and C. S. Ashley, J. Non-Cryst. Solids, 48, 47 (1982).
46. K. D. Keefer in Better Ceramics Through Chemistry II, edited by C. J. Brinker, D. E. Clark and D. R. Ulrich, (Mat. Res. Soc., Pittsburgh, PA, 1986).
47. D. W. Schaefer and K. D. Keefer, Phys. Rev. Lett. 33, 1383 (1984).
48. K. D. Budd, S. K. Dey, and D. A. Payne in Better Ceramics Through Chemistry II, edited by C. J. Brinker, D. E. Clark and D. R. Ulrich, (Mat. Res. Soc., Pittsburgh, PA, 1986).
49. H. Schmidt, private communication.
50. R. Nelson, J. D. F. Ramsay, J. L. Woodhead, J. A. Cairns, and J. A. A. Crossley, Thin Solid Films, 81, 329 (1981).
51. M. Silverberg, D. Farin, A. Ben-Shaul and D. Avnir, Proc. F³ Conference on Fragmentation, Form and Flow in Fractured Media, Ann. Isr. Phys. Soc., (1986), in press.
52. R. Botet and R. Jullien, Phys. Rev. Lett., 55, 1943 (1985).
53. P. M. Glaser and C. G. Pantano, J. Non-Cryst. Solids 63, 209 (1984).
54. C. J. Brinker and D. M. Haaland, J. Am. Ceram. Soc., 66, 758 (1983).
55. R. K. Brow and C. G. Pantano in Better Ceramics Through Chemistry, edited by C. J. Brinker, D. E. Clark and D. R. Ulrich, (Elsevier - NY, 1984), pg 361.
56. R. K. Brow, private communication.
57. T. A. Michalske and B. C. Bunker, J. Appl. Phys. 56, 2686 (1984).
58. C. J. Brinker, D. R. Tallant, E. P. Roth, and C. S. Ashley in Defects in Glasses, edited by F. L. Galeener, D. L. Griscom and M. J. Weber (Mat. Res. Soc., Pittsburgh, PA, 1986), pg 387.
59. G. Smith and C. G. Pantano, Appl. Surf. Sci., 9, 345 (1981).
60. C. G. Pantano, P. M. Glaser, and D. J. Armburst, in Ultrastructure Processing of Ceramics, Glasses and Composites, edited by L. L. Hench and D. R. Ulrich, (Wiley Interscience, New York, 1984), pg 161.

61. H. Hayashi, K. Kikuchi, Y. Yamaguchi, and T. Nakahama, Inst. Phys. Conf. Ser. 56: Chp. 5, 275 (1981).
62. R. K. Iler, The Chemistry of Silica, (John Wiley, NY, 1979).
63. W. Stöber, A. Fink, and E. Bohn, J. Coll. and Int. Sci., 26, 62 (1968).
64. M. Born and E. Wolf, Principles of Optics, (Pergamon, NY, 1975).
65. A. J. Hurd and C. J. Brinker, unpublished results.
66. D. M. Smith and C. J. Brinker, unpublished results.
67. P. M. Lenahan and C. J. Brinker, unpublished results.
68. K. J. Ward and C. J. Brinker, unpublished results.
69. R. B. Pettit and C. J. Brinker, unpublished results.
70. D. W. Schaefer and C. J. Brinker, unpublished results.

ORGANOMETALLIC COMPOUNDS AS STARTING MATERIALS FOR THE PREPARATION OF UNIFORM FINELY DISPERSED POWDERS

EGON MATIJEVIC'* AND PAOLA GHERARDI**
* Clarkson University, Dept. of Chemistry, Potsdam, NY 13676, USA
** Istituto G. Donegani, Via Fauser 4, 28100 Novara, Italy

INTRODUCTION

The interest in finely dispersed powders, consisting of particles uniform in size and shape, has increased dramatically in recent years. The main reason for this development is the recognition that such materials are essential if better products are to be achieved in many areas of high technology.

In this respect one may mention ceramic materials for specific uses, catalysts, pigments, electro-optical and magnetic devices, fillers, coatings, drug delivery carriers, etc. The specific properties of the particles depend on the intended use. Thus, the modal size requirements may range from tens of amstrongs to tens of micrometers, the desirable shapes include spheres, rods (whiskers), ellipsoids, platelets, oblate or prolate disks, etc. Finally, optical or magnetic characteristics also vary with applications.

In the past such uniform particles were reported in a variety of chemical compositions, organic and inorganic (elements or compounds). With the exception of polymer latexes, the obtained "monodispersed" colloids were subject to certain individual preparation procedures, with no general underlying scientific principles.

Over the past several years considerable progress has been made in the science of fine particles. Indeed, it is now possible to prepare a large variety of finely dispersed solids over a range of modal sizes and in many geometric shapes, including spheres, crystalline or amorphous (1-3). These powders may be of simple or mixed composition or they may be obtained as coated particles with layers of different chemical, physical and structural properties (4).

The important aspect of these studies is that such colloids can be obtained of organic (polymers), inorganic (oxides, hydrous oxides, sulphides, selenides, phosphates, carbonates, etc.), mixed inorganic/organic natures. In addition they are available in chemical compositions, which in the past seemed unattainable in well defined states.

It is the purpose of this paper to review several principles upon which the procedures for the generation of uniform particles have been developed, to illustrate a number of resulting solids, and to describe some of their properties. Special emphasis will be placed in the use of organometallic compounds in these processes.

TECHNIQUES

The procedures to be described can be classified into three major categories, each of which has many variations. These are:

1. Homogeneous precipitation
2. Phase transformation
3. Aerosols

In the first two methods, the chemistry of the systems has to be controlled according to certain principles and there is little possibility to a priori predict the actual size and shape of the particles. The reason for the inability to fully anticipate the results is in the complexity of chemical processes underlying the precipitation of such solids. However, once the principles are understood the set of conditions, that should yield uniform powders can be drastically narrowed down. Once such conditions are given, the variations allow for modification on particles natures. It has been repeatedly proved that the reproducibility of the developed procedures in individual cases is excellent.

The philosophy on the application of the third method, aerosol, is quite different.

R. M. Laine (ed.), Transformation of Organometallics into Common and Exotic Materials: Design and Activation, 279–288.
© 1988 by Martinus Nijhoff Publishers.

Relatively simple, straightformed chemical reactions are carried out in a physically defined state, i.e. in uniform liquid droplets.

1.   Homogeneous precipitation

In homogeneous precipitation one must control the chemical processes in solutions in order to generate particle forming solutes in a kinetically controlled manner so that one burst of nuclei occurs. The production of constituent species should then continue at a rate which assures their incorporation into existing nuclei without causing another critical supersaturation with subsequent secondary nucleation.

In such a procedure the decomposition of complexes is one of the convenient ways to release the necessary precursors to particle precipitation.

In principle the simplest process is the deprotonation of hydrated metal ions to form hydrolyzed intermediates in the production of hydrous (metal) oxides. Since with increasing temperature the hydrolysis is greatly enhanced, the formation of uniform particles can be accomplished by simply heating at low temperature (60-100°C, depending on the system) some metal salt solutions of appropriate initial pH. Indeed, the "forced hydrolysis" has been successfully used in the preparation of a number of metal (hydrous) oxides under relatively mild conditions. Thus, dispersions of narrow size distribution were obtained by heating solutions of aluminum (5, 6), chromium (III) (7, 8), iron (III) (9), titanium (IV) (10), thorium (11), etc. It must be noted that in all of these processes anions play an essential role and they may affect chemical composition, structure, shape and size of the particles, clearly pointing to the essence of complexation in solutions in which precipitation takes place.

The hydrolysis process can further be promoted by the addition into the reacting solutions of compounds that on heating release hydroxide ions in a controlled manner, such as is urea (Figure 1).

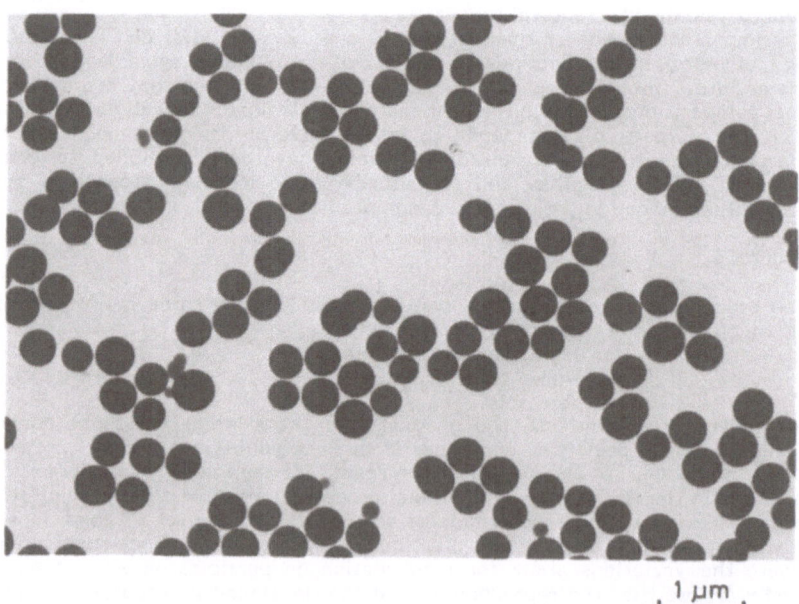

1 μm

Fig. 1. TEM of aluminum hydroxy sulfate obtained by aging at 100°C for 1 hour solutions of $Al_2(SO_4)_3$ in the presence of urea.

It is obvious that organometallic complexes which can decompose under proper conditions should be excellent starting materials for the purpose of formation of uniform colloids. A number of examples will be offered to show that such an approach is indeed feasible.

For example monodispersed copper oxide particles of different shapes were prepared by proper reaction of Fehling solutions (12) and monodispersed colloidal cubic cobalt oxide (Co_3O_4) from cobalt (II) acetate solutions (13).

Because of the extremely easy hydrolyzability of metal alkoxides, these organometallics are particularly suitable as starting compounds for the preparation of metal oxide and other kinds of colloidal particles. For example aluminum hydroxide was obtained by controlled decomposition of Al-sec-butoxide in highly acidic solutions (14). Similar process was used in a continuous reactor for the preparation of aluminum, titanium and zirconium oxide powders (15, 16).

Even more chemically complicated compounds, yet of great technological interest, can be obtained by similar processes. Figures 2a, b show spherical particles of narrow size distribution of barium titanate precursor prepared using Ti (IV)-isopropoxide and a barium compound as starting materials (17). These powders are amorphous and exhibit a peroxidic composition, which on calcination yield crystalline barium titanate.

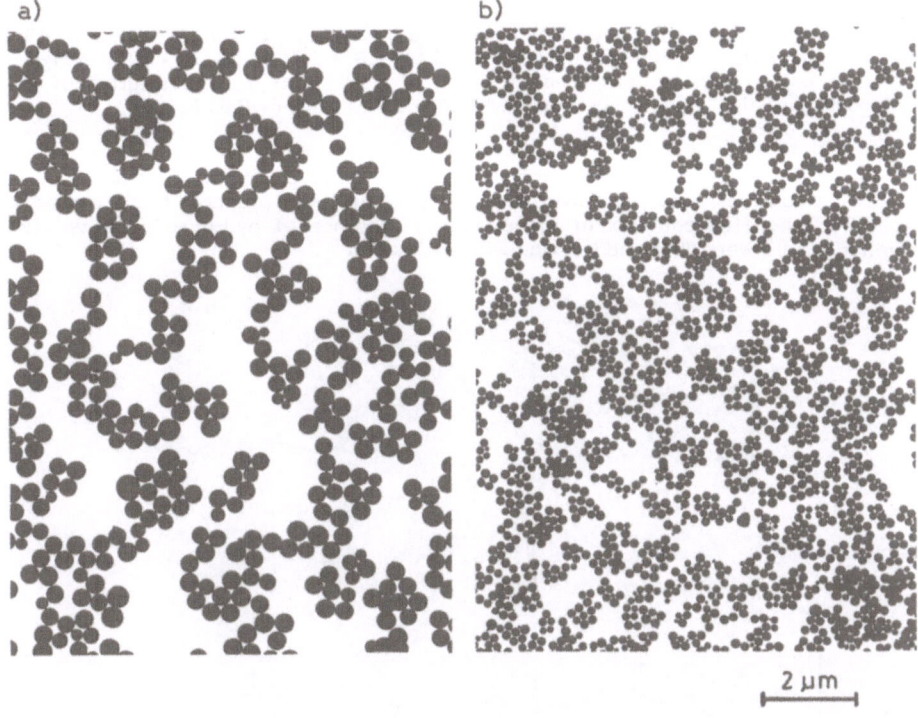

a)   b)

2 μm

Fig. 2. TEM of barium titanate peroxidix precursor obtained by aging at (a) 60°C for 2 hr, (b) 40°C for 2 hr a solution containing Ti(IV)iso-propoxide and Ba^{2+} species.

A variety of metal oxide particles in different shapes were synthesized by decomposing metal chelates, as thriethanolamine (TEA) complexes, at 250°C (18). Scanning electron micrographs in Figure 3 illustrate two such solids, a disk-like hematite and a rod-like zinc oxide (3, 18).

a)

b)

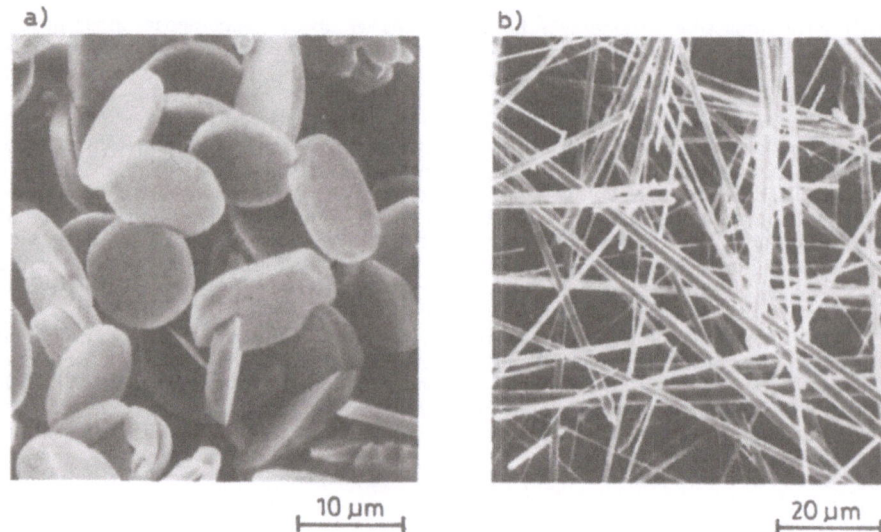

| 10 µm |

| 20 µm |

Fig. 3. SEM of: (a) hematite particles prepared by aging at 250°C for 1 hr a solution of the composition: 0.040 M Fe (NO$_3$)$_3$ + 0.20 M TEA + 1.20 M NaOH + 0.50 M H$_2$O$_2$; (b) Zinc oxide particles prepared by aging at 250°C for 2 hr a solution of the composition: 0.040 M Zn(NO$_3$)$_2$ + 0.20 M TEA + 1.2 M NaOH + 0.85 M N$_2$H$_4$.

Oxidizing or reducing agents can be added to modify the chemical composition of the resulting solids. Figure 4 shows metallic nichel particles prepared by decomposing at 250°C the TEA complex of nichel in the presence of hydrogen peroxide (19). Similar results were obtained working with hydrazine.

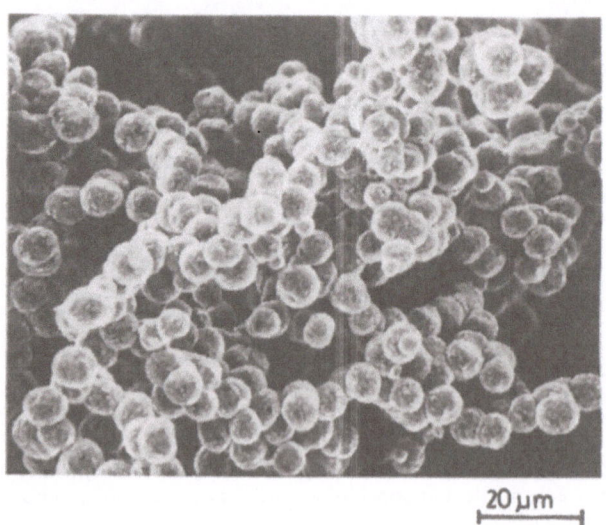

| 20 µm |

Fig. 4. SEM of nickel particles obtained by aging at 250°C for 2 hr a solution containing 0.040 M NiCl$_2$ + 0.20 M HEDTA + 1.2 M NaOH + 0.5 M H$_2$O$_2$.

2.   Phase transformation

Uniform dispersions can be achieved by precipitating a given solid phase and using it as a starting material for another powder of different chemical, morphological and structural characteristics. The latter may consist of changing a solid from amorphous to crystalline or transforming one kind of crystals into another kind. The common known procedure of this type is the sol-gel transformation. This process is one of many different varieties which can be clustered in the same category.

In view of the complexity of the processes involved it is not surprising that few common principles can be applied. For example, in some cases the new phase can form interiorly the precipitated matrix, while in some other cases the original solids will first dissolve and the new particles will then precipitate in the environment, the chemical composition of which will be determined by the composition of the solid precursor.

A few examples of colloidal dispersions obtained by phase transformation processes will be offered here.

Magnetite (20) and various ferrites (21-24) were obtained by first precipitating a ferrous hydroxide gel (a corresponding mixed metal hydroxide gel) with subsequent crystallization at elevated temperatures in the presence of a mild oxidizing agent. Under proper conditions the resulting particles are of spherical shape and of narrow size distribution (Figure 5).

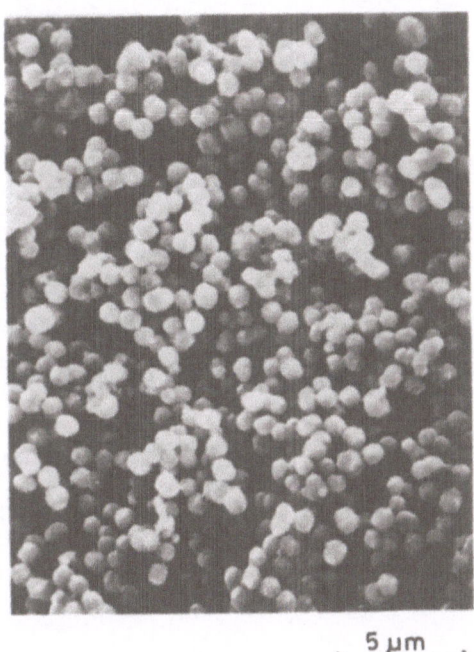

5 μm

Fig. 5. SEM of magnetite particles prepared by aging at 90°C for 4 hr a ferrous hydroxide gel containing an excess of $FeSO_4$ (5.0×10^{-3} M).

284

Exceedingly uniform hematite spindle-shaped particles were instead achieved by first precipitating ferric hydroxide in the presence of a small amount of phosphate ions and subsequent crystallization (25).

The length and the axial ratio of these particles could be altered by modifying the experimental conditions. By further phase transformation in a sequence of reduction/oxidation processes these hematite particles could be converted into magnetic γ-Fe$_2$O$_3$ dispersion without changing their morphology (26).

Chelating agents have been also used in the promotion of phase transformation. For example crystals of magnetite of well defined shapes could be obtained in aqueous dispersions of spherical hematite if these were autoclaved in the presence of triethanolamine (TEA) and a reducing agent (hydrazine) (18).

Figure 6 shows such a change in particle composition and morphology. The process was interrupted on purpose to see the transformation taking place. On continuation all of hematite spheres disappear and only magnetite particles are seen.

├──────────┤
10 μm

Fig. 6. Magnetite particles obtained by recrystallization, at 250°C for 1 hr, of a suspension of spherical α-Fe$_2$O$_3$ in the presence of 0.20 M TEA + 1.20 M NaOH + 0.85 M N$_2$H$_4$.

3.   Aerosols

The use of aerosols as a method for the preparation of well defined colloids represents a very promising yet unconventional technological approach to the problem. In principle, droplets of a reacting liquid are first prepared and then contacted with a vapour of another reactant. The droplets can be of a simple liquid obtained either by condensation in specially designed aerosols generators, or by nebulizing using different devices. In the latter case mixtures of liquids, solutions, or dispersion can be used as a starting phase.

The advantage of the aerosol technique is in its simplicity, versatility, and the ability to produce exceedingly pure products of known particles shape (always spheres) and of desidered size (predetermined by the one of the droplets). Furthermore, the technique lends itself equally well for the generation of organic, inorganic, simple or mixed powders. In the latter case the chemical composition is again adjustable by the ability to control the content of chemicals in the original liquid aerosol, in particular when this is produced by nebulization.

It is noteworthy that the aerosol technology offers the opportunity to prepare powders that cannot be obtained by other techniques. For example, organic latexes can be generated of polymers that cannot be produced by emulsion polymerization or other conventional methods.

In the aerosol technology organometallic precursors may be used as starting materials. Because of their great reactivity metal alkoxides are especially convenient for this purpose.

Figure 7a shows perfectly spherical particle of TiO_2 obtained by dispersing droplets of Ti (IV)-isopropoxide in air and reacting them with water vapour (27, 28). The so prepared powder is amorphous, but on calcination the particle crystallize to anatase and, at sufficiently high temperatures, to rutile.

a)   b)

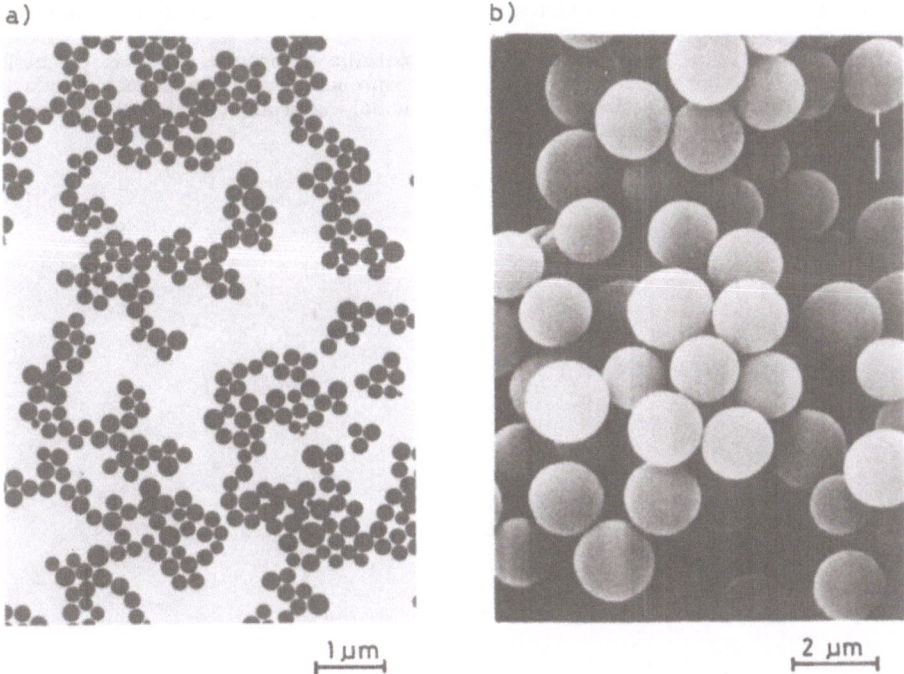

1 μm   2 μm

Fig. 7. (a) TEM of titanium dioxide (TiO_2) particles obtained by the hydrolysis with water vapor of titanium (IV) isopropoxide aerosol droplets; (b) SEM of polymer particles obtained by reaction in aerosol of droplets of hexamethylene diisocyanate with ethylenediamine.

The same procedure can be used to generate colloidal alumina (29) and mixed alumina/titania particles (30). Kinetic studies of hydrolysis in the aerosol state have shown that the process is exceedingly fast; the entire solidification is completed in less than 300 milliseconds (31).

An example of organic latex prepared by the aerosol technique is offered in Figure 7b, which represent polyurea particles prepared by reacting droplets of a diisocianate derivative with ethylenediamine (32). Other polymers were also generated by the same procedure (33, 34).

Indeed, it was possible to make such particles as larger as 30 µm in diameter (34), a size which cannot be obtained by emulsion polymerization and requires processing in gravity free atmosphere.

Mixed organic latexes with metal oxides (32), or metal oxide particles coated by organic layers were also prepared by modifications of the described aerosol technique (35).

CONCLUSIONS

The primary intention of this short review is to demonstrate that very well defined colloidal dispersions of a great variety of materials can now be prepared. It is impossible to offer all possible examples of materials obtained by the described techniques. For details of experimental procedures it is necessary to refer to the original referenced literature.

The results achieved so far clearly show that the underlying principles are of general applications assuming that proper chemical compounds are available in order for the reaction of generation of the heterogeneous phase to proceed at the required rates.

In view of the huge variety of organometallic compounds it is in principle possible to find such starting materials for the preparation of conceivable product. Indeed, the library of existing well defined colloidal materials is rapidly expanding to fill up many needs of modern technology.

REFERENCES

1) E. Matijevic': Monodispersed Colloids: Art and Science
Langmuir, 2, 12-20 (1986)
2) E.Matijevic': Production of Monodispersed Colloidal Particles
Ann. Rev. Materials Sci., 15, 483-516 (1985)
3) E. Matijevic': Monodispersed Metal (Hydrous) Oxides. A Fascinating Field of
Colloid Science
Acc. Chem. Res., 14, 22-29 (1981)
4) P. Gherardi and E. Matijevic': Interactions of Precipitated Hematite with
Preformed Colloidal Titania Dispersion
J. Colloid Interface Sci., 109, 57-76 (1986)
5) R. Brace and E.Matijevic': Aluminum Hydrous Oxide Sols. I. Spherical Particles
of Narrow Size Distribution
J. Inorg. Nucl. Chem., 35, 3691-3705 (1973)
6) W.B. Scott and E. Matijevic': Aluminum Hydrous Oxide Sols. III. Preparation of
Uniform Particles by Hydrolysis of Aluminum Chloride and Aluminum
Perchlorate
J. Colloid Interface Sci., 66, 447-454 (1978)
7) R. Demchak and E. Matijevic': Preparation and Particle Size Analysis of
Chromium Hydroxide Sols of Narrow Size Distribution.
J. Colloid Interface Sci., 31, 257-262 (1969)
8) E. Matijevic', A.D. Lindsay, S. Kratohvil, M.E. Jones, R.I. Larson and N.W.
Cayey: Characterization and Stability of Chromium Hydroxide Sols of Narrow
Size Distribution
J. Colloid Interface Sci., 36, 273-281 (1971)
9) E.Matijevic' and P. Scheiner: Ferric Hydrous Oxide Sols. III. Preparation of
Uniform Particles by Hydrolysis of Fe(III)-Chloride,-Nitrate, and-Perchlorate
Solutions.
J. Colloid Interface Sci., 63, 509-524 (1978)
10) E. Matijevic', M. Budnik, and L. Meites: Preparation and Mechanism of
Formation of Titanium Dioxide Hydrosols of Narrow Size Distribution
J. Colloid Interface Sci., 61, 302-311 (1977)
11) N.B. Milic' and E. Matijevic': Formation of Spherical Colloidal Thorium Basic
Sulfate Particles
J. Colloid Interface Sci., 85, 306-315 (1982)
12) P. McFadyen and E. Matijevic': Copper Hydrous Oxide Sols of Uniform Particle
Shape and Size
J. Colloid Interface Sci., 44, 95-106 (1973)
13) T. Sugimoto and E. Matijevic': Colloidal Cobalt Hydrous Oxides. Preparation of
Monodispersed Co_3O_4
J. Inorg. Nucl. Chem., 41, 165-172 (1979)
14) D.L. Catone and E. Matijevic': Aluminum Hydrous Oxide Sols.
II. Preparation of Uniform Spherical Particles by Hydrolysis of Al-sec-butoxide
J. Colloid Interface Sci., 48, 291-301 (1974)
15) B. Fegley, Jr. and E.A. Barringer: Synthesis, Characterization and Processing of
Monosized Ceramic Powders.
Better Ceramics Through Chemistry. Materials Research Society, 32, 187-197
(1984)
16) B.E. Novich: Continuous Processing of Ceramic Oxide Powders
Ceramics Processing Research Laboratory Report ≠ Q1 M.I.T. Cambridge
17) P. Gherardi and E. Matijevic': Preparation of Uniform Spherical Colloidal Barium
Titanate Particles
In preparation.
18) R.S. Sapieszko and E. Matijevic': Preparation of Well Defined Colloidal Particles
by Thermal Decomposition of Metal Chelates. I. Iron Oxides
J. Colloid Interface Sci., 74, 405-422 (1980)
19) R.S. Sapieszko and E. Matijevic': Preparation of Well Defined Colloidal Particles
by Thermal Decomposition of Metal Chelates. II. Cobalt and Nickel
Corrosion, 36, 522-530 (1980)

288

20) T. Sugimoto and E. Matijevic': Formation of Uniform Spherical Magnetite Particles by Crystallization from Ferrous Hydroxide Gels
J. Colloid Interface Sci., 74, 227-243 (1980)

21) A.E. Regazzoni and E. Matijevic': Formation of Spherical Colloidal Nickel Ferrite Particles as Model Corrosion Products.
Corrosion, 38, 212-218 (1982)

22) H. Tamura and E. Matijevic': Precipitation of Cobalt Ferrites.
J. Colloid Interface Sci., 90, 100-109 (1982)

23) A.E. Regazzoni and E. Matijevic': Formation of Uniform Colloidal Mixed Cobalt-Nickel Ferrite Particles.
Colloids Surf., 6, 189-201 (1983)

24) E. Matijevic', C.M. Simpson, N. Amin and S. Arajs: Preparation and Magnetic Properties of Well Defined Colloidal Chromium Ferrites.
Colloids Surf. In press

25) M. Osaki, S. Kratohvil and E. Matijevic': Formation of Monodispersed Spindle-type Hematite Particles.
J. Colloid Interface Sci., 102, 146-151 (1984)

26) M. Osaki and E. Matijevic': Preparation and Magnetic Properties of Monodispersed Spindle-type γ-Fe$_2$O$_3$ Particles.
J. Colloid Interface Sci., 107, 199-203 (1985)

27) M. Visca and E. Matijevic': Preparation of Uniform Colloidal Dispersions by Chemical Reaction in Aerosols. I. Spherical Particles of Titanium Dioxide.
J. Colloid Interface Sci., 68, 308-319 (1979)

28) E. Matijevic' and M. Visca: Titanium Dioxide.
US Pat. 4241042

29) B.J. Ingebrethsen and E. Matijevic': Preparation of Uniform Colloidal Dispersions by Chemical Reactions in Aerosols. 2. Spherical Particles of Aluminum Hydrous Oxide.
J. Aerosol Sci., 11, 271-280 (1980)

30) B.J. Ingebrethsen, E. Matijevic' and R.E. Partch: Preparation of Uniform Colloidal Dispersions by Chemical Reactions in Aerosols. III. Mixed Titania/Alumina Colloidal Spheres.
J. Colloid Interface Sci., 95, 228-239 (1983)

31) B.J. Ingebrethsen and E. Matijevic': Kinetics of Hydrolysis of Metal Alkoxide Aerosol Droplets in the Presence of Water Vapor.
J. Colloid Interface Sci., 100, 1-16 (1984)

32) R.E. Partch, K. Nakamura, K.J. Wolfe and E. Matijevic': Preparation of Polymer Colloids by Chemical Reactions in Aerosols. III. Polyurea and Mixed Polyurea-Metal Oxide Particles.
J. Colloid Interface Sci., 105, 560-569 (1985)

33) R. Partch, E. Matijevic', A.W. Hodgson and B.E. Aiken: Preparation of Polymer Colloids by Chemical Reactions in Aerosols. I. Poly(p-tertiary-butylstyrene).
J. Polymer Sci., Polym. Chem. Ed., 21, 961-967 (1983)

34) K. Nakamura, R.E. Partch and E. Matijevic': Preparation of Polymer Colloids by Chemical Reactions in Aerosols. II. Large Particles.
J. Colloid Interface Sci., 99, 118-127 (1984)

35) F. Mayville, R.E. Partch and E. Matijevic': Submitted.